Field Trips in the Southern Rocky Mountains, USA

Edited by

Eric P. Nelson
Department of Geology and Geological Engineering
Colorado School of Mines
Golden, Colorado 80401
USA

and

Eric A. Erslev
Department of Geosciences
Colorado State University
Fort Collins, Colorado 80523
USA

THE
GEOLOGICAL
SOCIETY
OF AMERICA

Field Guide 5

3300 Penrose Place, P.O. Box 9140 ▪ Boulder, Colorado 80301-9140 USA

2004

Published by The Geological Society of America, Inc.
3300 Penrose Place, P.O. Box 9140, Boulder, Colorado 80301-9140, USA
www.geosociety.org

Printed in U.S.A.

GSA Books Science Editor: Abhijit Basu

Library of Congress Cataloging-in-Publication Data

Field trips in the southern Rocky Mountains, USA / edited by Eric P. Nelson and Eric A. Erslev.
   p. cm. -- (Field guide ; 5)
 Includes bibliographical references.
  ISBN 0-8137-0005-1 (pbk.)
   1. Geology--Colorado--Guidebooks. 2. Geology--New Mexico--Guidebooks. 3. Geology--Rocky Mountains Region--Guidebooks. 4. Geology, Stratigraphic--Guidebooks. 5. Colorado--Guidebooks. 6. New Mexico--Guidebooks. 7. Rocky Mountains Region--Guidebooks. I. Nelson, Eric P., 1951- II. Erslev, Eric A. III. Field guide (Geological Society of America) ; 5.

QE91.F43 2004
557.88-dc22

                                  2004058060

Cover: Digital elevation model of the Colorado Front Range between Golden and Ft. Collins created from a mosaic of 7.5 min U.S. Geological Survey topographic quadrangles using Natural Scene Designer; data were gridded at 30 m (courtesy of Karl Mueller, University of Colorado, Boulder). The eastern margin of the Front Range near Boulder, shown in purple and blue, is formed of east-vergent folds and thrust faults. As the range front steps eastward farther to the north, structures are smaller and are west vergent above reactivated NW-trending basement faults. The central part of the region (from north to south) shows gently east-dipping surfaces of Quaternary age that are now deeply incised by river networks. The elevation of these surfaces once marked regional base levels along the base of the Front Range, which has now been significantly lowered by removal of easily eroded Cretaceous and Tertiary strata. Regions of higher elevation show glacial cirques and valleys sculpted by alpine glaciers. The western edge of the Front Range is marked by yellow and green where west-vergent thrust faults emplace basement rocks over late Cretaceous and Tertiary strata. This region also marks the headwaters of the Colorado River that drains the Front Range to the west.

10 9 8 7 6 5 4 3 2 1

Contents

Preface v

1. *Navajo sand sea of near-equatorial Pangea: Tropical westerlies, slumps, and giant stromatolites* 1
   David Loope, Len Eisenberg, and Erik Waiss

2. *Strike-slip tectonics and thermochronology of northern New Mexico: A field guide to critical exposures in the southern Sangre de Cristo Mountains* 15
   Eric A. Erslev, Seth D. Fankhauser, Matthew T. Heizler, Robert E. Sanders, and Steven M. Cather

3. *Structural implications of underground coal mining in the Mesaverde Group in the Somerset Coal Field, Delta and Gunnison Counties, Colorado* 41
   Christopher J. Carroll, Eric Robeck, Greg Hunt, and Wendell Koontz

4. *West Bijou Site Cretaceous-Tertiary boundary, Denver Basin, Colorado* 59
   Richard S. Barclay and Kirk R. Johnson

5. *Buried Paleo-Indian landscapes and sites on the High Plains of northwestern Kansas* 69
   Rolfe D. Mandel, Jack L. Hofman, Steven Holen, and Jeanette M. Blackmar

6. *The Colorado Front Range—Anatomy of a Laramide Uplift* 89
   Karl S. Kellogg, Bruce Bryant, and John C. Reed Jr.

7. *Continental accretion, Colorado style: Proterozoic island arcs and backarcs of the Central Front Range* 109
   Lisa Rae Fisher and Thomas R. Fisher

8. *Eco-geo hike along the Dakota hogback, north of Boulder, Colorado* 131
   Peter W. Birkeland, E.E. Larson, C.S.V. Barclay III, E. Evanoff, and J. Pitlick

9. *The South Cañon Number 1 Coal Mine fire: Glenwood Springs, Colorado* 143
   Glenn B. Stracher, Steven Renner, Gary Colaizzi, and Tammy P. Taylor

10. *Field guide to the paleontology and volcanic setting of the Florissant fossil beds, Colorado* 151
    Herbert W. Meyer, Steven W. Veatch, and Amanda Cook

11. *Paleoceanographic events and faunal crises recorded in the Upper Cambrian and Lower Ordovician of west Texas and southern New Mexico* 167
    John F. Taylor, Paul M. Myrow, Robert L. Ripperdan, James D. Loch, and Raymond L. Ethington

12. *The consequences of living with geology: A model field trip for the general public*
 (second edition) 185
 David M. Abbott Jr. and David C. Noe

13. *Surface and underground geology of the world-class Henderson molybdenum*
 porphyry mine, Colorado 207
 James R. Shannon, Eric P. Nelson, and Robert J. Golden Jr.

14. *Walking with dinosaurs (and other extinct animals) along Colorado's Front Range:*
 A field trip to Paleozoic and Mesozoic terrestrial localities 219
 Joanna L. Wright

Preface

The theme of the 2004 Geological Society of America Annual Meeting and Exposition is "Geoscience in a Changing World," a theme that covers both new and traditional areas of the earth sciences. Traditionally, field trips are a very important part of the GSA Annual Meeting, both for disseminating the results of current research as well as for educational purposes. Education is an important goal implicit in the mission of GSA, and in the broader earth science community, and students in our changing world come from academic, industry, governmental, and political arenas as well as interested laypersons. The Front Range of the Rocky Mountains and the High Plains preserve an outstanding record of geological processes from Precambrian through Quaternary times, and thus serve as an excellent educational exhibit. With its energy and mineral resources, geological hazards, water issues, geoarcheological sites, and famous dinosaur fossil sites, the Front Range and adjacent High Plains region provide ample opportunities for field trips focusing on our changing world.

The chapters in this field guide all contain technical content as well as a field trip log describing field trip routes and stops. Of the 25 field trips offered at the meeting, 14 are described in this guidebook and are presented in the order that the trips were listed and organized at the meeting. In keeping with the theme of the meeting, the guidebook chapters cover a wide variety of geoscience disciplines, with chapters on tectonics (Precambrian and Laramide), stratigraphy and paleoenvironments (for example, early Paleozoic environments, Jurassic eolian environments, the K-T boundary, the famous Oligocene Florissant fossil beds), economic deposits (coal and molybdenum), geological hazards, and geoarchaeology. Chapter topics are summarized below.

Three chapters discuss tectonics. Fisher and Fisher (chapter 7) present field evidence in the Front Range for the existence and paleoenvironments of Proterozoic island arc and backarc rock packages that formed during the accretion of the North American Craton. Although most basement rocks in the Front Range have undergone high *T*-low *P* amphibolite facies metamorphism under anatectic conditions, the rocks examined on this trip are slightly lower grade and without anatectic melting. Kellogg and coauthors (chapter 6), who have extensive past and recent mapping experience in the Front Range, present a spectacular transect across the Front Range to explore Precambrian through Rio Grande Rift tectonics, with a special focus on the controversy surrounding interpretations of Laramide structure along the flanks of the range. Erslev and coauthors (chapter 2) examine the flanks of the southern Sangre de Cristo arch in New Mexico and address its history of uplift and the large (>35 km) dextral offsets of Precambrian lithologic belts on N-striking fault zones that have been attributed to Proterozoic, Ancestral Rocky Mountain, and Laramide events by recent authors. This field trip tests these conflicting hypotheses by combining new and old mapping with new thermochronological techniques.

Six chapters discuss stratigraphy and paleoenvironmental interpretations. Taylor and coauthors (chapter 11) present a revised, and more useful, lithostratigraphy for Lower Paleozoic strata in New Mexico and west Texas based on new biostratigraphic (trilobite and conodonts) and chemostratigraphic (carbon isotope) data. Wright (chapter 14) examines terrestrial sedimentary units exposed along the east flank of the Front Range, with a focus on sedimentological and trace fossil evidence for paleoenvironmental interpretations during erosion of Ancestral and Laramide mountains. Loope and coauthors (chapter 1) present new research results on eolian deposits of the Early Jurassic Navajo Sandstone in Utah that suggest that pluvial episodes punctuated the history of dune sea development during this otherwise arid period. Spectacular outcrops of preserved dune sand features are illustrated. Barclay and Johnson (chapter 4) present extensive data from the West Bijou K-T boundary site (only discovered in August 2000) that constrain the position of this important boundary preserved in a terrestrial environment. This is a great example of a multidisciplinary study directed at understanding the importance of the K-T boundary and, in general, of impact events in earth history. Birkeland and coauthors (chapter 8) offer a trip using public and human-powered transportation that gives an overview of the Front Range Phanerozoic stratigraphy and Laramide magmatism while focusing on the development of Quaternary geomorphology and surficial deposits in the Boulder area. Meyer and coauthors (chapter 10) illustrate how Oligocene volcanic activity and related landforms in the southern Front Range influenced sedimentation of the Florissant Formation, which houses a spectacularly diverse, world-famous flora and insect fauna consisting of more than 1,700 described species. This fossil site continues to provide significant data for interpretations of paleoenvironments and the evolution of life.

Two chapters cover aspects of economic geology. Carroll and coauthors (chapter 3) present recent results of structural studies on faults, fractures, and coal cleats in important coal mines within the Mesaverde Group in the Somerset Coal Field in westernmost Colorado. The results have important implications for mine planning in the area and for the structural evolution of the Mesaverde Group during the Eocene and Neogene. Shannon and coauthors (chapter 13) present previously unpublished mapping data on veins and hydrothermal alteration associated with the peripheral zones of the world-famous Urad-Henderson porphyry Mo deposits, and suggest that many features of the deposit and the peripheral zone were structurally controlled by old Precambrian shear zones. The geology and operations of this world-class mine are also illustrated.

Two chapters cover important and current aspects of geological hazards. Abbott and Noe (chapter 12) present an excellent and updated version of this previously-run field trip that examines the geological hazards present along the heavily populated east flank of the Front Range. This trip is an important model for future trips on the consequences of living with geology that could be offered to the general public, service clubs, politicians, and school groups anywhere in the world. One of the biggest challenges for geoscientists is appropriately disseminating the results and importance of geoscientific studies to society in general. Stracher and coauthors (chapter 9) present new, exciting data and methods of study on the composition and environmental effects of gases and mineral condensates related to an underground coal fire in western Colorado. This is an important study because such underground fires, combined with recent drought conditions in the west, have led to surface fires with drastic effects on floral and faunal, including human, habitats. This study also illustrates how geoscience and fire science are being combined to address associated hazards.

In the chapter on geoarchaeology, Mandel and coauthors (chapter 5) present recent research on the High Plains of northwestern Kansas, where soil-stratigraphic and geomorphological studies have shed light on the buried landscapes that harbor the material record of what may be the earliest humans in the region. This is an excellent illustration of how classical geological techniques can aid in archeological interpretations of past civilizations and how these civilizations may have interacted with the physical environment.

We hope that these guidebook chapters will illustrate to current and future readers how geoscience is changing and how it is addressing issues in a changing world.

Eric P. Nelson
Eric A. Erslev

Geological Society of America
Field Guide 5
2004

Navajo sand sea of near-equatorial Pangea: Tropical westerlies, slumps, and giant stromatolites

David Loope
Department of Geosciences, University of Nebraska, Lincoln, Nebraska 68588-0340, USA

Len Eisenberg
Consultant, 223 Granite Street, Ashland, Oregon 97520, USA

Erik Waiss
Department of Geosciences, University of Nebraska, Lincoln, Nebraska 68588-0340, USA

ABSTRACT

Studies of Quaternary dune fields during pluvial intervals have shown that, given appropriate shifts in climate, eolian sand seas can become verdant landscapes with widespread surface water. The Early Jurassic Navajo Sandstone of the western interior United States, although it appears superficially to have been an immense, arid, active dune sea during the entirety of its depositional history, contains subtle evidence of pluvial episodes. In situ fossil tree stumps, abundant dinosaur tracks, bioturbated strata up to 20 m thick, and large stromatolite-bearing, mass flow–capped interdune deposits bear witness to long-lived pluvial episodes. The Navajo accumulated within the tropics; dunes were swept by cross-equatorial winds in winter and watered by monsoonal rains in summer. Most of the evidence for pluvial conditions occurs around the southern and eastern margins of the Navajo dune sea, in the lower half of the formation.

Keywords: Navajo Sandstone, pluvial, tracks, burrows, stromatolites, interdune, eolian.

INTRODUCTION AND PURPOSE

During the Quaternary, active eolian depositional environments have alternated with pluvial/fluvial environments. For example, Kocurek et al. (1991) and Lancaster et al. (2002) show that in the past 25,000 yr in the western Sahara desert of North Africa, several thousand-year-long episodes of active dune migration alternated with humid episodes of a similar length. During the humid episodes, soils formed on stabilized dunes, and lakes and attendant animal life were common. Shifts of a similar sort and duration have been documented in the Nebraska Sand Hills (Loope et al., 1995; Mason et al., 1997) and in the Namib erg of southwest Africa (Svendsen et al., 2003). In Quaternary studies,

the climate fluctuations affecting sand seas are linked to the same Milankovitch forcing that controlled the expansion and contraction of high-latitude glaciers. Shifts in eolian depositional regimes have also been recognized in pre-Quaternary eolian systems; examples include the late Paleozoic Cedar Mesa Sandstone in southeastern Utah (Loope, 1985) and the Middle Jurassic Page Sandstone of south-central Utah and adjacent Arizona (Blakey et al., 1996). In both of these sequences, intertonguing of eolian with fluvial or shallow marine strata can be correlated to bounding surfaces within the eolian section that record cessation of eolian activity, deflation to a planar surface, high water tables, and plant colonization.

These sorts of changes within eolian settings can be recognized in the rock record by the presence of widespread mud-

Loope, D., Eisenberg, L., and Waiss, E., Navajo sand sea of near-equatorial Pangea: Tropical westerlies, slumps, and giant stromatolites, *in* Nelson, E.P. and Erslev, E.A., eds., Field trips in the southern Rocky Mountains, USA: Geological Society of America Field Guide 5, p. 1–13. For permission to copy, contact editing@geosociety.org. © 2004 Geological Society of America cracks, corrugated surfaces, paleosols, lacustrine or fluvial deposits, trampled horizons, and bioturbated zones. Such features have been attributed to changes either in climate, sea level, or tectonic configuration, or combinations of these factors (Loope, 1985; Kocurek, 1988; Kocurek and Havholm, 1993).

In contrast to many eolian units, the Early Jurassic Navajo Sandstone appears deficient in these sorts of obvious features. To casual observers, the Navajo, because of its striking uniform lithology, can appear to have accumulated in an immense, hyperarid sand sea during the entire duration of its deposition. This character also makes direct stratigraphic correlation between Navajo outcrops problematic (Blakey, 1994). An uncommon exception to the Navajo's uniform lithology occurs in the form of thin (<2 m), widely scattered lenses of interbedded sandstone and carbonate (Gregory, 1917; Peterson and Pipiringos, 1979). These lenses have been ascribed to deposition in short-lived interdune ponds created by heavy rainstorms or temporary rises in the local water table, as well as possibly being indicative of a wetter climate and dune stabilization (Gilland, 1979; Kocurek, 1988; Winkler et al., 1991).

This paper first reviews evidence of abundant water in the Navajo Sandstone and then, in the section describing specific field trip sites, describes features that imply long-term shifts in the erg's climatic regime. Although often subtle, this evidence shows that, as some workers have long suspected, there was much more water and variety in the Navajo erg than its gross geologic character would suggest.

EVIDENCE OF ABUNDANT WATER IN THE NAVAJO SANDSTONE

Plant and animal fossils are rare in the Navajo, but reported finds include rare or single occurrences of sphenophytes (horsetails), ferns, and other plants, as well as pollen and spores, ostracodes and brachiopods, a crocodylomorph, herbivorous and carnivorous dinosaurs, an herbivorous tritylodontid synapsid, unidentified burrows, scorpion tracks, tracks of dinosaurs (*Grallator* and *Eubrontes*), prosauropds (*Otozoum*), and possibly mammal-like tritylodont tracks (*Brasilichnium*), and even "dino-turbated" (trampled) horizons (Picard, 1977; Gilland, 1979; Winkler et al., 1991; Lockley, 1991; Stokes, 1991; Lockley and Hunt, 1994). One of us (Eisenberg) has observed theropod dinosaur tracks and trampled horizons in interdune strata in the Navajo Sandstone at Comb Ridge in southeastern Utah. Calcareous rhizoliths have been noted at one locality in the Navajo Sandstone (Loope, 1988), and both of the authors observed rhizoliths as clasts in a flood deposit in the Navajo Sandstone at Comb Ridge in southeastern Utah. All of these finds occur in or are associated with interdune strata and suggest the presence of abundant, probably surface water. What is not known from these finds is if water was present due to rare intense rainfall events or floods in an eolian setting or whether the water was present because of a shift to a pluvial or fluvial environment. At each of these fossil sites, it is possible that an interdune low filled with water for a few months or years, was visited or briefly colonized by flora and fauna, and then was abandoned when the anomalous water disappeared.

As best as can be determined with the meager fossil evidence, it appears that the faunal assemblage present in the Navajo Sandstone is a continuation of that present in the underlying fluvial Kayenta Formation (Winkler et al., 1991; Lockley and Hunt, 1994; Lockley et al., 1998). This suggests that when the westward-flowing Kayenta streams were buried by southward-moving Navajo dunes (Peterson, 1994), animals and plants would enter the eolian environment where appropriate environments persisted or temporarily reappeared.

The sparse Navajo fossils are commonly associated with interdune carbonate rocks, and these strata suggest that water was present for more than just a few years. Interdune carbonates are commonly laminated and in part could be deposited by microbial action. If this is the case, based on modern rates of microbial deposition of laminated carbonate (Chivas et al., 1990; Rasmussen et al., 1993; Moore and Burne, 1994), a 2-m-thick interdune carbonate could represent as much as several thousand years of aqueous deposition. It should be noted, however, that these beds commonly also include significant amounts of sand and could be primarily evaporitic rather than microbial in origin, such that the likely time of deposition could be measured in decades rather than hundreds of centuries (Lancaster and Teller, 1988; Langford and Chan, 1989). Navajo interdune carbonate also does not by itself suggest stabilization of dunes. Downwind portions of wide flats between migrating dunes can remain open for many hundreds or thousands of years (Hummel and Kocurek, 1984), plenty of time for a 2-m-thick interdune carbonate to form.

None of the evidence discussed above points directly to extended, extensive pluvial episodes interrupting a basically eolian regime in the Navajo erg. All of the features mentioned above could have been formed in isolated and short-term ponds or lakes resulting from unusual rainfall events or temporary and local groundwater intersection of an interdune surface. More direct indications of long-lasting humid episodes in the Navajo Sandstone have been recognized at the following localities: (1) near the Arizona/Utah border at Coyote Buttes; (2) in the region around Capitol Reef National Park; and (3) Canyonlands National Park at Horseshoe Canyon.

STOP-BY-STOP GEOLOGY OF THE NAVAJO SANDSTONE

The trip begins and ends in Grand Junction, Colorado. Full-day hikes on Days 2 (Coyote Buttes) and 4 (Cottonwood and Five Mile Canyons) form the core of this field trip. These are strenuous, off-trail hikes over steep slickrock exposures. Day 1 has only two short, near-the-van stops—the first is at a rest area on I-70 near the west margin of the San Rafael Swell, and the second is along the east entrance road in Zion National Park. Day 3 is mainly a travel day, but, if weather permits, we will walk through a short (<2 km) "slot canyon" (Willis Creek Gorge) that cuts through the Navajo outcrop below the cliffs of Bryce Canyon National Park. Day 5 features a 2–3 h hike that will take us below the rim of Horseshoe Canyon.

Geologic Context of the Navajo Erg, San Rafael Swell

Day 1, Stop 1: Ghost Rock Viewpoint, westbound I-70 (west of Green River, Milepost 120; Fig. 1).

General Geologic Setting

Throughout the late Paleozoic and early Mesozoic, large areas of the western interior United States were covered by dune seas as eolian environments alternated with periods of fluvial, pluvial, or shallow marine deposition. As much as 2500 m of mostly sand-dominated sediment was deposited (Hintze, 1988). The most prominent in outcrop and the thickest and most widespread of the eolian units is the Navajo Sandstone. It is nearly 700 m thick in south-central Utah (Peterson and Pipiringos, 1979), and with its stratigraphic equivalents, extends across most of Utah and into Arizona, Nevada, Colorado, Wyoming, and Idaho (Fig. 2) (Blakey et al., 1988). It is immediately underlain by a fluvial sandstone (Kayenta Formation) and overlain by the shallow marine, partly evaporitic Carmel Formation, or, in limited areas, by a laterally equivalent eolian unit, the Page Sandstone (Blakey, 1994). A tremendous amount of fine sand accumulated in eolian and other environments during this time, but where it all came from is not known for certain. At least for the Jurassic, evidence points to transcontinental transport via westward-flowing rivers from eastern and possibly southern North America (Blakey, 1994; Dickinson and Gehrels, 2003; Rahl et al., 2003). The Ancestral Rockies, on the eastern margin of the Navajo erg, were also a source of sediment (Peterson, 1994; Blakey, 1994; Dickinson and Gehrels, 2003). Sand that reached the sea on the western margin of the continent was recycled by wind and coastal currents and returned to the interior to feed eolian systems. When for one reason or another the eolian system shut down, sand was deposited in fluvial, pluvial, or shallow marine environments (Blakey, 1994).

Paleowind Directions Based on Eolian Cross-Strata

For more than 40 years, geologists have attempted to use the dip directions of cross-strata within ancient eolian sandstones to better understand how continental positions and planetary winds have changed through time. Eolian strata from the western United States clearly show that, from Pennsylvanian through Early Jurassic time, winds have been directed generally southward (= northerly winds), and this observation has long been used to show that the region was in the northern hemisphere during this interval. In their classic paper, Opdyke and Runcorn (1960) noted that "…between about 20° N and 20° S, *except in the monsoon belt of India*, trade winds blow with more or less constancy depending on the topography" (italics added; see "Tropical Westerlies" below). They argued that because cross-strata within Permian-Pennsylvanian rocks (Tensleep, Weber, and Casper Sandstones of Wyoming and northern Utah) dip toward the southwest, the western United States was within the trade wind belt at that time. They concluded that the agreement of paleoclimatic and paleomagnetic data require a major northward drift of the continent

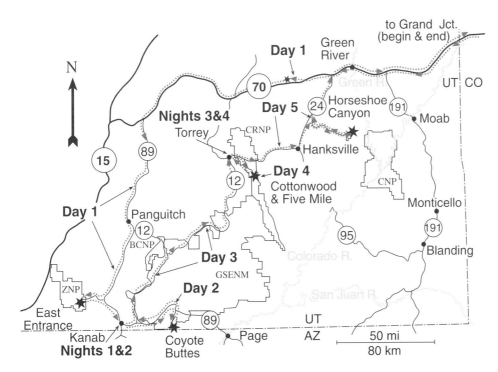

Figure 1. Map showing the southernmost one third of Utah, with field trip route and main geologic stops (stars). ZNP—Zion National Park; BCNP—Bryce Canyon National Park; GSENM—Grand Staircase–Escalante National Monument; CRNP—Capitol Reef National Park; CNP—Canyonlands National Park.

relative to the equator since deposition of the strata. Beginning with the work of Runcorn (1963) and Poole (1963), the abundant cross-strata in southern Utah with dips toward the southeast (Early Permian through Early Jurassic) were interpreted as products of northwesterly winds blowing just north of the "wheel round" (where mid-latitude northwesterly winds blowing around zones of high pressure [anticyclones] turn to become northeasterly trade winds; today's "wheel rounds" are generally ~20°–30° from the equator). Using the paleogeographic reconstruction of the Paleomap Project (see Scotese, 2004) with the Four Corners at ~18°N, maps by Parrish and Peterson (1988) also explained Navajo paleowinds in terms of flow around an anticyclone at the northern edge of the trade wind belt.

Paleogeography Based on Paleomagnetism of Plateau Strata

Using a different paleogeography that is based on paleomagnetic work on Plateau strata by two independent teams, Loope et al. (2004) introduced a very different interpretation. The Plateau paleomagnetic data place the Four Corners 10° farther south, at ~9° N during the Early Jurassic. In this paleogeographic context, the northwesterly winds that drove the Navajo dunes were clearly too close to the equator to be related to mid-latitude westerlies that blew around the northern limb of an anticyclone.

Tropical Westerlies and Monsoon Rains

Loope et al. (2004) interpreted the winds as tropical westerlies—winds that today form a belt that lies equatorward (not poleward) of the trade winds. This interpretation is supported by the presence of southwest-dipping Navajo cross-strata at the north margin of the Colorado Plateau (along the flanks of the Uinta Mountains), which can be interpreted as products of northwesterly trade winds. Cross-equatorial winds (tropical westerlies) are today developed only in monsoonal systems in which very low pressure develops in the summer hemisphere and draws air across the equator from the winter hemisphere. This circulation is today restricted to the monsoonal belts of the eastern hemisphere (Fig. 3). Because the Coriolis force is zero at the equator, the winds can turn sharply toward the low pressure zone, as seen on the east coast of Africa in June, July, and August, when the southern hemisphere trade winds become northwesterlies before they reach the equator. In continental settings, tropical westerlies can penetrate as much as 30° from the equator into the summer hemisphere (northern India in July), but probably can't develop more than 10° from the equator in the winter hemisphere. In the simulation of Early Jurassic atmospheric circulation by Chandler et al. (1992), winter winds in the tropics of western Pangea become westerlies at ~10° north of the equator.

Lawver et al. (2002) show the Four Corners at ~10° during the Early Jurassic (200 Ma), which thereby fits the Navajo paleowind pattern and the Plateau paleomagnetic data better than the Paleomap reconstruction (Scotese, 2004). Dip directions of eolian cross-strata have thus again proved useful for choosing between competing paleogeographic reconstructions. The tropical westerlies recorded by the Navajo cross-strata represent some

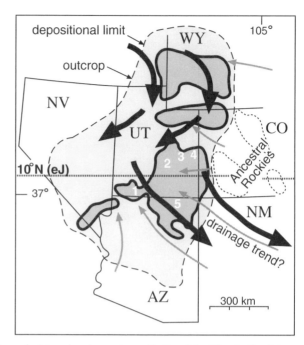

Figure 2. Map showing outcrop limits of the Navajo Sandstone and its stratigraphic equivalents, possible depositional limits of the Navajo erg, paleowinds (bold arrows), hypothetical fluvial sand delivery systems (thin arrows), position of the Ancestral Rockies, and early Jurassic paleolatitude. Southernmost winds are tropical westerlies. Map elements modified from Blakey et al. (1988), Marzolf (1988), Peterson (1994), Blakey (1994), and Loope et al. (2004). Noted localities: 1—bioturbated strata and dinosaur tracks at Coyote Buttes (Day 2; Loope and Rowe, 2003); 2—stromatolite-bearing interdune deposits at Capitol Reef National Park (Day 4; Eisenberg, 2003a); 3—interdune carbonates with 15 m of depositional relief (Day 5); 4—in situ fossil tree stumps near Moab, Utah (not visited; Stokes, 1991); 5—dinosaur and tritylodontid synapsid fossils from Eggshell Arch, Arizona (not visited; Winkler et al., 1991).

of the last vestiges of an atmospheric circulation that persisted over what is now the Colorado Plateau for ~100 m.y., having become established in the Early Permian.

Because they depend on development of a strong low pressure zone across the equator, tropical westerlies can only be sustained for ~3 months per year (in this northern hemisphere case, December, January, and February). Thus, southeast-dipping cross-strata provide a record of environmental conditions representing only one quarter of the year. From Coyote Buttes (on the Utah/Arizona border; see Day 2), Loope et al. (2001) reported slump deposits within large-scale cross-strata that contain annual depositional cycles. The position of the slumps within the cycles show that slumping did not take place during the main season of dune-driving winds. The slumps were interpreted as products of mass wasting during summer (June, July, August) monsoonal rains. This interpretation is consistent with Marzolf's (1983) observation that the features indicating surface water or shallow groundwater within the Navajo are associated with cross-strata dipping to the southeast.

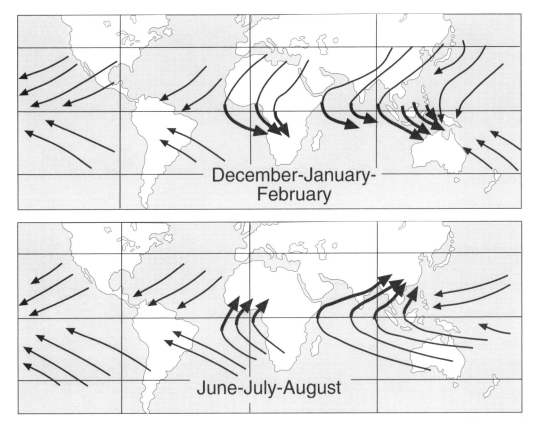

Figure 3. Modern surface winds, showing tropical westerlies (bold arrows) associated with the monsoon systems of the eastern hemisphere (from Webster, 1987). Loope et al. (2004) used these winds as analogs for northwesterly Navajo paleowinds.

Dry Erg Sequence with Annual Depositional Cycles

Day 1, Stop 2: Zion National Park (Utah Highway 9, 2.7 km [1.7 mi] west of east entrance gate).

General Setting

The east entrance to Zion National Park provides spectacular views of Navajo cross-strata. Here there are very few indications of surface water or near-surface groundwater, and those that are present are restricted to the lowest part of the formation (Marzolf, 1983). Thick sets of south- and southwest-dipping cross-strata are piled one upon another without interruption (Fig. 4). Except at the formation's base, interdunes are represented only by erosional surfaces. The site of this stop was recently used to demonstrate the utility of subsurface imaging of bedrock by ground penetrating radar (Jol et al., 2003). Recent studies of diagenesis of the Navajo have shown that outcrops on Laramide structural highs have been bleached by hydrocarbons (Beitler et al., 2003) and that iron-oxide–cemented concretions ("Moki marbles") are close analogs to recently described features in Martian rocks (Chan et al., 2004).

Cyclic Crossbedding

Hunter and Rubin (1983) recognized two forms of cyclic crossbedding in the Navajo Sandstone. At Stop 2, we will examine their example of "concordant cyclic crossbedding"; the other form they recognized, "compound cyclic crossbedding," is well developed at Coyote Buttes (Day 2). Within the 11-m-thick set of SSW-dipping foresets prominently displayed at Stop 2, the increments of dune advance (depositional cycles) that Hunter and Rubin (1983) attributed to fluctuating winds average ~0.3 m (also see Chan and Archer, 1999). Because many dune fields experience daily and annual fluctuations in wind speed and direction, Hunter and Rubin (1983) argued persuasively that the strong depositional cyclicity seen at this site represents annual increments of dune migration. Given the large size of the dunes, the distance of dune migration is too great for the cycles to represent a *daily* advance, but is an appropriate scale for an *annual* increment of dune advance. In comparison, compound cyclic crossbedding at Kanab Canyon and Coyote Buttes records the yearly advances of southeastward-migrating dunes that are two to three times greater than those displayed here (Hunter and Rubin, 1983; Loope et al., 2001).

Pluvial Conditions I: West Margin of Paria Plateau at Coyote Buttes

Day 2; hike starts from Wire Pass trailhead, House Rock road (13.3 km [8. 3 mi] south of highway 89). *Important note*:

Public access to Coyote Buttes is strictly controlled; contact the Kanab Field Office, Bureau of Land Management, Kanab, UT 84741, USA; (435) 644-4600.

General

Coyote Buttes may be the best single place on the Colorado Plateau to see the physical and biogenic sedimentary structures of eolian sandstones. The biogenic structures are present in profusion within three distinct stratigraphic intervals (Fig. 5). Paleowinds were from the northwest during deposition of the entire exposed interval.

Eolian Stratification Types

Hunter (1977) showed that three processes occurring on dune lee slopes can explain the stratification found in small eolian dunes: (1) fallout of grains that are transported over the crest (grainfall strata); (2) avalanching of dry sand when grainfall oversteepens the lee slope (grainflow); and (3) saltation that leads to the climb of wind ripples.

Grainflows in the Navajo intertongue with wind-ripple strata at the toe of the slip-face and are bounded by thin laminations that have been called pin stripes (Fig. 6). Navajo outcrops in southwestern and south-central Utah provide exceptional opportunities to examine eolian grainflows. Due to post-depositional compaction of these loosely packed deposits, few grainflows within Permian through Jurassic eolian sandstones on the Colorado Plateau dip more steeply than 25° (Hunter, 1981). Some of the Navajo grainflows are very thick; one grainflow near the east entrance to Zion National Park measures 17 cm (Loope, 2004). Unlike the modern grainflows Hunter described from several localities that have a narrow, tongue-like shape, the Navajo grainflows typically extend several meters parallel to strike.

At Coyote Buttes, the dip-parallel cross-sectional area of one measured grainflow is over 10,000 cm². This area would have been even greater before compaction. If this grainflow represented a single mass movement, a huge amount of sand would have been stored on the upper lee face before it was triggered. Field and modeling studies of eolian grainflows indicate that nearly all grainfall takes place within 2 m of the dune crest (McDonald and Anderson, 1995; Nickling et al., 2002). A more likely possibility is that grainflows originating from near the crest of the Navajo dunes thinned and stopped before reaching the toe of the very long lee slope. The thick grainflows found in the basal cross-strata probably formed by failure of an over-steepened mid-slope.

Concave-up scour surfaces filled by thick grainflows are common within Navajo cross-strata. These scours formed by scarp retreat during avalanching (Hunter, 1977) and demonstrate that numerous grainflows were initiated far below the dune crest, on the lower lee slope.

Soft-Sediment Deformation

With porosities approaching 50%, grainflow deposits that lie below the water table are especially vulnerable to liquefaction (Hunter, 1981). The abundance of grainflow stratification at

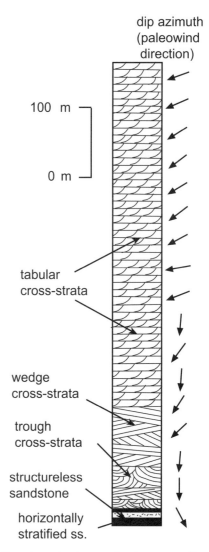

Figure 4. Sedimentary structures and paleocurrent directions for the Navajo Sandstone from Zion Canyon (from Marzolf, 1983).

Coyote Buttes thus probably contributed to the large volume of soft-sediment deformation that can be observed there. Horowitz (1982) concluded that the dune cross-strata most vulnerable to liquefaction lie a few meters below the water table and directly beneath an interdune surface. The higher overburden stress under dunes prevents liquefaction, leading to lateral squeezing and partial dune collapse (Horowitz, 1982).

Marzolf (1983) called attention to the connection between soft-sediment deformation and shallow groundwater and showed that the lower part of the Navajo Sandstone contains more indicators of shallow groundwater than the upper part (and was dominated by northwesterly [monsoonal] paleowinds rather than northeasterly trade winds; see below).

From an exposure a few km south of this site, Loope et al. (2001) described a 6-m-thick set of cross-strata with 36 annual

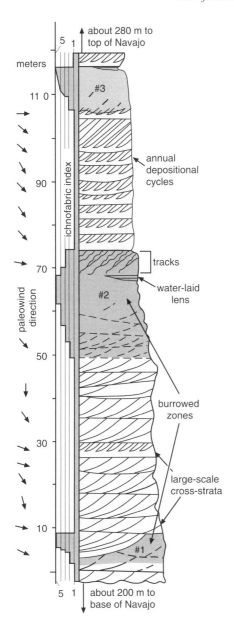

Figure 6. Interbedded grainflows (steep, dark) and wind-ripple strata (light, thin bedded) at the base of a thick set of Navajo Sandstone cross-strata at Coyote Buttes. Pin stripes (the light-colored, thin laminae between grainflow strata) are contiguous with wind ripple deposits and are composed of very fine sand and coarse silt; grainflows are composed of medium sand.

Figure 5. Stratigraphic section at Coyote Buttes showing distribution of burrows and tracks. Ichnofabric index varies from 1 (no bioturbation) to 5 (complete bioturbation). Dip directions of cross-strata show that the dominant, northwesterly wind direction did not change during deposition (compare with Fig. 4; modified from Loope and Rowe, 2003).

Figure 7. Thin slumps at Coyote Buttes. Down-dip sliding due to rainfall events generated thrust faults (upper photo) and normal faults (lower photo) on dune slip face. Only a few slumps are exposed in the northern part of Coyote Buttes; much greater numbers are present a few km to the south.

depositional cycles, 24 of which contained slump deposits. They attributed 20 of the slumps to mass movements due to heavy summer rain events. Slumps are separable from other forms of soft sediment deformation by (1) thrust faults indicating down-dip sliding; (2) smooth truncation surfaces at their tops that are conformably overlain by undeformed strata; and (3) breccia sheets at their distal (down-dip) margins. Only a few small slumps are present at this site (Coyote Buttes North; Fig. 7).

Dune Types and Annual Depositional Cycles

Nearly all of the Navajo cross-strata at Coyote Buttes were produced by the migration of large dunes that did not have superimposed bedforms on the lower portions of their lee slopes. The simple style of cross-stratification and the dominance of grainflow strata are probably related to a wind regime that was dominated by very strong flow from a single direction. Compound cyclic crossbedding (Hunter and Rubin, 1983) is developed in a large portion of the exposed rocks. Cycles consist of couplets, each composed of a thick grainflow-dominated portion and a thinner, wedge-shaped accumulation of wind-ripple strata (Fig. 8). Cycle thicknesses indicate a dune migration rate of about a meter per year—a rate achieved by cross-equatorial winds that blew only from December through February (Loope et al., 2004). Loope et al. (2001) and Loope et al. (2004) hypothesize that summer rainfall on the Navajo erg came from air masses moving north from near the equator. No direct evidence of south-to-north winds, however, has been described yet from exposures at Coyote Buttes or, to our knowledge, from Lower Jurassic eolian sandstones on the Colorado Plateau. In their study of the Lower Jurassic Wingate Sandstone in northeastern Arizona, however, Clemmensen and Blakey (1989) called upon paleowinds from the southwest to explain the dynamics of inferred oblique bedforms.

Biogenic Structures

Intense burrowing is present in both interdune and dune strata at three stratigraphic levels at Coyote Buttes (Figs. 5 and 9). The middle burrowed interval is more than 20 m thick. The large populations of burrowers (probably insects) were likely sustained by plant growth in interdune areas, but no plant remains or traces have been identified. In the upper part of the middle burrowed interval, two types of vertebrate tracks (*Grallator* and *Brasilichnium*) are present within a well exposed, thick set of cross-strata. The trampled cross-strata contain meter-scale depositional cycles, and more than 160 m of dune migration is exposed. The high density of *Grallator* tracks (Fig. 10), their pervasive distribution, and their persistence during a long distance of dune migration suggest a resident (non-migratory) population that was continually present in the dune field for at least 100 yr (Loope and Rowe, 2003). The great thickness of the burrowed zones suggest that the pluvial intervals were long-lived (at least thousands, and more likely, tens of thousands, of years).

It is surprising that lithologic evidence of water, in the form of interdune carbonate, paleosols, or rooted horizons, is sparse, but lens-shaped massive sandstones, 1–2 m thick, associated with the biotic horizons, are probably the remnants of sandy interdunes, intensely bioturbated by plants and animals. Loope and Rowe (2003) interpreted the absence of interdune carbonate to rainfall recharge rather than fluvial or subsurface input. Although the measurable duration of the dune migration for the trampled cross-strata zone is one hundred years or more, it is likely that the pluvial episode lasted much longer. This view is supported by both the vertical extent (up to 25 m) and the lateral extent of the principal burrowed zone (at least 115 km²).

Figure 8. Compound cyclic cross-strata at Coyote Buttes interpreted as annual depositional cycles. Couplets are composed of grainflow-dominated, steep cross-strata interbedded down-dip with wedges of wind-ripple laminae. Grainflows were generated by dominant, northwesterly flow (winter); wind ripple deposits were banked against the lee face during the remaining nine months.

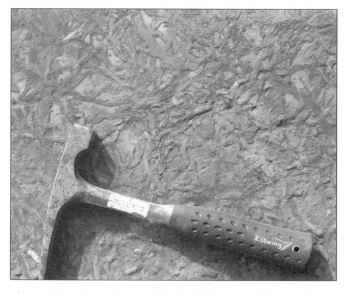

Figure 9. *Planolites* burrows in flat-bedded, interdune sandstone, Coyote Buttes. Most of the burrows at Coyote Buttes are within eolian cross-strata.

The outcrops described in Loope and Rowe (2003) lie along the SE-NW drainage trend described by Blakey (1994) (Fig. 2). They are also in the same area noted by Loope et al. (2001, 2004) to have been subject to intense monsoonal rainfall. It therefore seems likely that adequate surface water to support

Figure 10. Theropod dinosaur tracks (*Grallator*) on the upper surface of angle-of-repose grainflow strata, Coyote Buttes.

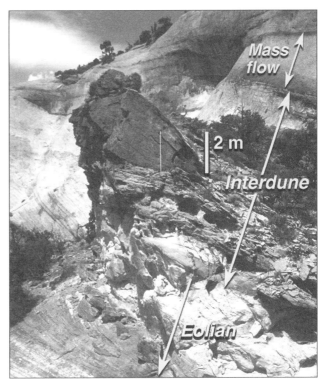

Figure 11. A 5-m-high stromatolite at South Cottonwood site, Capitol Reef National Park (from Eisenberg, 2003a).

the biotic activities described was provided by rainfall. Additionally, substantial subsurface flow probably contributed to a higher groundwater table in this area, which in turn would have made it easier for seasonal rainfall recharge to fill numerous interdune lows with ponds and lakes.

Pluvial Conditions II: Capitol Reef National Park, Cottonwood Canyon (Day 4)

Giant Stromatolites in Interdune Carbonates

At Capitol Reef National Park, stromatolites up to 5 m high in the growth direction occur on paleonorth (paleosouth-facing) interdune margins (Fig. 11). Based on modern rates of stromatolite growth (0.5–1 m/k.y.), it would have taken at least a few thousand yr and perhaps much longer than 10,000 yr to build such large microbial growth structures (Eisenberg, 2003a). This is strong evidence that a relatively continuous, abundant supply of surface water was present in the interdunes where stromatolites grew. Because the stromatolites occur on paleonorth interdune margins, and because primary sand transporting winds in the Navajo erg blew from the north or northwest (Peterson, 1988), dunes surrounding the stromatolite-bearing interdune deposits must have been stable during stromatolite growth. If not stable, dune migration would have buried any nascent stromatolites within a few years. Stromatolites at Capitol Reef grew

preferentially on the slopes of paleonorth (south-facing) interdune margins because of the greater light energy received there (Eisenberg, 2003a). Stromatolites and thick interdune sections also occur at scattered localities east and south of Capitol Reef. At these sites, stromatolites also commonly occur on paleosouth-facing interdune margins, although at some localities small stromatolites occur in central or even paleonorth-facing interdune margins. But even these indicate an episode of abundant water and dune stability. The interdune basins in which they lie are only a few hundred meters across. At a typical Navajo dune migration rate of 0.5 m to 3 m/yr (Hummel and Kocurek, 1984), these would have been buried by advancing dunes in roughly 500 yr or less, and a 2-m-high stromatolite would have taken a thousand or more years to grow.

In addition to the stromatolites, the unusual thickness (as much as 20 m) of associated interdune strata suggests a period of dune stability (Eisenberg, 2003a). These interdune outcrops all occur in steep-sided, bowl-shaped erosive depressions, and all occur between one third and one quarter of the way up in the Navajo Sandstone. In contrast, typical interdune deposits in both modern and ancient eolian strata, including the Navajo Sandstone, are planar and are commonly less than 3 m thick (Ahlbrandt and Fryberger, 1981; Hummel and Kocurek, 1984; Lancaster and Teller, 1988; Langford and Chan, 1989; Verlander, 1995).

Mass Flows

At several Capitol Reef sites and at a few locations northeastward toward Canyonlands National Park, stromatolite-bearing, thick interdune sections are buried beneath mass flow deposits, which can reach a preserved thickness of nearly 30 m and contain clasts of ripped-up interdune carbonate up to 12 m long (Eisenberg, 2003b). These unique flood deposits confirm that large volumes of surface water were present at these sites. These mass flow–capped sites probably formed during a single depositional episode of widespread and abundant surface water in the Navajo erg and can be traced across at least 1400 km². A handful of stromatolite-bearing, thick interdune sites occur south of the Capitol Reef–Canyonlands trend. They lack capping mass flows but also occur stratigraphically one third to one quarter of the way up the Navajo Sandstone. They may be part of the same depositional episode as the mass flow–capped sites or be part of a series of abundant water episodes. Added together, the two site types can be traced across 3200 km² of central Utah.

Pluvial Conditions III: Canyonlands National Park, south of Green River

Our hike starts at Horseshoe Canyon parking area (Day 5).

Fossil Wood

Fossil tree stumps are good indicators of the availability of abundant water during deposition of the Navajo Sandstone. More than a dozen sites around Moab, Utah, near the eastern edge of the Navajo erg, have been identified as containing in situ fossil coniferous tree stumps as well as pieces of large logs. In situ stumps exceed 25 cm in diameter, and some fossil logs are close to a meter in diameter (Fig. 12). Fossil wood occurs in typical Navajo interdune strata, that is, in lens-shaped deposits a few meters thick and a few hundred meters wide (Loope, 1979; Stokes, 1991; F. Barnes, 2001, personal commun.). At one site near Moab, dozens, if not hundreds, of in situ fossil tree stumps are scattered across a sandy carbonate bed that caps a few-meters-thick interdune deposit at least one kilometer long and 500 m wide.

In situ fossil wood indicates that near-surface groundwater was continuously abundant in the Navajo around Moab for at least decades if not longer. If water had not been available continuously, the trees would not have grown so large or in such numbers. This also implies that surface water was probably locally common during tree-growing episodes. This notion is supported by the observation that, locally, perhaps up to one fourth of the lower Navajo Sandstone in the Moab area consists of interdune strata (Verlander, 1995). Dunes need not have been completely stable during the growth of these trees: For example, at Coral Pink Sand Dunes State Park in Utah, numerous large conifers grow in an area of partially stabilized eolian dunes (Ford and Gillman, 2000). The exact stratigraphic position of each of these Navajo fossil wood deposits is unknown, but ongoing work by one of us (Eisenberg) is aimed at determining if they

Figure 12. Silicified wood at Horseshoe Canyon site, Canyonlands National Park. Staff is 2 m long.

occur throughout the Navajo stratigraphic section and represent multiple episodes or if they are restricted to one, or a very few, particular horizons.

The only known in situ fossil wood in the Navajo west of the Green River occurs at two localities near Horseshoe Canyon, ~55 km northwest of Moab, and at one site near the Dirty Devil River, ~90 km southwest of Moab (Stokes, 1991; Eisenberg, 2003, unpublished field notes). All three sites are in carbonate-rich interdune deposits. It is possible that drainage from the Ancestral Rockies, following a westward-flowing trend established in the underlying, fluvial Kayenta Formation (Blakey, 1994; Peterson, 1994), continued to bring relatively abundant subsurface and perhaps ephemeral surface water into the Navajo erg around what is now Moab (Stokes, 1991; Verlander, 1995) and on to the west-southwest. Another possibility is that trees were widespread, but conditions suitable for preservation of the wood were restricted to what is now east-central Utah.

Relief on Carbonate Bed

The thin carbonate bed at the base of the stromatolites shows depositional relief of over 15 m (Fig. 13). This implies that water depth in the interdune during deposition was at least 15 m. In an active dune field, infiltration of rainfall into the substrate would be nearly 100% because runoff would be nil. If the dunes were stabilized by soil crusts (and there is no evidence for stabilization by vascular plants), evapotranspiration from the dune surface would be very low. If, for example, potential evaporation was 300 cm and annual rainfall was 50 cm, lakes could have persisted in small, closed basins between large, sinuous-crested dunes. On

Figure 13. Foreshortened (telephoto) view of a single carbonate bed (X) at Horseshoe Canyon, emphasizing depositional relief of ~15 m. Several stromatolites lie just below the left X.

the other hand, if a dune field with broad, flat interdunes was stabilized, perennial lakes would be much less likely because evaporation would have exceeded infiltration over a large portion of the landscape.

DISTRIBUTION OF LOCALITIES AND POSSIBLE CORRELATIONS

The evidence noted above for abundant water occurs mostly along the presumed eastern or southern margin of the Navajo erg (Fig. 2). Paleogeographic and paleoclimatic reconstructions of the Navajo erg (Blakey, 1994; Peterson, 1994; Parrish and Peterson, 1988) suggest that these areas may have been subject to significantly greater input of rain and groundwater than more interior regions of the erg. On the southern side, Blakey (1994) interprets the presence of a long-standing northwest-flowing, long-distance drainage system to account for the intertonguing of fluvial (Kayenta) and eolian (Navajo) strata in that area in the lower part of the Navajo Sandstone. Even after the area was overwhelmed by eolian sediment that moved in from the north, groundwater would likely have continued to be input into the dune system. This could have led to the more common presence of lakes and ponds and plant and animal life in that area.

Similarly, evidence presented by Peterson (1994), Dickinson and Gehrels (2003), and Rahl et al. (2003) suggests that the Ancestral Rockies may have been a source of water on the eastern margin of the Navajo erg. Lastly, westward-flowing rivers that headed in eastern North America and found their way by various routes to the sea west of the Navajo erg may have, at times, brought significant amounts of water to the Navajo erg on its eastern margin.

In addition to possible groundwater or surface-water inputs, Loope et al. (2001) show that heavy monsoonal rains occurred in the Navajo erg, at least on its southern side. During the northern hemisphere summer, the intertropical convergence zone would have migrated northward to cover the Colorado Plateau (Loope et al., 2004). If moist air were available, this convergence would have generated the lift needed for precipitation. The specific source of the moisture for summer precipitation during pluvial intervals and the efficacy of the coastal mountains along Pangea's western margin in generating a rain shadow during the dry intervals are presently unknown, but climate modeling may shed light on these questions.

The rough similarity in stratigraphic position between outcrops along the Arizona/Utah border that contain evidence of pluvial episodes and those in the vicinity of Capitol Reef National Park invites the question: Are they part of the same depositional episode? If so, it would mean that dunes were stable in one part of the Navajo erg while still migrating in another. This would not be unexpected, especially in an erg as large as the Navajo. In the Coral Pink Sand Dunes of Utah, part of the dune field is stable, part is actively migrating, and part is in the process of transforming from one state to the other (Ford and Gillman, 2000).

Ultimately, the question is probably unanswerable, even if the stratigraphic position of the Arizona/Utah border outcrops were to be more precisely determined. At the minimum, however, the stratigraphic similarity suggests there were several pluvial intervals of undetermined duration during Navajo deposition when wet interdunes and interdune lakes were common and supported a diverse flora and fauna.

CONCLUSIONS

The erg preserved as the Navajo Sandstone and its stratigraphic equivalents was one of the largest known to have existed on our planet. Although it was generally an intensely arid environment of actively migrating large dune forms, at times water was locally abundant at the surface. Stratigraphic intervals rich in dinosaur tracks and insect burrows, extensive areas of in situ tree stumps, and scattered sites bearing large stromatolites, thick interdune sections, and flood deposits attest to these wetter conditions. Most evidence for wetter conditions occurs toward the southern and eastern margin of the Navajo erg, where monsoonal rains and fluvial input were most likely. Ongoing work will seek new field evidence for the number and duration of pluvial episodes during Navajo Sandstone deposition and, through the use of global and regional climate models, attempt to delineate changes in atmospheric circulation during pluvial episodes as well as ascertain the source of the moist air.

ACKNOWLEDGMENTS

Eric Nelson reviewed the manuscript and made many helpful suggestions. Part of the work reported here was supported by a grant from the National Science Foundation (EAR-02-07893).

REFERENCES CITED

Ahlbrandt, T.S., and Fryberger, S.G., 1981, Sedimentary features and significance of interdune deposits, in Ethridge, F.G., and Flores, R.M., eds., Recent and ancient nonmarine depositional environments: Models for exploration: Society for Sedimentary Geology (SEPM) Special Publication 31, p. 293–314.
Beitler, B., Chan, M.A., and Parry, W.T., 2003, Bleaching of Jurassic Navajo Sandstone on Colorado Plateau Laramide highs: Evidence of exhumed hydrocarbon supergiants?: Geology, v. 31, p. 1041–1044, doi: 10.1130/G19794.1.
Blakey, R.C., 1994, Paleogeographic and tectonic controls on some Lower and Middle Jurassic erg deposits, Colorado Plateau, in Caputo, M.V., Peterson, J.A., and Francyzk, K.J., eds., Mesozoic Systems of the Rocky Mountain Region, USA: Special Publication of the Rocky Mountain Section of the Society for Sedimentary Geology (SEPM), p. 273–298.
Blakey, R.C., Peterson, F., and Kocurek, G., 1988, Synthesis of late Paleozoic and Mesozoic eolian deposits of the Western Interior United States: Sedimentary Geology, v. 56, p. 3–125, doi: 10.1016/0037-0738(88)90050-4.

Blakey, R.C., Havholm, K.G., and Jones, L.S., 1996, Stratigraphic analysis of eolian interactions with marine and fluvial deposits, Middle Jurassic Page Sandstone and Carmel Formation, Colorado Plateau, U.S.A.: Journal of Sedimentary Research, Section B: Stratigraphy and Global Studies, v. 66, no. 2, p. 324–342.
Chan, M.A., and Archer, A.W., 1999, Spectral analysis of eolian foreset periodicities: implications for Jurassic decadal-scale paleoclimatic oscillators: Paleoclimates, v. 3, p. 239–255.
Chan, M.A., Beitler, B., and Parry, W.T., 2004, A possible terrestrial analogue for haematite concretions on Mars: Nature, v. 429, p. 731–734, doi: 10.1038/NATURE02600.
Chandler, M., Rind, D., and Ruedy, R., 1992, Pangean climate during the Early Jurassic: GCM simulations and the sedimentary record of paleoclimate: Geological Society of America Bulletin, v. 104, p. 543–559, doi: 10.1130/0016-7606(1992)1042.3.CO;2.
Chivas, A.R., Torgersen, T., and Polach, H.A., 1990, Growth rates and Holocene development of stromatolites from Shark Bay, Western Australia: Australian Journal of Earth Sciences, v. 37, p. 113–121.
Clemmensen, L.B., and Blakey, R.C., 1989, Erg deposits in the Lower Jurassic Wingate Sandstone, northeastern Arizona: Sedimentology, v. 36, p. 449–470.
Dickinson, W.R., and Gehrels, G.R., 2003, U-Pb ages of detrital zircons from Permian and Jurassic eolian sandstones of the Colorado Plateau, USA: Paleogeographic implications: Sedimentary Geology, v. 163, p. 29–66, doi: 10.1016/S0037-0738(03)00158-1.
Eisenberg, L.I., 2003a, Giant stromatolites and a supersurface in the Navajo Sandstone, Capitol Reef National Park, Utah: Geology, v. 31, p. 111–114.
Eisenberg, L.I., 2003b, Stromatolites, floods and regional correlation in the Navajo Sandstone: American Association of Petroleum Geologists, Abstracts with Programs, AAPG Annual Convention, May, 2003, Salt Lake City, Utah, p. A49.
Ford, R.L., and Gillman, S.L., 2000, Geology of Coral Pink Sand Dunes State Park, Kane County, Utah, in Sprinkel, D.A., Chidsey, T.C. and Anderson, P.B., eds., Geology of Utah's Parks and Monuments: Salt Lake City, Utah Geological Association Publication 28, p. 365–389.
Gilland, J.K., 1979, Paleoenvironment of a carbonate lens in the lower Navajo Sandstone near Moab, Utah: Utah Geology, v. 6, p. 28–37.
Gregory, H.E., 1917, Geology of the Navajo country: A reconnaissance of parts of Arizona, New Mexico, and Utah: U.S. Geological Survey Professional Paper 93, 161 p.
Hintze, L.F., 1988, Geologic history of Utah: Brigham Young University Geology Studies Special Publication 7, 202 p.
Horowitz, D.H., 1982, Geometry and origin of large-scale deformation structures in some ancient wind-blown sand deposits: Sedimentology, v. 29, p. 155–180.
Hummel, G., and Kocurek, G., 1984, Interdune areas of the back-island dune field, North Padre Island, Texas: Sedimentary Geology, v. 39, p. 1–26, doi: 10.1016/0037-0738(84)90022-8.
Hunter, R.E., 1977, Basic types of stratification in small eolian dunes: Sedimentology, v. 24, p. 361–387.
Hunter, R.E., 1981, Stratification styles in eolian sandstones: Some Pennsylvanian to Jurassic examples from the western interior, U.S.A., in Ethridge, F.G., and Flores, R.M., eds., Recent and ancient nonmarine depositional environments: Models for exploration: Society for Sedimentary Geology (SEPM) Special Publication 31, p. 315–329.
Hunter, R.E., and Rubin, D.M., 1983, Interpreting cyclic crossbedding, with an example from the Navajo Sandstone, in Brookfied, M.E. and Ahlbrandt, T.S. eds., Eolian Sediments and Processes: Amsterdam, Elsevier, p. 429–454.
Jol, H.M., Bristow, C.S., Smith, D.G., Junck, M.B., and Putnam, P., 2003, Stratigraphic imaging of the Navajo Sandstone using ground-penetrating radar: The Leading Edge, v. 22, p. 882–887, doi: 10.1190/1.1614162.
Kocurek, G., 1988, First-order and super bounding surfaces in eolian sequences—Bounding surfaces revisited, in Kocurek, G., ed., Late Paleozoic and Mesozoic Eolian deposits of the Western Interior of the United States: Sedimentary Geology, v. 56, p. 193–206.
Kocurek, G., Deynoux, M., Havholm, K., and Blakey, R., 1991, Amalgamated accumulations resulting from climatic and eustatic changes, Akchar Erg, Mauritania: Sedimentology, v. 38, p. 751–772.
Kocurek, G., and Havholm, K.G., 1993, Eolian sequence stratigraphy—A conceptual framework, in Weimer, P. and Posamentier, H., eds., Siliciclastic sequence stratigraphy: American Association of Petroleum Geologists Memoir 58, p. 393–409.

Lancaster, N., and Teller, J.T., 1988, Interdune deposits of the Namib sand sea: Sedimentary Geology, v. 55, p. 91–107, doi: 10.1016/0037-0738(88)90091-7.

Lancaster, N., Kocurek, G., Singhvi, A., Pandey, V., Deynoux, M., Ghienne, J.F., and Lo, K., 2002, Late Pleistocene and Holocene dune activity and wind regimes in the western Sahara Desert of Mauritania: Geology, v. 30, p. 991–994.

Langford, R.P., and Chan, M.A., 1989, Fluvial-aeolian interactions, Part II, ancient systems: Sedimentology, v. 36, p. 1037–1051.

Lawver, L.A., Dalziel, I.W., Gahagan, L.M., Martin, K.M., and Campbell, D., 2002, Plates 2002: Atlas of Plate Reconstructions (750 Ma to present day): University of Texas Institute for Geophysics, http://www.ig.utexas.edu/research/projects/plates/plates.htm.

Lockley, M.G., 1991, Tracking dinosaurs, a new look at an ancient world: New York, Cambridge University Press, 238 p.

Lockley, M.G., and Hunt, A.P., 1994, Review of Mesozoic vertebrate ichnofaunas of the western interior United States: Evidence and implications of a superior track record, *in* Caputo, M.V., Peterson, J.A. and Francyzk, K.J., eds., Mesozoic Systems of the Rocky Mountain Region, USA: Special Publication of the Rocky Mountain Section of the Society for Sedimentary Geology (SEPM), p. 95–108.

Lockley, M.G., Hunt, A.P., Meyer, C., Rainforth, E.C., and Schultz, R.J., 1998, A survey of fossil footprint sites at Glen Canyon National Recreation Area (Western USA): A case study in documentation of trace fossils resources at a national preserve: Ichnos, v. 5, p. 177–211.

Loope, D.B., 1979, Fossil wood and probable root casts in the Navajo Sandstone: Geological Society of America Abstracts with Programs, v. 11, no. 6, p. 278.

Loope, D.B., 1985, Episodic deposition and preservation of eolian sands: a Late Paleozoic example from southeastern Utah: Geology, v. 13, p. 73–76.

Loope, D.B., 1988, Rhizoliths in ancient eolianites, *in* Kocurek, G., ed., Late Paleozoic and Mesozoic eolian deposits of the Western Interior of the United States: Sedimentary Geology, v. 56, no. 1–4, p. 301–314.

Loope, D.B., 2004, Origin, disruption, and reconstitution of pin stripes: Lower Jurassic Navajo Sandstone: Geological Society of America Abstracts with Programs, v. 36, no. 5 (in press).

Loope, D.B., Swinehart, J.B., and Mason, J.P., 1995, Dune-dammed paleovalleys of the Nebraska Sand Hills: Intrinsic versus climatic controls on the accumulation of lake and marsh sediments: Geological Society of America Bulletin, v. 107, p. 396–406.

Loope, D.B., and Rowe, C.M., 2003, Long-lived pluvial episodes during deposition of the Navajo Sandstone: Journal of Geology, v. 111, p. 223–232, doi: 10.1086/345843.

Loope, D.B., Rowe, C.M., and Joeckel, R.M., 2001, Annual monsoon rains recorded by Jurassic dunes: Nature, v. 412, p. 64–66, doi: 10.1038/35083554.

Loope, D.B., Steiner, M.B., Rowe, C.M., and Lancaster, N., 2004, Tropical westerlies over Pangaean sand seas: Sedimentology, v. 51, p. 315–322.

Marzolf, J.E., 1983, Changing wind and hydrologic regimes during deposition of the Navajo and Aztec Sandstones, Jurassic (?), southwestern United States, *in* Brookfield, M.E. and Ahlbrandt, T.S., eds., Eolian sediments and processes: Amsterdam, Elsevier, Developments in Sedimentology, v. 38, p. 635–660.

Marzolf, J.E., 1988, Controls on late Paleozoic and early Mesozoic eolian deposition of the western United States, *in* Kocurek, G., ed., Late Paleozoic and Mesozoic eolian deposits of the Western Interior of the United States: Sedimentary Geology, v. 56, p. 167–191.

Mason, J.P., Swinehart, J.B., and Loope, D.B., 1997, Holocene history of lacustrine and marsh sediments in a dune-blocked drainage, Southwestern Nebraska Sand Hills, U.S.A.: Journal of Paleolimnology, v. 17, p. 67–83.

McDonald, R.R., and Anderson, R.S., 1995, Experimental verification of an eolian saltation and leeside deposition models: Sedimentology, v. 42, p. 39–56.

Moore, L.S., and Burne, R.V., 1994, The modern thrombolites of Lake Clifton, Western Australia, *in* Sarfati, J.B., and Monty, C., eds., Phanerozoic Stromatolites II: Dordrecht, Kluwer, p. 3–29.

Nickling, W.G., McKenna Neuman, C., and Lancaster, N., 2002, Grainfall processes in the lee of transverse dunes, Silver Peak, Nevada: Sedimentology, v. 49, p. 191–209, doi: 10.1046/J.1365-3091.2002.00443.X.

Opdyke, N.D., and Runcorn, S.K., 1960, Wind direction in the western United States in the Late Paleozoic: Geological Society of America Bulletin, v. 71, p. 959–971.

Parrish, J.T., and Peterson, F., 1988, Wind directions predicted from global circulation models and wind directions determined from eolian sandstones of the western United States—A comparison: Sedimentary Geology, v. 56, p. 261–282, doi: 10.1016/0037-0738(88)90056-5.

Peterson, F., 1988, Pennsylvanian to Jurassic eolian transportation systems in the western United States: Sedimentary Geology, v. 56, p. 207–260, doi: 10.1016/0037-0738(88)90055-3.

Peterson, F., 1994, Sand dunes, sabkhas, streams and shallow seas: Jurassic paleogeography in the southern part of the Western Interior Basin, *in* Caputo, M.V., Peterson, J.A. and Francyzk, K.J., eds., Mesozoic Systems of the Rocky Mountain Region, USA: Special Publication of the Rocky Mountain Section of the Society for Sedimentary Geology (SEPM), p. 233–270.

Peterson, F., and Pipiringos, G.N., 1979, Stratigraphic relations of the Navajo Sandstone to Middle Jurassic formations, southern Utah and northern Arizona: U.S. Geological Survey Professional Paper 1035-B, 43 p.

Picard, M.D., 1977, Stratigraphic analysis of the Navajo Sandstone: a discussion: Journal of Sedimentary Petrology, v. 47, p. 475–483.

Poole, F.G., 1963, Palaeowinds in the western United States, *in* Nairn, A.E.M., ed., Problems in Palaeoclimatology: London, Interscience Publishers, p. 394–405.

Rahl, J.M., Reiners, P.W., Campbell, I.H., Nicolescu, S., and Allen, C.M., 2003, Combined single-grain (U-Th)/He and U/Pb dating of detrital zircons from the Navajo Sandstone, Utah: Geology, v. 31, p. 761–764, doi: 10.1130/G19653.1.

Rasmussen, K.A., Macintyre, I.G., and Prufert, L., 1993, Modern stromatolite reefs fringing a brackish coastline, Chetumal Bay, Belize: Geology, v. 21, p. 199–202, doi: 10.1130/0091-7613(1993)0212.3.CO;2.

Runcorn, S.K., 1963, Palaeowind directions and palaeomagnetic latitudes, *in* Nairn, A.E.M., ed., Problems in Palaeoclimatology: London, Interscience Publishers, p. 409–419.

Scotese, C.R., 2004, Paleomap Project: http://www.scotese.com/jurassic.htm.

Stokes, W.L., 1991, Petrified mini-forests of the Navajo Sandstone, east-central Utah: Survey Notes, Utah Geological and Mineral Survey, v. 25, no. 1, p. 14–19.

Svendsen, J., Stollhofen, H., Krapf, C.B.E., and Stanistreet, I.G., 2003, Mass and hyperconcentrated flow deposits record dune damming and catastrophic breakthrough of ephemeral rivers, Skeleton Coast Erg, Namibia: Sedimentary Geology, v. 160, p. 7–31, doi: 10.1016/S0037-0738(02)00334-2.

Verlander, J.E., 1995, Basin scale stratigraphy of the Navajo Sandstone, southern Utah, USA [Ph.D. thesis]: University of Oxford, 159 p.

Webster, P.J., 1987, The elementary monsoon, *in* Fein, J.S. and Stephens, P.L., eds., Monsoons: New York, John Wiley & Sons, p. 3–32.

Winkler, D.A., Jacobs, L.L., Congleton, J.D., and Downs, W.R., 1991, Life in a sand sea: Biota from Jurassic interdunes: Geology, v. 19, p. 889–892, doi: 10.1130/0091-7613(1991)0192.3.CO;2.

Printed in the USA

Geological Society of America
Field Guide 5
2004

Strike-slip tectonics and thermochronology of northern New Mexico: A field guide to critical exposures in the southern Sangre de Cristo Mountains

Eric A. Erslev
Seth D. Fankhauser
Department of Geosciences, Colorado State University, Fort Collins, Colorado 80523, USA

Matthew T. Heizler
Robert E. Sanders
Steven M. Cather
New Mexico Bureau of Geology and Mineral Resources, New Mexico Tech, 801 Leroy Place, Socorro, New Mexico 87801, USA

ABSTRACT

The history of fault initiation and reactivation in the southern Rocky Mountains remains highly debated, as does the region's exhumation history. Nowhere has the evidence been more contested than in the southern Sangre de Cristo Mountains, where major, 30+ km dextral separations of basement rocks and their aeromagnetic anomalies have been attributed to Proterozoic, Ancestral Rocky Mountain and Laramide orogenies. Since the sum of these dextral separations is in the range of 100 km, unambiguous determination of the age(s) of faulting would have major implications to Rocky Mountain tectonics. Likewise, the history of exhumation and stabilization of the western North American craton provides an important example of continental lithospheric evolution.

This field trip will start by visiting excellent exposures of spectacularly brecciated yet indurated basement rocks and flanking Paleozoic sedimentary rocks along the Picuris-Pecos fault system, which has 37 km of dextral separation of Proterozoic contacts. Hypotheses for the age(s) of slip will be examined in light of stratigraphic and fault relationships, thin section petrography and isotopic analyses. The region's history of fault reactivation and associated K-metasomatism will be discussed by combining thermochronology, largely based on new $^{40}Ar/^{39}Ar$ K-feldspar analyses, with recent seismic data across the Laramide front of the Sangre de Cristo Mountains. The regional tectonic implications of new geologic mapping, fault analyses, $^{40}Ar/^{39}Ar$ thermochronology and seismic studies will be discussed on the outcrop, with a full examination of all hypotheses.

Keywords: strike-slip faulting, New Mexico, thermochronology, Laramide, Picuris-Pecos, reactivation.

Erslev, E.A., Fankhauser, S.D., Heizler, M., Sanders, R., and Cather, S.M., 2004, Strike-slip tectonics and thermochronology of northern New Mexico: A field guide to critical exposures in the southern Sangre de Cristo Mountains, *in* Nelson, E.P. and Erslev, E.A., eds., Field trips in the southern Rocky Mountains, USA: Geological Society of America Field Guide 5, p. 15–40. For permission to copy, contact editing@geosociety.org. © 2004 Geological Society of America.

INTRODUCTION

The processes of craton stabilization and reactivation are crucial to our understanding of continental lithospheres. The Rocky Mountain province of the western United States provides an important example of lithospheric processes because excellent exposures, coupled with extensive academic research and voluminous subsurface data from hydrocarbon exploration, have made the Rockies one of the best documented examples of craton development and subsequent deformation. But despite a huge volume of data and analyses on the Rockies, many basic questions remain about its tectonic history.

This field trip will examine the geologic and isotopic relationships in the southern Sangre de Cristo Mountains (Fig. 1).

This is a key area in the debates about cratonic stabilization and fault reactivation because the Proterozoic basement has been reactivated by later Proterozoic, Pennsylvanian Ancestral Rocky Mountain, Cretaceous to Eocene Laramide, and Neogene Rio Grande Rift deformation. The area also exposes a 37-km dextral separation of Proterozoic lithologic belts (Montgomery, 1963) on the NNE-striking Picuris-Pecos fault system, the largest, unambiguous strike-slip fault in the southern Rocky Mountains.

New field and petrographic observations along the Picuris-Pecos fault system (Fankhauser and Erslev, 2004) resulted in widely differing interpretations that prompted a storm of debate and additional work. Ongoing mapping and petrographic analyses by S.D. Fankhauser, as part of his ongoing M.S. thesis research, and E.A. Erslev, Fankhauser's advisor, has stimulated

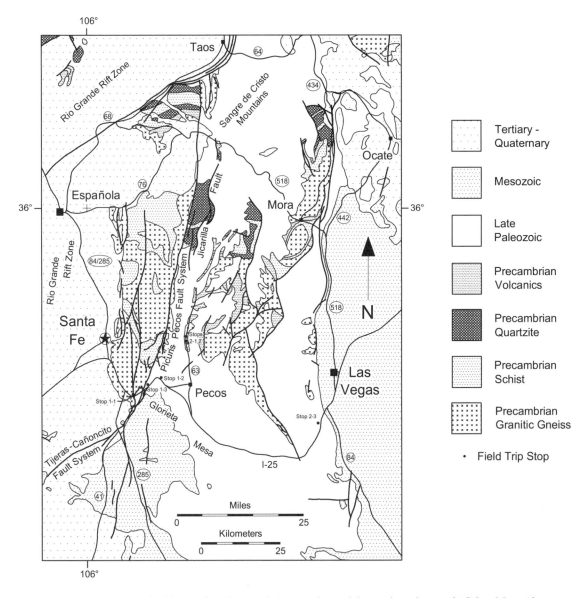

Figure 1. Regional geologic map of north-central New Mexico and the southern Sangre de Cristo Mountains.

independent and simultaneous remapping by S.M. Cather and A.S. Read as well as independent petrographic analyses, detailed stratigraphic age and $^{40}Ar/^{39}Ar$ age determinations, and the initiation of paleomagnetic, fluid inclusion, and fission-track age analyses. This field trip will visit these contested exposures at Deer Creek canyon and explore the history of Rocky Mountain exhumation and metasomatism revealed by the thermochronology of the Proterozoic basement. The inclusion of geoscientists from both sides of this debate as field trip leaders and the rapidly increasing amount of data on the exposures should result in a dynamic field trip. This research-in-progress will undoubtedly have major implications to Rocky Mountain tectonic and resource accumulation models as well as basic insights on thermochronologic methodology as applied to the stabilization of continental lithosphere.

PICURIS-PECOS FAULT SYSTEM AND ROCKY MOUNTAIN STRIKE-SLIP DEBATES

Over the past 25 years, the tectonic significance of north-striking, dextral strike-slip faulting in New Mexico has been one of the most contentious topics in Rocky Mountain geology. The southern Sangre de Cristo Mountains loom large in this debate because they expose the Picuris-Pecos fault system, which has the largest strike-slip separation in the Rockies (Fig. 1). Despite the clear dextral separation of Proterozoic units across this fault system, studies of the timing of strike-slip faulting based on regional stratigraphic relationships (Woodward et al., 1997, 1999; Yin and Ingersoll, 1997; Cather, 1999; Cather and Lucas, 2004a) and structural patterns (Baltz, 1967; Karlstrom and Daniel, 1993; Baltz and Myers, 1999; Erslev, 2001; Fankhauser and Erslev, 2004) have failed to reach a consensus.

The Picuris-Pecos fault system is exposed between Taos and Lamy, New Mexico. South of Taos, the Picuris-Pecos Fault emerges from the eastern edge of the Neogene Rio Grande Rift and crosses the Proterozoic rocks of the Picuris Mountains. The fault is then covered for ~20 km (Bauer and Ralser, 1995) as it crosses the Rio Pueblo, Rio Santa Barbara, and Rio Chiquito valleys before emerging to traverse the eastern flank of the Santa Fe Range, the western spine of the southern Sangre de Cristo Mountains (Fig. 1). South of Taos, north-south strike-slip faults cut Oligocene and Miocene volcanic and clastic rocks (Bauer et al., 1999; Erslev, 2001; McDonald and Nielsen, 2004), recording a clear post-Laramide signature. However, a larger pre-rift offset is clearly indicated by the 37 km dextral separation between E-W–striking belts of alternating metasedimentary and granitic rocks with distinctive, correlative ductile folds of Proterozoic age (Montgomery, 1963). Grambling et al. (1989) determined relatively uniform metamorphic equilibration conditions of 3.5–4.5 kb and 500 °C in Proterozoic rocks from exposures on both sides of the fault, indicating that the net slip on the Picuris-Pecos fault system was dominated by strike slip, not dip slip. Using truncated aeromagnetic anomalies, Cordell and Keller (1984) and Karlstrom and Daniel (1993) extended the zone of

postulated dextral strike-slip to parallel, less well-exposed faults, including one paralleling the eastern margin of the southern Sangre de Cristo Mountains.

Montgomery (1963) and Bauer (1988) reported continuous folding in the metasedimentary rocks of the Picuris Mountains and inferred that the apparent ductile nature of this folding indicated slip during late Proterozoic metamorphic conditions. Along the Picuris-Pecos fault system in the southern end of the Sangre de Cristo Mountains (Fig. 1), domains of Proterozoic granitic and metasedimentary rocks are juxtaposed across the fault or have been mapped as being in fault contact with Paleozoic strata. Montgomery (1963) reported that local zones of indurated breccia form large, north-striking ribs that stand above the local topography and are located up to a kilometer away from fault contacts between Paleozoic and Proterozoic lithologies. At the southern end of Proterozoic exposures, between Glorieta and Cañoncito just northwest of Interstate I-25 (Fig. 2), north-striking zones of brecciated granitic rocks have been mapped by Budding (1972), Booth (1976), and Ilg et al. (1997), who show the contact between Paleozoic sedimentary strata and crystalline rocks as a high-angle fault striking N20E.

The questioning of a Proterozoic age for strike slip on the Picuris-Pecos fault system was initiated by Chapin and Cather (1981), who suggested that 26 km of the dextral separation on the Picuris-Pecos fault system may be Laramide in age. Karlstrom and Daniel (1993) and Daniel et al. (1995) suggested that folding adjacent to the Picuris-Pecos fault system could have been generated by distributed brittle deformation. Karlstrom and Daniel (1993) inferred that the entire 37 km of dextral separation was probably Laramide, in agreement with Chapin and Cather (1981).

Bauer and Ralser (1995) re-emphasized the continuity of apparent dextral drag folding adjacent to the northern Picuris-Pecos fault system and agreed with Montgomery (1963) that this continuity indicates ductile, and thus probably Proterozoic, deformation. In addition, they pointed out evidence for Proterozoic faulting, including minimally offset Phanerozoic strata overlying major basement fault zones that parallel the Picuris-Pecos fault system. In their conclusions, Bauer and Ralser (1995) suggested that the dextral separations on the Picuris-Pecos fault system might have been caused by the combination of ~26 km of transpressive Laramide slip superimposed on ~11 km of Proterozoic slip. In contrast, McDonald and Nielsen (2004) hypothesized that the Picuris-Pecos fault system was segmented by four NW- to NE-trending strike-slip faults during the Laramide orogeny, ruling out latest Laramide motion on the Picuris-Pecos fault system.

The concept of large, Laramide strike-slip motions was applied to the entire Laramide foreland by Chapin and Cather (1981) who proposed a stage of late Laramide, NE-trending shortening with major dextral displacements paralleling the Neogene Rio Grande Rift and a series of narrow, en echelon, axial basins of Eocene age. Chapin and Cather (1981, p. 190) and Cather (2004, p. 239) proposed that northward yielding of the Colorado Plateau may have begun in the early Laramide, as shown by the

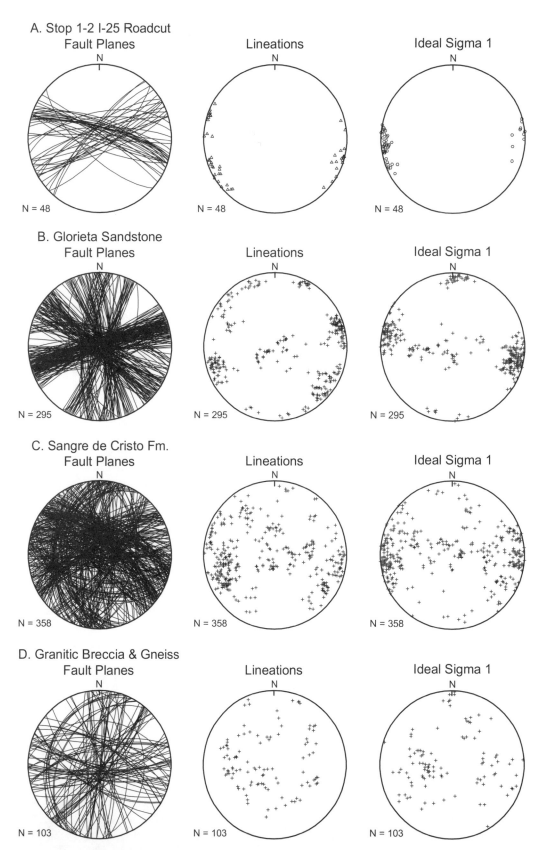

Figure 2. Minor fault analyses from (A) stop 1–2, (B) the Glorieta Sandstone between Cañoncito and Glorieta, (C) the Sangre de Cristo Formation near Deer Creek, and (D) slickensided granitic gneiss and breccia. Ideal sigma one (σ_1) orientations are calculated using a 25° angle between the σ_1 orientation and the slickenline as detailed in Compton (1966) and Erslev (2001).

inception of crustal shortening in the Late Cretaceous north of the plateau. Chapin (1983) proposed that dextral strike-slip on north-striking faults dominated late stage Laramide deformation, citing numerous candidates for strike-slip flower structures and attempting to quantify the amount of slip using separations of Proterozoic rock units and their aeromagnetic signatures across faults like the Picuris-Pecos fault system. Karlstrom and Daniel (1993) and Daniel et al. (1995) summed the dextral separations of Proterozoic rock units and magnetic anomalies and estimated from 50 to 170 km of dextral slip in a zone of spaced strike-slip faults paralleling the Rio Grande Rift. They hypothesized that dextral slip may have connected Laramide extensional collapse in Arizona with foreland shortening in Wyoming. Cather (1999) reexamined these separations, and concluded that they indicated between 85 and 110 km of dextral strike slip.

In contrast, Wawrzyniec et al. (2002) analyzed minor fault populations and paleomagnetic evidence for limited vertical axis rotations near major faults and concluded that the Laramide-age deformation is characterized by a lack of slip partitioning, an interpretation that is consistent with an overall pattern of oblique-slip faults and modest block rotation. As a result, they estimated that larger estimates of dextral Laramide displacement are excessive and they calculated a minimum of 12–19 km of northward displacement of the Colorado Plateau. Larger slip-estimates would be expected to, at some scale, produce much larger rotations (Wawrzyniec, 2004, personal commun.). Cather and Harrison (2002) and Cather (2004) noted that the 85-km lower slip limit of Cather (1999) was based on a process-of-elimination argument that utilized the analysis of Beck and Chapin (1994), who regarded Ancestral Rocky Mountain deformation in New Mexico to be sinistral. Subsequent work, however, has shown areas of significant dextral Pennsylvanian deformation in New Mexico (Baltz and Myers, 1999; Cather, 2000), causing Cather and Harrison (2002) and Cather (2004) to abandon the earlier 85-km minimum dextral strike-slip estimate.

Analyses of separations of Phanerozoic isopach contours have been variously used to suggest 33–110 km (Cather, 1999) and 5–20 km (Woodward et al., 1997) of Laramide dextral strike slip based on Mesozoic isopachs, and ~100 km (Cather and Harrison, 2002) of combined Ancestral Rocky Mountain and Laramide dextral slip in southern New Mexico based on offsets of lower Paleozoic isopachs. Similarly, Laramide sedimentary basin geometries have been used to suggest both substantial (Chapin and Cather, 1981; Chapin, 1983) to negligible (Yin and Ingersoll, 1997) Laramide dextral strike slip. On the basis of their interpretation of continuous Mesozoic isopachs, Woodward et al. (1997) concluded that "major right slip estimated on the basis of metamorphically defined piecing lines in Proterozoic rocks must predate the Laramide orogeny and is probably of Precambrian and/or late Paleozoic age." Woodward et al. (1999) added an assumption that Proterozoic deformation would probably be ductile and hypothesized that the major dextral offsets of basement rocks were generated during the Pennsylvanian Ancestral Rocky Mountain Orogeny. Baltz (1999, p. 163) agreed that the

Ancestral Rocky Mountain Orogeny was caused by northeast-southwest compression that could have generated dextral strike slip on north-striking faults but he concluded that the total Pennsylvanian dextral slip need not exceed 5 km.

Erslev (2001) and Fankhauser and Erslev (2004) measured minor faults in Mesozoic and Paleozoic rocks adjoining the southern Picuris-Pecos fault system to help determine the Phanerozoic slip history of the region (Fig. 2). Three major conjugate fault sets were found in Permian and younger rocks, including an early, presumably Laramide, slip system with east-west–directed thrust faults and ENE-striking dextral and ESE-striking sinistral faults. These are overprinted by sets of generally north-striking normal faults and NNW-striking dextral and NNE-striking sinistral faults. The latter faults also cut Neogene rocks, indicating a post-Laramide origin, perhaps associated with Rio Grande rifting. The fact that no major sets of dextral faults parallel the N- to NNE-striking Picuris Pecos fault system can be used to suggest a lack of major post-Permian strike slip on the fault system (Erslev, 2001) or slip partitioning, where the strike-slip component is partitioned on a weak major fault and minor faults in adjoining strata only represent unresolved fault-perpendicular shortening (Cather, 2004). The question of whether these adjoining minor faults are representative of the overall Phanerozoic slip on the Picuris-Pecos fault system partially prompted the current research in Deer Creek canyon, which exposes a mapped trace of the Picuris-Pecos fault system.

Although there is agreement that the Picuris-Pecos Fault is a major strike-slip structure, it has hosted well-documented Phanerozoic dip separations as well. As first noted by Miller et al. (1963), east-down separation on the Picuris-Pecos Fault formed part of the basin boundary between the Uncompahgre uplift and the Taos Trough during the Pennsylvanian, and was again reactivated in a east-down sense during the Laramide orogeny. Kelley and Chapin (1995) documented Laramide uplift and apatite fission-track cooling (74–44 Ma) at Santa Fe Baldy just west of the Picuris-Pecos Fault. In the Oligocene and Miocene, subsidence and sedimentation occurred along part of the Picuris-Pecos fault system south of Taos (Bauer et al., 1999; McDonald and Nielsen, 2004). Extension along the Rio Grande Rift reactivated parts of the fault system in a west-down sense, forming major basin-bounding faults along the east side of the Peñasco embayment and the San Luis Basin (Sangre de Cristo Fault). Smith (2004, p. 351–352) describes sedimentological evidence from the Espanola Basin that suggests sediment dispersal patterns may have been disrupted by west-up deformation along the Picuris-Pecos Fault.

In summary, the larger estimates of Laramide dextral slip in the southern Rocky Mountains (100–170 km by Karlstrom and Daniel, 1993; 85–110 km by Cather, 1999) are partially based on the assumption that offsets of magnetic anomalies are Laramide in age. This interpretation hinges on whether the dextral separation of Proterozoic units across the Picuris-Pecos fault system is Laramide. But the age of this Picuris-Pecos separation is still in question considering that while Chapin and

Cather (1981), Chapin (1983), Karlstrom and Daniel (1993), Daniel et al. (1995), and Cather (1999) attributed the separations mostly to the Laramide deformation, Woodward et al. (1999) regarded them as Pennsylvanian in origin, and Montgomery (1953, 1963), Sutherland (1963), Baltz and Myers (1999, p. 15) and Fankhauser and Erslev (2004) attributed them mostly to Proterozoic deformation. This field trip will visit an excellent exposure of the Picuris-Pecos Fault in Deer Creek canyon to test these hypotheses (Fig. 3). Be forewarned, however, that opinions on previous field trips to this area were often quite split, with some stating that all features could be generated by Laramide deformation and others seeing little evidence of Phanerozoic deformation other than the development of fault-propagation folds and local cross-cutting faults.

THERMOCHRONOLOGY OVERVIEW

^{40}Ar/^{39}Ar thermochronology studies are contributing detailed information about the intensity of 1450–1350 Ma tectonism and subsequent exhumation history (1350–600 Ma) for much of the New Mexico and Colorado region (Heizler et al., 1997, Karlstrom et al., 1997, Marcoline et al., 1999; Shaw et al., 2004). Our record of the overall exhumation history that eventually brought mid-crustal (10–20 km) rocks to the surface and preserved them under the Great Unconformity is almost entirely provided by thermochronologic data. In very important areas where limited exposures of Proterozoic sedimentary rocks occur in the southwestern United States, the basement exhumation path appears variable in time and space. For instance, in central and southern Arizona,

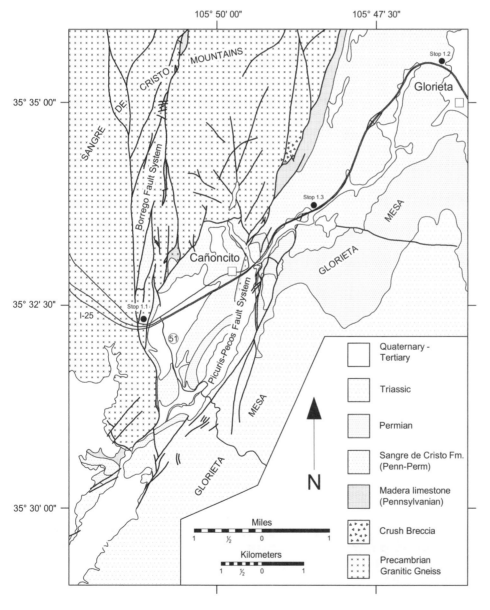

Figure 3. Regional geologic map of the Glorieta 7.5 min geologic quadrangle map (Ilg et al., 1997) and thesis map of field area (Booth, 1976), with modified fault and stratigraphic contacts from Fankhauser and Erslev (2004).

1420 Ma basement is exhumed and overlain by the 1330 Ma Pioneer Shale (Stewart et al., 2001). In the Grand Canyon area, exhumation was more protracted as evidenced by the nonconformity between basement and the 1255 Ma Bass Formation (Timmons et al., 2005). In most of the Rocky Mountain region, including the field trip locations within the southern Sangre de Cristo Mountains, no Proterozoic sedimentary rocks are preserved, and therefore, insights about the billion year lacuna at the basal Paleozoic unconformity with basement requires thermochronological data. As will be explored on this field trip, thermochronology studies are providing a wealth of information about the spatial and temporal variability of the exhumation history, delineating timing of movement on discrete structures, timing of hydrothermal systems and information on the age of brecciation.

Paleoproterozoic rocks of southern Colorado and all of New Mexico almost ubiquitously yield hornblende, muscovite, and biotite ^{40}Ar/^{39}Ar dates that are ca. 1450 Ma or younger (Karlstrom et al., 1997; Shaw et al., 2004). These data indicate regionally high temperatures of ~500 °C during ca. 1400 Ma metamorphism that are independent of the spatial distribution of 1400 Ma plutons. A compilation of hornblende and mica ^{40}Ar/^{39}Ar dates from the southern Sangre de Cristo Mountains is shown on Figure 4A and 4B. Figure 4B is a modified histogram of the age data that is constructed from the cumulative probability of age and errors assigned to each sample (cf. Deino and Potts, 1992). When the Y-axis is assigned a numerical value, the area under each curve will equal 1; however, in this case the relative probabilities for each age population for individual minerals are compiled. Essentially, age results with high uncertainty have a low probability relative to more precise age data. As shown on Figure 4B, hornblende ages are between 1460 and 1370 Ma and are interpreted to indicate either complete argon loss due to reheating or record new hornblende crystal growth during ca. 1400 Ma metamorphism. Mica ages are younger due to lower argon closure temperatures (~300–350 °C for biotite and ~350–400 °C for muscovite) and also are more scattered. Muscovites generally indicate cooling below ~375 °C by ca. 1350 Ma; however, some muscovites (presumably with lower argon retentivities) record ages as young as ca. 1270 Ma. Biotites yield the most variable dates (1410–1090 Ma) and indicate long-term crustal residence at temperatures of ~300 °C until ca. 1100 Ma.

K-feldspar ^{40}Ar/^{39}Ar data can provide a detailed record of the cooling history between ~300 °C and 150 °C. The K-feldspar results can be interpreted in the context of the multiple diffusion domain (MDD) model that views individual samples as having many discrete diffusion length-scales that are related to their complex microstructures (e.g., Lovera et al., 1989). This model can provide a more detailed thermal history analysis compared to the mica and amphibole data in that a record of the cooling history path can be retrieved rather than a single time-temperature point. This modeling technique assumes that argon is transported by thermally activated diffusion and that the same diffusion boundaries that provided pathways for argon transport in nature are those that act in the laboratory. There is significant controversy over the MDD model because factors such as low temperature recrystallization can severely impact the accuracy of model thermal histories (cf. McDougall and Harrison, 1999). Also, the models are not unique, as reheating can cause argon loss that cannot be quantitatively defined. However, in the cases presented here, the point at which accelerated cooling initiates is robust for either a slow-cooling or a reheating model.

In parts of the southern Sangre de Cristo Mountains (Pecos Valley) K-feldspar also occurs as a metasomatic phase or as discrete vein material that can be directly dated by the ^{40}Ar/^{39}Ar method. Figure 4C summarizes the regional thermal histories that are derived from K-feldspar analyses collected across the southern Sangre de Cristo Mountains. These thermal history curves show the mean and 90% confidence windows about solutions that provide model age spectra that match the measured age spectra (see Quidelleur et al., 1997). In this case, the darker central windows represent the mean time and temperature data whereas the lighter surrounding windows represent the 90% confidence levels about the mean values. In general, K-feldspar samples west of the Montezuma Fault record cooling from ~300–150 °C between 1000 and 800 Ma whereas samples in the subsurface east of the fault appear to have cooled significantly later. Many K-feldspars from New Mexico and Colorado record a post–1000 Ma cooling history similar to the samples west of the Montezuma Fault that we suggest represents a period where significant exhumation occurred. Much of the Rocky Mountain region differs from Arizona in that the majority of basement exhumation appears to have initiated late in the 1100 Ma Grenville orogeny rather than prior to 1200 Ma. During the field conference we will explore the geological and thermochronological constraints for the evolution of the southern Sangre de Cristo Mountains. Specifically, we will discuss the timing constraints for the breccias exposed in Deer Creek canyon and Pecos Valley, for the hydrothermal activity in Pecos Valley and for the initiation and reactivation of the Montezuma Fault. All argon data and methods for this paper can be found at http://geoinfo.nmt.edu/labs/argon/DataRep/DataRep.html.

ROAD LOGS AND FIELD TRIP STOP DESCRIPTIONS

These road logs start east of Santa Fe, New Mexico, where the St. Francis entrance ramp (exit 282B) intersects northbound I-25.

Day 1: Picuris-Pecos Fault System

Go east 12 mi (19 km) on I-25N through fractured and altered outcrops of Proterozoic crystalline rocks to exit 294, Cañoncito at Apache Canyon. Take the first left under I-25, and another left on the Old Las Vegas Highway. Go l.5 mi (2.4 km) and park on the right.

STOP 1-1: Overview of the Picuris-Pecos Fault System from the Borrego Fault

In the distance to the east, relatively undeformed Paleozoic and Mesozoic strata are exposed in Glorieta Mesa. A complex zone of faults, folds, and tilted strata defining Glorieta Mesa's western margin marks the mapped continuation (Booth, 1976; Ilg

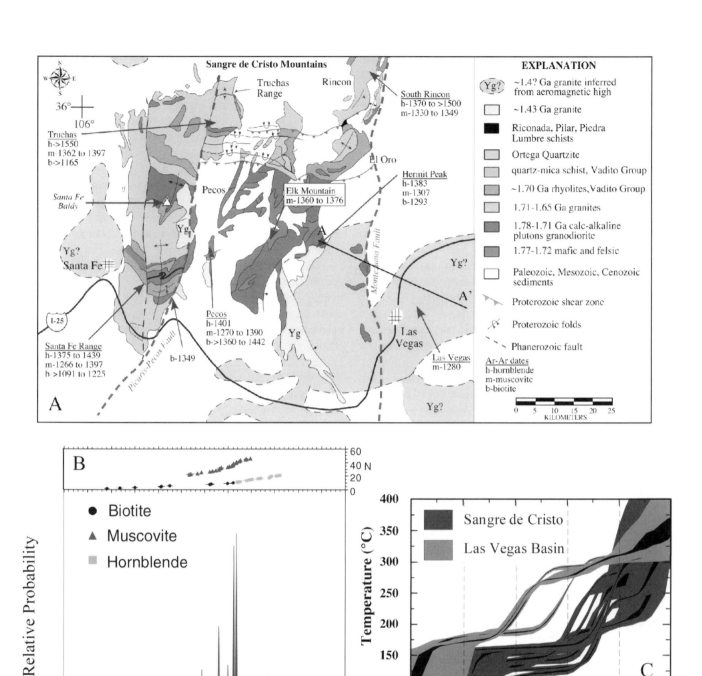

Figure 4. (A) Generalized geological map of the southern Sangre de Cristo Mountains Proterozoic rocks adopted from Karlstrom et al. (2004). (B) Probability diagram of the distribution of hornblende and mica ^{40}Ar/^{39}Ar dates from the region. The age distribution reflects the variable closure temperatures with hornblende recording the oldest apparent ages followed by muscovite and biotite. (C) Compilation of thermal histories derived from ^{40}Ar/^{39}Ar K-feldspar analyses. Although there is some variability, the samples west of the Montezuma Fault are quite similar and indicate initiation of cooling earlier than samples east of the Montezuma Fault.

et al., 1997) of the N- to NNE-trending Picuris-Pecos fault system from basement exposures to the north (Fig. 3). To the south, this zone of faulting in Triassic to Cretaceous units is cut by the undeformed, 26.55 ± 0.30 Ma Galisteo dike (Peters and McIntosh, 1999; Erslev, 2001). Farther south, the zone of deformation is buried by the Miocene fill of the Estancia Basin.

The western end of the road cut at this stop shows the juxtaposition of Proterozoic basement lithologies with clastic rocks of the Pennsylvanian-Permian Sangre de Cristo Formation across the Borrego Fault. The Borrego Fault is here characterized by unconsolidated gouge and breccia, which is typical of many Laramide and Rio Grande Rift fault zones. To the north, however, the Borrego Fault exhibits high-standing ridges of indurated fault breccia (P.W. Bauer, 2004, personal commun.). The Borrego Fault has had a complex Phanerozoic history, showing fault-bounded horses of Paleozoic rocks along its strike (Fig. 3). The basement lithologies mapped by Ilg et al. (1997) and Moench et al. (1988) show both dextral and sinistral separations. Paleozoic inliers along the Borrego Fault are mostly limestone of the Pennsylvanian Madera Formation (P.W. Bauer, 2004, personal commun.). Given the small areal extent of these inliers, it is quite unlikely that limestone (as opposed to clastics) would have been deposited in them had they been actively subsiding during the Pennsylvanian. The structural development of these inliers thus probably occurred during the subsequent Laramide and/or Rio Grande Rift deformations.

Ilg et al. (1997) mapped contacts between the Sangre de Cristo Formation and the basement rocks as either high-angle faults or low-angle, roughly bedding-parallel detachments. Locally, thin zones of Pennsylvanian Madera Formation separate basement and the Sangre de Cristo Formation. Baltz and Myers (1999) considered the Sangre de Cristo Formation to directly overlie the basement locally in this area due to west-side-up displacements on the Picuris-Pecos Fault to the east during the Pennsylvanian Ancestral Rocky Mountain orogeny.

To the south of I-25, the Picuris-Pecos fault system is expressed as an area of complicated faulting and folding (Booth, 1976; Ilg et al., 1997). Baltz and Myers (1999, p. 15) stated that the only place where Laramide strike-slip movement on the extension of the Deer Creek Fault could have linked to the possibly dextral, northeast-striking Tijeras-Cañoncito fault system was covered by alluvium along the Galisteo Creek near Lamy. They discounted this possibility, however, stating the following:

Booth (1976) showed that, in at least one place, steeply dipping beds of Triassic and Jurassic rocks exposed on terrace margins on opposite sides of Galisteo Creek project into each other along strike. Therefore, it is unlikely that a fault with major strike-slip components exists beneath the valley of Galisteo Creek.

New field mapping and reexamination of the maps of Booth (1976) and Ilg et al. (1997) actually suggest a left-lateral separation of 0.4 km of a monoclinal fault-propagation fold across

Galisteo Creek. This separation is compatible with late stage, NNE-striking left-lateral minor faults seen in Neogene and older rocks throughout the region. If these correlations are correct and a strand of the Picuris-Pecos fault system underlies Galisteo Creek, there is no net, post-Permian dextral slip on this strand. Alternatively, there could be Laramide dextral slip on the Picuris-Pecos fault system if (1) the monoclines on each side of Galisteo Creek are not correlative, (2) sinistral slip during the Neogene exceeded the earlier Laramide dextral slip, or (3) the main strands of the Picuris-Pecos fault system are at the base of Glorieta Mesa farther to the east (where Booth [1976] shows faults trending more directly toward Deer Creek) and these faults accommodated most of the Laramide strike slip. Where exposed, however, the highly altered fault zones on the flank of Glorieta Mesa juxtapose Permian and Triassic rocks that lack large numbers of NNE-striking dextral minor faults.

Recent geological mapping by Lisenbee (1999; 2000) has documented that a system of faults strikes southward from the complex intersection of the Picuris-Pecos, Borrego, and Tijeras-Cañoncito fault systems near Lamy until it disappears beneath the Miocene fill of the Estancia Basin ~25 km to the south. Along this southward extension of the Picuris-Pecos fault system, Cather and Lucas (2004a, 2004b) report fault juxtaposition of differing measured sections of Upper Cretaceous Dakota Sandstone that span a normal-fault–dominated, dextral step-over 25 km south of Deer Creek canyon, which they interpret to have required at least several km of dextral slip during the Laramide.

Driving instructions: Return to I-25N, travel northeast 5 miles (8 km) to exit 299, Glorieta/Pecos. Cross over I-25, turn left to merge on to I-25S heading west and go 0.8 mi. (1.3 km) west before parking on the side of I-25S at the crest of the hill.

STOP 1-2: Minor Faulting in Sangre de Cristo Formation Sandstone

One of the tools used to determine the type of major faulting is the study of minor faults. This outcrop shows an excellent conjugate set of strike-slip faults consistent with east-west shortening and compression (Fig. 2A). This set of faults can be seen throughout the area in Paleocene and older rocks. As a result, Erslev (2001) interpreted these as forming during the Laramide orogeny.

These faults can be interpreted as either reflecting the regional slip system, and thus the Laramide slip on the Picuris-Pecos fault system (Erslev, 2001), or as resulting from slip partitioning next to a major strike-slip fault where components of dip slip and strike slip are spatially separated adjacent to a fault (Cather, 2004). Both behaviors exist—many Laramide strike-slip systems are characterized by extensive strike-slip faulting that includes a set of faults that parallel the slip on the major fault. Likewise, however, minor faults and other indicators of horizontal compression associated with many oblique orogens commonly show horizontal shortening and compression perpendicular to major strike-slip faults, such as has been described for parts of the San Andreas fault system (Mount and Suppe, 1987; Tavarnelli, 1998).

Driving instructions: Carefully merge back into I-25S (going west) and travel 3 miles (4.8 km) before pulling off into dirt parking area beside the highway at Deer Creek.

STOP 1-3: Picuris-Pecos Fault System at Deer Creek Canyon

Note: This stop will involve the rest of the day, so bring extra clothes, water, and food.

The stop will consist of a 1 mi (1.5 km) hike through the Permian and Pennsylvanian rocks that abut the Deer Creek Fault, which is the southern extension of the Picuris-Pecos fault system, followed by another 1 mi (1.5 km) of light scrambling to examine the crush breccia and contacts within the immediate vicinity of the Paleozoic-crystalline rock contacts (Fig. 5). The following features will be examined:

Folded Sangre de Cristo and Madera formations. Deer Creek intersects I-25 within the Permian (Wolfcampian) Sangre de Cristo Formation, which consists of interbedded, reddish-brown to purplish mudstones, shales, arkosic sandstones and conglomerates deposited by a fluvial-deltaic system (Baltz and Myers, 1999). Beneath the Sangre de Cristo Formation, carbonate units to the northwest in Deer Creek canyon and east of the Picuris-Pecos fault system were mapped as the Pennsylvanian Madera Formation by Budding (1972), Booth (1976), and Ilg et al. (1997). However, recent work has shown middle Mississippian carbonate rocks to be present within, and possibly above, the breccias of the fault zone itself, as will be discussed later. The dominant Pennsylvanian lithologies consist of alternating layers of gray crystalline limestone, gray fossiliferous limestone, yellow to brown, fine- to coarse-grained arkose and quartz arenite, and gray, green, maroon, and brown mudstone. Some fossiliferous limestone and arkosic beds near the base of the sequence contain small (1–5 mm) angular grains of brick-red microcline that resemble the microcline in the adjoining crystalline rocks.

Bedding dips steepen from 10° to 37° in the Sangre de Cristo Formation as the Picuris-Pecos fault system is approached from the southeast. The bedding dips in the underlying limestone-dominated sequence increase westward until overturned attitudes are attained in the east limb of an oblique, asymmetric anticline with a northerly trend. This fold contains nearly flat-lying and locally west-dipping upright strata in close proximity to the crystalline rocks. Paleozoic strata immediately adjacent to the crystalline rocks exhibit a variety of attitudes, locally with dips as low as 17° to the southeast. Stereonet plots (Fig. 5) of bedding orientations indicate south-plunging fold axes, indicating that deeper structural levels are exposed to the north. This observation is consistent with regional map patterns of Paleozoic rocks and with the general southward plunge of the southern end of the Sangre de Cristo Mountains.

Minor thrust and strike-slip faults in Permian and Mesozoic strata record two stages of horizontal shortening (Fig. 2C). Thrust and strike-slip faults indicate an early, probably Laramide, east-west shortening and are cut by strike-slip and normal faults indicating a later, possibly Rio Grande Rift–related, transtensional deformation characterized by both east-west extension and north-south shortening. A cross section roughly parallel to Deer Creek immediately north of the interstate (Fig. 6) can be interpreted as either a dip-slip fault-propagation fold or a transpressive flower structure.

Indurated crush breccias dominated by granite clasts. The crystalline rocks west of the limestone-dominated sequence consist of fully lithified crush breccias and protocataclasites of granitic gneiss (here abbreviated as breccias) with irregular enclaves of mafic rocks, granitic gneiss, and attenuated quartz veins. These brittlely deformed rocks grade into less cataclasized areas of granitic gneiss to the west. Bright brick-red microcline is the dominant feldspar in the granitic units. The Proterozoic granitic gneisses display high-angle (>68°) foliations with ductile fabrics including microcline augen and S-C fabrics. They are crosscut by breccia zones and faults that generally increase in frequency as the Paleozoic strata are approached.

The indurated granitic breccias and cataclasites follow the trend of the Picuris-Pecos fault system (Ilg et al., 1997) and form rounded, resistant outcrops rising up to 50 m above Deer Creek. The cataclastic zone adjacent to the contact is as wide as 300 m and commonly anastomoses, with subsidiary zones wrapping around domains of less deformed granitic gneiss. Foliation directions in the gneissic clasts are highly variable, with offset clasts commonly showing both sinistral and dextral separations, but no ductile drag on clast margins. Some of these coherent breccia exposures are cross-cut by cm-wide zones of unlithified gouge that appear to have been generated by later brittle deformation. Quartz veins, locally containing large, pegmatitic brick-red microcline, form coherent yet attenuated bands within the breccia. Their brecciated textures, both in outcrop and in thin section, show that the veins have experienced penetrative cataclasis along with the granitic breccias. Mafic enclaves within the breccias locally show a thorough mixing of dark-green mafic rock fragments with clasts of brick-red microcline and quartz.

In thin section, angular granitic breccia clasts with clear ductile fabrics are surrounded by cataclasite with no evidence of cementation or mm-scale vugs or vug mineralization. Mineralogically, the granitic breccias are dominated by tartan-twinned microcline, plagioclase, quartz, and chlorite or biotite. The mafic minerals in most breccia samples are dominated by relatively clean books of chlorite in apparent equilibrium with cataclasized microcline and strained quartz. Two thin sections contain slightly deformed sheets and smaller, planar laths of brown biotite showing only minor retrogression to chlorite. This biotite may or may not have been in stable equilibrium during cataclasis. Thin sections of mafic crush breccia show pervasive secondary chlorite overprinting diabasic textures with felted plagioclase laths altered to fine-grained white mica aggregates and clear, probably albitic plagioclase. One thin section from an outcrop several meters from the Paleozoic sedimentary rocks showed younger fractures filled with very fine cataclasite and microcline partially altered to clay minerals.

Fankhauser and Erslev (2004) interpreted the unusual (from a Laramide perspective) induration, void-free nature of the

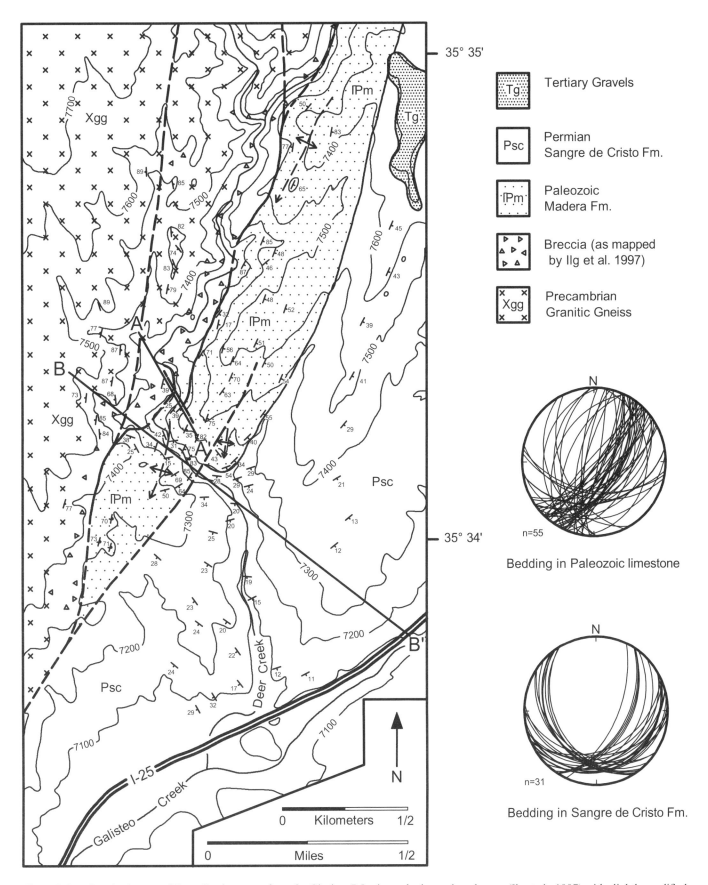

Figure 5. Local geologic map of Deer Creek canyon from the Glorieta 7.5 min geologic quadrangle map (Ilg et al., 1997) with slightly modified contacts. New bedding and foliation orientations are plotted on the map and in the inset stereonets.

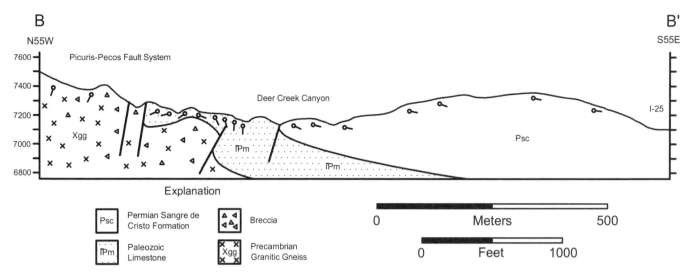

Figure 6. Cross section B–B′ paralleling Deer Creek canyon (see line of section in Fig. 5). Note the more uniformly tilted bedding in the Sangre de Cristo Formation relative to the folded and faulted Pennsylvanian Madera (?) Formation.

brecciated rocks, and possible low-grade metamorphic mineral assemblages of the breccias as indicating higher pressures and temperatures than those experienced by nearby Paleozoic rocks. The only comparable thickness of brecciated rocks attributed to Laramide deformation that we have seen in the Rocky Mountains is in the hanging wall of the Laramide Williams Range Thrust along the western flank of the Front Range in central Colorado (Kellogg, 1999). In contrast to the breccias along Deer Creek, these breccias are not resistant to erosion, typically underlying grassy meadows, and are highly susceptible to Holocene landsliding (Kellogg, 1999). Other Laramide basement fault zones in the Front Range, Teton Range, Owl Creek Mountains, Wind River Range, Bighorn Mountains, Beartooth Mountains, Madison Range, and San Juan Mountains are characterized by extensive alteration of feldspars to clay (Erslev et al., 1988; Yonkee and Mitra, 1993; Erslev and Rogers, 1993). This alteration results in the fault zones forming topographic lows and allows them to be readily excavated with hand tools. While it is possible that Laramide breccias were locally indurated by Tertiary mineralizing fluids, one would expect those same fluids to indurate the immediately adjacent grus and shaly layers at the base of the Paleozoic limestones at Deer Creek canyon.

In contrast, one of the authors of this paper, S.M. Cather, regards the lack of megascopic voids in the breccias to require burial depths of only 1–2 km. He notes that chlorite similar to that in the breccias is also present in Pennsylvanian clastic rocks east of the Picuris-Pecos Fault, and thus its presence does not require metamorphic conditions, but perhaps only normal basinal diagenesis or possible hydrothermal activity along the fault. Additionally, Cather notes that the indurating agent in the breccias (silica, as reported elsewhere along the fault by Miller et al., 1963, p. 16) may have spatial rather than temporal significance. Although breccias are well developed in places on the Picuris-Pecos Fault and other faults throughout the Sangre de Cristo Mountains, these breccias generally are indurated only near the latitude of Santa Fe. North or south of this latitude, with few exceptions, breccias are typically unindurated and form valleys. It is possible that the localization of induration reflects the distribution of fossil hydrothermal systems, which in turn may be related to proximity to Phanerozoic (Tertiary or Cambrian) hypabyssal intrusions. An example may be the late Eocene (35 Ma) volcanic or shallow intrusive rocks recently documented by Melis et al. (2000) in the Pecos River valley, ~20 km northeast of Deer Creek.

Breccia thermochronology in Deer Creek canyon. $^{40}Ar/^{39}Ar$ age spectrum analyses on six K-feldspars and one biotite from Proterozoic clasts within the Deer Creek canyon breccia are shown in Figure 7. The samples are all from within 2 km of the same location and represent coarse-grained quartz–K-feldspar veins or pegmatites (KSP1, 2, 3, 5), a biotite–K-feldspar granitic gneiss (KSP4), and a fairly altered medium grained quartz–K-feldspar clast with significant internal brecciation (KSP6). A biotite separate from KSP4 granite-gneiss yields a somewhat complex spectrum, but has a fairly flat portion that gives a weighted mean age of 1349 ± 4 Ma (Fig. 7D). As compared to the compilation in Figure 4, this age is similar to many biotites from the area and indicates cooling to ~350 °C between 1400 and 1350 Ma.

Like most K-feldspars from the southern Sangre de Cristo Mountains, K-feldspar spectra from Deer Creek canyon reveal very steep initial age gradients with minimum ages between ca. 300–500 Ma climbing to ca. 800–900 Ma over the first few percent of ^{39}Ar released. In general the spectra show a gentler gradient for the remainder of the ^{39}Ar released that are dominated by ages between ca. 900 and 1000 Ma. Several of the spectra record a drop in apparent age between ~30% and 60% ^{39}Ar released, after which ages once again rise. The MDD modeling technique

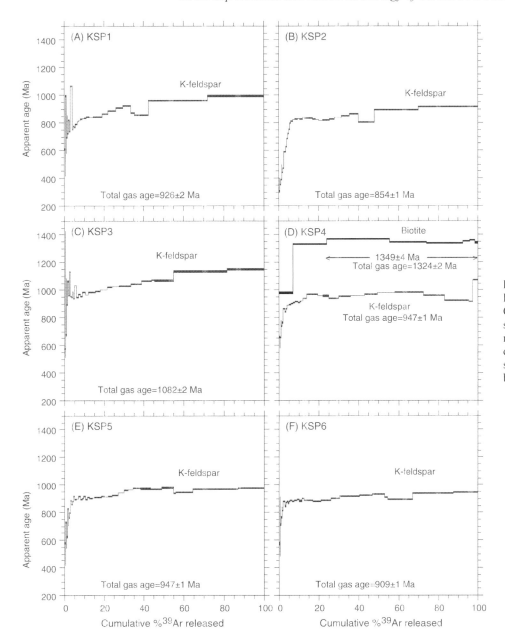

Figure 7. (A–F) ^{40}Ar/^{39}Ar age spectra for K-feldspars from breccia clasts in Deer Creek canyon. K-feldspar spectra are somewhat complex, but not atypical of most spectra from the Southern Sangre de Cristo Mountains. (D) Biotite age spectrum suggests cooling of KSP4 clast below ~350 °C by ca. 1350 Ma.

does not predict this age drop and therefore most of the samples from Deer Creek canyon are not appropriate for modeling.

The initial Paleozoic ages recorded by some samples (KSP2; Fig. 7B) could indicate that the Deer Creek canyon region reached ~150 °C during this time. This initial part of the spectrum could also indicate alteration and associated minor argon loss during residence near the surface or flow of fluids within the fault zone. For instance, several generations of chlorite are observed with one being clearly post Proterozoic and therefore perhaps minor Paleozoic argon loss can be related to fluids that precipitated these chlorites. It is important to stress that the Phanerozoic signal in Deer Creek canyon K-feldspar spectra indicates that the area as a whole has not been significantly heated to greater

than ~150–200 °C since ca. 800–900 Ma. The least complex age spectrum is given by KSP3 (Fig. 7C) and MDD modeling results are shown in Figure 8. This thermal history is like most of the thermal histories from the region (Fig. 4C) and suggests that this clast remained below ~175 °C since 900 Ma.

Although determining the timing of brecciation at Deer Creek canyon is not straightforward, it is important to emphasize that the thermochronology of brecciated samples is similar to most of the results from throughout the southern Sangre de Cristo region. This suggests that either brecciation predated 900 Ma or that brecciation occurred at low enough temperatures so that the feldspars were largely unaffected by the deformation. For low-temperature breccia formation (~150 °C) without the growth of new K-feldspar, the brecciation itself would not be readily recorded by the $^{40}Ar/^{39}Ar$ system since argon loss is primarily a thermally activated process. However, if brecciation occurred at elevated temperatures (e.g., >200 °C), K-feldspar argon data could record a minimum age for the event. Therefore, in the case of Deer Creek canyon, if the primary induration of the breccias occurred at relatively high temperature following brecciation, the K-feldspar thermochronology would support that brecciation occurred at or prior to 900 Ma. Unfortunately, we are uncertain about the temperature of brecciation, annealing or cementation. Well-preserved, pristine chlorite and perhaps secondary biotite in the breccia matrix could support lower greenschist grade conditions for breccia formation. This, coupled with the argon data would support a Proterozoic brecciation history. Alternatively, if the brecciation occurred at relatively low temperatures, the argon thermochronology would not provide a measure of the age of brecciation.

Tabular, lensoidal, and finely infilled carbonate rocks within the granitic crush breccias. Breccia outcrops were examined carefully for Paleozoic rock fragments as a means of determining whether the breccia contains Phanerozoic fault-bounded clasts and thus is Phanerozoic in age. In general, brecciation is much more pervasive in Proterozoic rocks than in adjoining Paleozoic strata, suggesting that pre-Mississippian faulting did occur on the Picuris-Pecos fault system. On both sides of Deer Creek, however, narrow (up to 30 cm), tabular and irregular, subequant carbonate-rich bands and pods were observed in the granitic breccias. Their contacts with the crystalline breccias are sharp; the largest carbonate band, 25 m long and up to a meter wide, strikes NW-SE with a high-angle dip.

Thin sections of the carbonate bands and pods show fragments of crystalline rock, which are commonly composites of Proterozoic gneiss cut by cataclastic bands. These breccia clasts are truncated by recrystallized microspar carbonate similar in texture to that seen in Mississippian dedolomitized carbonate units. Brachiopod and ostracod shell fragments as well as scarce relict dolomite confirm a middle Mississippian age for the material in the pods and bands (D. Ulmer-Scholle, 2004, personal commun.). Regionally, this recrystallization probably occurred in the mid-Mississippian due to the influx of fresh water into a hypersaline carbonate sequence (D. Ulmer-Scholle, 2004, personal commun.). Cross-cutting stylolites, locally brecciated chert and carbonate, and calcite veins suggest varying degrees of post-recrystallization deformation. In the samples that we have thin sectioned, the carbonate pods and bands, while deformed, have not been disjointed to the same degree as the surrounding crystalline breccias and gneisses. However, the degree of brecciation within the Mississippian carbonate beds seems to increase to the east.

Three lenticular limestone pods ~10–20 cm in length are exposed in the trail next to Deer Creek and macroscopically resemble clasts in a fault breccia. Adjoining outcrops of breccia reveal complex networks of thin (<1 cm), commonly anastomosing bands of fine-grained carbonate cut by carbonate veins. A thin section of one of these bands revealed apparent injection fea-

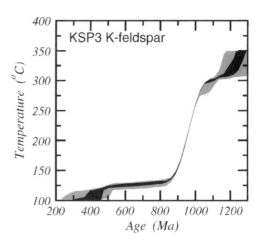

Figure 8. Thermal history derived from multiple diffusion domain modeling of KSP3 K-feldspar. This thermal history is much like those derived from nonbrecciated rocks from throughout the range and shows overall post-Grenville cooling from ~300–150 °C.

tures, including clasts of granitic breccia surrounded by recrystallized microspar limestone, and no evidence of bedding. Clear dedolomite textures strongly suggest a Mississippian age for the carbonate within the microspar limestone pods and thin bands.

The authors of this paper agree that the thin, anastomosing bands represent clastic dikes in preexisting (pre-Mississippian) breccia. The lack of brecciated limestone fragments in the thinnest carbonate bands combined with their mm-scale widths suggest that these carbonate bands were injected as fine-grained sediment into "Neptunian" clastic dikes filled from above. Alternatively, these three pods along the trail can be considered to be fault breccia clasts. This would indicate that brecciation postdates lithification and dedolomitization of the Mississippian limestone, which was probably completed by the beginning of the Pennsylvanian (D. Ulmer-Scholle, 2004, personal commun.). This interpretation suggests that the pods are pieces of brecciated Mississippian bedrock mixed into the granitic breccias during Pennsylvanian and/or Laramide shearing. Fankhauser and Erslev (2004) have advanced the hypothesis that the pods are an attenuated, Mississippian clastic dike much like those in the adjoining outcrops except that faulting associated with stratal folding attenuated and separated the pods during post-brecciation reactivation of the fault zone. This would make the major breccia-forming event at this outcrop pre-Mississippian in age. While the aspect ratio of the lensoidal carbonate pods described above is suggestive of a fault origin, Fankhauser and Erslev (2004) have suggested that the 25 × 1 m aspect ratio of the largest major carbonate band and its NW-SE strike are not consistent with an origin as a breccia clast in the N20E-striking Picuris-Pecos fault system. S.M. Cather has countered that the apparent lack of subhorizontal and meniscate bedding that would be expected within an aggrading, steeply dipping fracture, however, is difficult to reconcile with a fissure-fill origin.

Contacts between Paleozoic limestones and granitic crush breccias. Montgomery (1963) and Sutherland (1963) indicated that the linear trace of the southern Picuris-Pecos fault system was due to the near-vertical dip of the Phanerozoic faults separating the crystalline rocks from the Paleozoic strata. As a result, it may be expected that the linear mapped contact between the crystalline and Paleozoic rocks in Deer Creek canyon is a high-angle fault, and that Paleozoic beds would increase in dip toward the fault, as is typical along fault-related folds (Erslev and Rogers, 1993). This relationship is visible in several areas along the contact on the eastern side of Deer Creek, but in several places, upright carbonate strata overlie crystalline rocks with shallow eastward dips (32°, 39° and 25°). In these areas, Fankhauser and Erslev (2004) reported that the basal limestone is separated from indurated breccias by a zone of shale, weathered grus and/or green-gray altered (?) basement rocks. They hypothesized that these may represent an unconformity, locally modified by slip on this weak interface related post-Mississippian deformation. Other coauthors (Cather, Sanders) consider these contacts to be faults with major east-down stratigraphic separation, based on the local presence of brecciated and sheared Pennsylvanian strata, slickenlines, missing Paleozoic section, and the truncation of limestone beds by the contact at angles incompatible with simple depositional onlap.

Several of these poorly indurated intervals were excavated, showing a range of deformational and perhaps depositional features (Fankhauser and Erslev, 2004). Overlying limestones contain clastic detritus and fossils (including fenestrate bryozoans and crinoids) suggesting a Pennsylvanian age (D. Ulmer-Scholle, 2004, personal commun.). One locality, however, contains relict patches of dolomite that might be indicative of a Mississippian age. Underlying the basal limestone at one locality (Trench T1 in Fankhauser and Erslev, 2004) are black and green shales containing nodules with a chloritic matrix surrounding thoroughly altered fragments of metamorphic rock with aligned, lensoidal quartz grains and no remaining altered microcline. A sharp, moderately discordant and locally faulted contact separates nodule-rich shales and granitic grus, which then grades into indurated breccia. According to Fankhauser and Erslev (2004), the feldspathic grus appears to show excellent cataclastic and brecciated fabrics, indicating that grus formation postdated the brittle deformation responsible for the brecciated rocks.

These younger-over-older contacts on the east side of Deer Creek can be interpreted as fossil nonconformities with minor slip due to regional folding (Fankhauser and Erslev, 2004) or as major faults, perhaps concealing hundreds of meters of Pennsylvanian and Mississippian section. According to S.M. Cather and A.S. Read, ongoing mapping indicates that younger-over-older (normal) faults are not uncommon near Deer Creek. They hypothesize that these normal faults may be related to the presence of a dextral bend in the Picuris-Pecos fault nearby to the south of Deer Creek (Ilg et al., 1997), which would have produced extension during Laramide dextral slip.

On a ridge crest just west of Deer Creek, an isolated keel-like slab of limestone overlies crystalline breccia. The limestone contains relict patches of brown-weathering dolomite, dedolomite textures including pervasive matrix recrystallization to microspar calcite, brachiopod spines, thick-walled ostracod fragments (locally replaced by chert) and possible replacements of evaporite minerals. In combination, these are highly diagnostic of the Mississippian section of the Sangre de Cristo Mountains, most probably the Espiritu Santo Formation or Macho member of the Tererro Formation (Scholle and Ulmer-Scholle, 2003; D. Ulmer-Scholle, 2004, personal commun.). Bedding-parallel stylolites, abundant calcite twins, and calcite veins cut the basal limestone but do not, in the samples that have been thin sectioned, brecciate the rocks into disjointed clasts except along a cross-cutting fault that bounds the edge of the outcrop.

On the southern, eastern, and western sides of this sedimentary inlier, the stylolitic limestone bedding is separated from the underlying breccia by a thin zone of poorly indurated granular material on top of gray-green, feldspar-poor rock. At 40 cm depth below the limestone, traces of pink feldspar are first seen, and by 67 cm depth, the brick-red microcline of the indurated, brecciated rocks is visible. Along the northern side of the outcrop, the bedding is folded into an asymmetrical syncline that is bounded by an unconsolidated gouge zone cutting the underlying crystalline breccia. If the bedding orientation (N15E-39SE) is extended to the southeast across Deer Creek parallel to this bounding fault, it intersects below Pennsylvanian limestone outcrops of the same general attitude on the eastern side of the stream valley (Fig. 9). Unfortunately, the contacts between the Pennsylvanian, Mississippian, and crystalline rocks are not exposed directly across the creek due to alluvial sediments and talus in the canyon. However, to the northeast, across the fault bounding the northeastern side of the keel, massive, medium-grained green-to-red rock underlies a gray carbonate unit with relict patches of dolomite. These relationships were initially interpreted as upper Pennsylvanian sandstone conformably underlying upper Pennsylvanian limestone. Thin sections of rocks collected by E.A. Erslev, however, show little K-feldspar (a commonly used indicator of late Pennsylvanian rocks in the area) in the sandstone and relict patches of dolomite in the overlying limestone. This suggests to Erslev that these units could be Mississippian or early Pennsylvanian in age.

This hypothesis is in contrast to the conclusions of previous mappers (Budding, 1972; Booth, 1976; Ilg et al., 1997) and because of the importance of the lithologic identity and inferred age of the carbonate and clastic rocks exposed east of Deer Creek, the opinion of Dana Ulmer-Scholle and Adam S. Read in early 2004 was sought. Ulmer-Scholle has much experience with Mississippian carbonate diagenesis and stratigraphy in northern New Mexico (Ulmer and Laury, 1984; Ulmer, 1992), and Read has mapped several quadrangles in the region that contain widespread exposures of Upper Paleozoic strata (e.g., Read et al., 1998; Read and Rawling, 2002). In their opinion, the mixed carbonate-clastic sequences in contact with fault breccias east of Deer Creek are Madera Formation. Moreover, based on the presence of interbeds of reddish, oxidized clastic rocks that resemble the transitionally overlying Sangre de Cristo Formation (Perm-

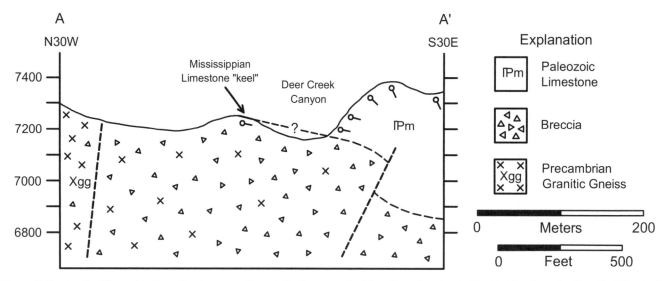

Figure 9. Section A–A′ crossing Deer Creek canyon through the limestone keel (see line of section in Fig. 5). Note how the dip of bedding in the keel can be extrapolated across the canyon to limestone exposures immediately to the southeast.

ian, Wolfcampian), Ulmer-Scholle and Read inferred that the exposures in question are probably the upper Madera Formation (middle or late Pennsylvanian)—the opinion strongly supported by co-author Cather. Anticipated paleontological dating will help delineate between these hypotheses. The implication of this last hypothesis is that the lower and middle part of the Madera Formation and the entire Pennsylvanian Sandia Formation and underlying Mississippian beds lie unexposed in the subsurface east of Deer Creek. If this hypothesis is correct, and the isopach thickness of the Pennsylvanian section at Deer Creek is ~300 m thick (Baltz and Myers, 1999, their Figure 72), then it predicts that Mississippian strata probably lie a few hundred meters in the subsurface to the east of Deer Creek.

Fankhauser and Erslev (2004) agree that a fault is possible along the east side of Deer Creek, but, for the following reasons, still entertain the possibility of an unconformable relationship:

1. The younger-over-older relationship of the carbonate rocks over the crystalline rocks combined with the fact that these contacts are subparallel to bedding indicates that kilometers of slip on a low-angle (relative to bedding) normal fault would be needed to remove the missing Paleozoic section as proposed above.

2. The low amount of deformation in the carbonate rocks overlying the crystalline rocks is more consistent with the deformation associated with a folded unconformity than a major Phanerozoic fault. Laramide faults with kilometer-scale slip are characterized by tens of meters of cataclasized and highly attenuated sedimentary rocks (e.g., Erslev and Rogers, 1993).

3. Adjacent minor faults and folds dominantly indicate lateral shortening, not the extension that would be expected next to a major, younger-over-older fault with substantial normal slip.

4. Currently, our only thin section of the basal limestone on the east side of Deer Creek contains relict dolomitic patches that are partially replaced by calcite. According to D. Ulmer-Scholle

(2004, personal commun.), fauna in the thin section are suggestive, but non-definitive, of a Pennsylvanian age whereas dedolomitization textures indicated by the relict dolomite are suggestive of a Mississippian age.

5. The above estimation of Pennsylvanian-Mississippian section thickness at Deer Creek is an extrapolation from areas outside Deer Creek. As pointed out by Fankhauser and Erslev (2004), 10 km to the NNE along the Picuris-Pecos Fault, Moench et al. (1988) mapped major lateral changes in the Paleozoic rocks nonconformably overlying the Precambrian basement. In this area, undifferentiated Pennsylvanian strata are mapped as overlying Proterozoic basement south of a NNW-striking cross fault. North of this fault, the Pennsylvanian undifferentiated strata wedge out over a distance of one km, at which point the Permian Sangre de Cristo Formation overlies basement. Two more kilometers to the northeast, Mississippian Terrero and Espiritu Santo formations overlie basement, and are themselves overlain by the Permian Sangre de Cristo Formation with no intervening Madera Formation. While the accuracy of this mapping can be questioned, it is clear from Baltz and Myers (1999, their Figure 72) that Paleozoic unit thicknesses vary dramatically over very short distances in this area.

The northwest-striking fault bounding the keel may also provide limits on the age of the brecciation. It parallels a well-mixed tabular body with carbonate clastic material mixed with fragments of breccia. These breccia chips are locally aligned with the margin of the dike, suggesting fault shear during sediment injection may have aligned the fragments. According to Fankhauser and Erslev (2004), across the canyon to the east the contact between the Paleozoic strata and breccia makes a large step, suggesting either a fault or a fold. This observation is disputed by S.M. Cather, who, during recent mapping (Cather, S.M., and Read, A.S., mapping in progress), observed no such offset. All workers agree that the Pennsylvanian limestone on the ridge above this locality shows no fold or fault offset. This suggests that if the keel fault is present east of Deer Creek, as suggested by Fankhauser and Erslev (2004), the fault might be Mississippian or early Pennsylvanian in age, which would match the age of the Mississippian sediment in the sub-parallel carbonate bands that these authors interpret as clastic dikes. If this fault is an Ancestral Rocky Mountain fault analogous to that mapped by Beck and Chapin (1994), and if it crosses the breccia-Paleozoic limestone contact, then it provides a pin line that doesn't allow major Laramide strike slip on this contact. Alternatively, the fault may die out before it crosses Deer Creek, or terminate at the hypothesized strand of the Picuris-Pecos Fault (see above discussion) that, if the contact is a fault, may define the crystalline rock-Pennsylvanian contact just east of Deer Creek.

Driving instructions: Return to Santa Fe on I-25 south.

DAY 2: Geologic Overview, Thermochronology, K-Metasomatism, and Fault Reactivation in the Pecos River Valley and Southeastern Front of the Sangre de Cristo Mountains

Overview

The Pecos River Valley trends north-northeast, on strike with the Picuris-Pecos Fault that is ~5 km to the west. Proterozoic basement rocks beneath the Mississippian-Proterozoic unconformity consist of amphibolite, tonalite, granodiorite and other granitic lithologies that record crustal accretion and subsequent metamorphism and plutonism in northern New Mexico (e.g., Condie, 1982). The oldest rocks (1720 Ma zircon age; Stacey et al., 1976) in the area are metaigneous and metasediments that comprise the Jones metavolcanic suite (Fig. 10). Other rock units are the Windy Bridge Tonalite (1718 Ma zircon age; Bowring and Condie, 1982), Indian Creek Granite (1650 Ma; zircon age, Bowring and Condie, 1982), Macho Creek Granite (1480 Ma zircon age, Bowring and Condie, 1982) and the Pecos Granodiorite (age unknown).

$^{40}Ar/^{39}Ar$ dates for hornblende and mica indicate this area was subject to high temperatures (>500 °C) during the ca 1400 Ma thermotectonic events that affected much of the southwestern United States (Melis, 2001). As mentioned, $^{40}Ar/^{39}Ar$ K-feldspar age spectra and MDD models for rocks throughout the Sangre de Cristo Mountains support accelerated basement cooling from 300 to 150 °C between 1000 and 800 Ma, but in the Pecos Valley, K-feldspar yield more diversity and complex age spectra reflecting both Neoproterozoic exhumation and punctuated K-metasomatic alteration.

Zones of brecciated basement rock are present throughout the Pecos Valley. Breccia textures vary from clast-supported and vuggy with silica cement to matrix-supported, intensely metasomatized outcrops. K-metasomatism has affected most Proterozoic rocks exposed in the Pecos River valley indiscriminate of host rock lithology. Amphibolite near Dalton Canyon, tonalite at Windy Bridge, and tonalite and granodiorite east of Brush Ranch

are pervasively altered (Fig. 10). K-feldspar and epidote veinlets with K-feldspar+hematite selvages crosscut the basement, and primary plutonic sodic plagioclase has been selectively replaced by secondary K-feldspar plus epidote. Zones of intense alteration are localized in breccia zones in mylonitic Windy Bridge tonalite and appear to trend ~NNE near Brush Ranch. Alteration halos extend away from these breccias for tens to hundreds of meters and grade to less altered rock. Breccias consist of angular to sub-angular fragments of the host tonalite that range in size between 1 and 10 cm in diameter. Lithic fragments are matrix-supported, rotated and lack megascopic open spaces. Both the lithic fragments and supporting matrix have been metasomatized. Mylonites in this area are also altered to K-feldspar, epidote and chlorite and barely preserve their original texture.

The extent of K-metasomatism is difficult to quantify due to limited Proterozoic exposure in the Pecos Valley. Additionally, to the west, the Indian Creek and Macho Creek granites dominate the crystalline rocks and contain large amounts of primary pink microcline, limiting the usefulness of using the occurrence of pink K-feldspar in mapping metasomatism. Basement rocks directly underlying the Proterozoic-Paleozoic contact are sporadically metasomatized and crosscut by epidote veinlets, but typically do not represent the most altered samples in the region. Overlying clastic and carbonate rocks show no signs of non-diagenetic alteration or brecciation.

Driving instructions: From where the St. Francis entrance ramp (exit 282B) intersects northbound I-25, go east on I-25 North 17 mi (27 km) to exit 299 (Glorieta/Pecos). Cross the overpass, turn right onto NM-50, and proceed 5.9 mi (9.4 km) to junction of roads NM 223 and NM 63. Turn left (north) on NM 63 and travel 9.5 mi (15.2 km) to Windy Bridge. Park in the Windy Bridge picnic area on the west side of the road.

STOP 2-1. Windy Bridge, Pecos Valley: K-metasomatism

The Windy Bridge area in the Pecos Valley provides an opportunity to view a variety of K-feldspar occurrences that are hosted by the 1720 Ma Windy Bridge tonalite (Fig. 11A). Here K-feldspar occurs within a coarse-grained K-feldspar/albite pegmatite dike that grades into a finer grained K-feldspar/quartz rock (Fig. 11B and 11C; samples 007d, 007e). Secondary K-feldspar is also disseminated throughout the tonalite (sample 007c) and along discrete fractures (sample 007b) (Fig. 11D and 11E and Fig. 11F and11G, respectively). K-feldspars in samples 007d and 007e appear to be primary phenocryst phases within the dike whereas samples 007c and 007b have K-feldspar replacement of plagioclase. Based on the backscattered electron microprobe image of sample 007b (Fig. 11G) showing oscillatory zoning, this K-feldspar is apparently precipitated from an aqueous fluid.

$^{40}Ar/^{39}Ar$ dating of individual K-feldspar occurrences yields remarkably variable results (Fig. 11H). Samples 007d and 007e are similar and have age spectra with initial ages between ca. 750 and 850 Ma that climb to between ca. 1100 and 1300 Ma (Fig. 11H). Sample 007c has a much flatter age spectrum with ages primarily between 800 and 850 Ma that are similar to the

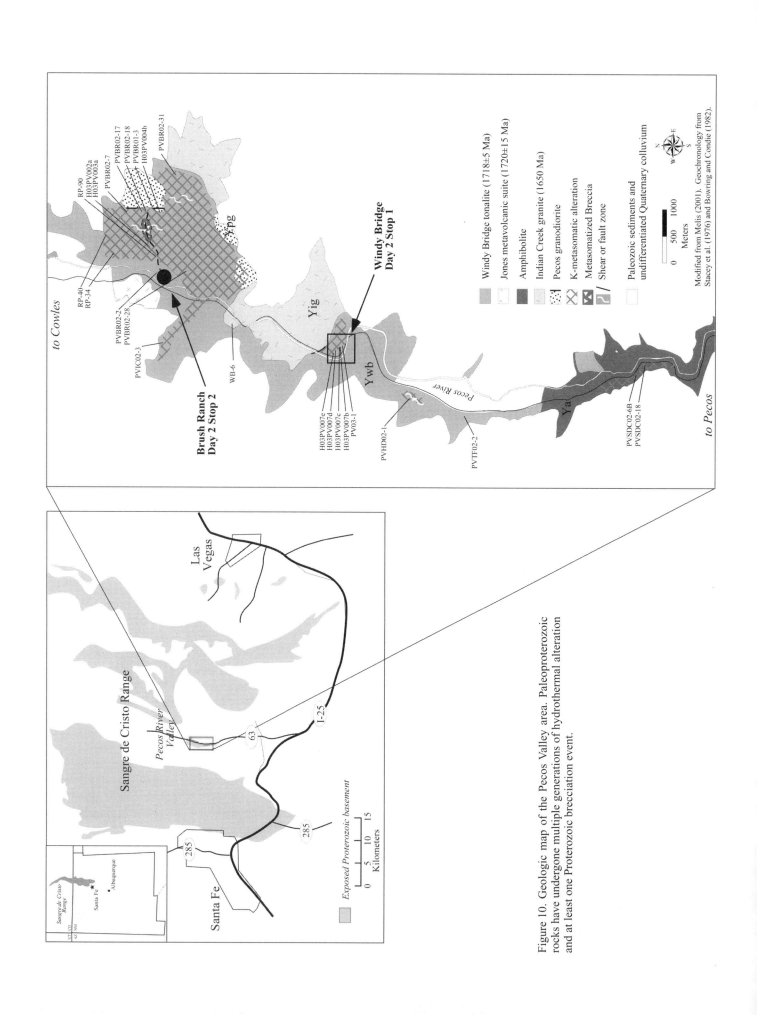

Figure 10. Geologic map of the Pecos Valley area. Paleoproterozoic rocks have undergone multiple generations of hydrothermal alteration and at least one Proterozoic brecciation event.

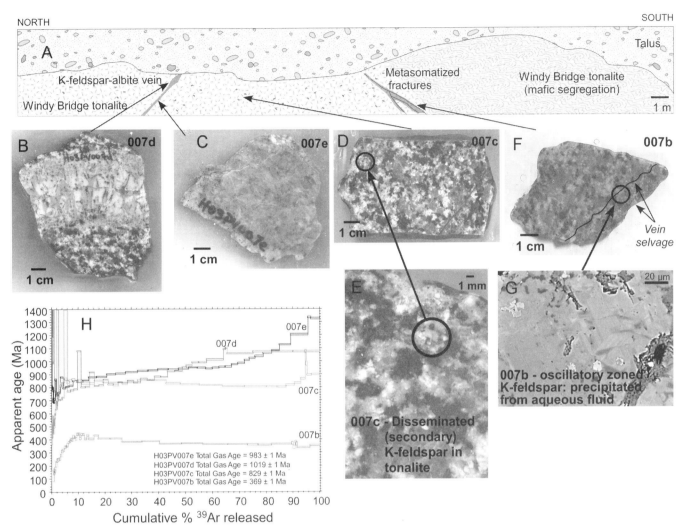

Figure 11. (A) Outcrop map at Windy Bridge location in the Pecos Valley. Windy Bridge tonalite has been intruded by pegmatite (B, C), has undergone disseminated secondary K-feldspar alteration (D, E), and has discrete factures containing secondary K-feldspar (F, G). Multiple K-feldspar occurrences provide the opportunity to delineate the timing of important events within the region. (H) Age spectra for individual K-feldspar occurrences. Samples 007d and 007e are similar to most basement feldspars from the Southern Sangre de Cristo range, whereas 007c appears to record a discrete K-feldspar growth episode at ca. 800–850 Ma. Many secondary K-feldspars within the Pecos Valley record spectra similar to 007c and indicate that much of the K-metasomatism in the region is Proterozoic. Sample 007b is much younger and records a discrete fluid event that occurred between 350 and 400 Ma.

initial parts of the spectra obtained from samples 007d and 007e (Fig. 11H). The fracture-controlled K-feldspar from sample 007b is much younger than the other K-feldspars and yields crystal growth ages between ca. 350 and 400 Ma for the majority of the spectrum. Samples 007c, 007d, and 007e represent much of the range of the K-feldspar results that are observed throughout the southern Sangre de Cristo area. Samples 007d and 007e represent the regionally pervasive cooling patterns that suggest basement temperatures were ~250–300 °C between 1100 and 1000 Ma and significant cooling to ~150 °C occurred by 800–900 Ma (Fig. 4C). The ca. 800–850 Ma ages given by the secondary K-feldspar (007c) is part of a wide occurrence of similar ages recorded by several metasomatic K-feldspars from the Pecos Valley (see

below). Because samples 007d and 007e indicate that the basement was relatively cool by ca. 800 Ma, we can suggest that the ca. 800 Ma ages recorded by sample 007c represent the actual age of the metasomatic event that precipitated this sample. By similar rationale, we suggest that sample 007b K-feldspar grew between 350 and 400 Ma as fluids moved through discrete fractures.

A highly quantitative evaluation of the Windy Bridge K-feldspars is hampered somewhat by age spectrum complexity and difficulty in determining what the local metasomatic effects are on the primary K-feldspars. For instance, the fairly flat parts of the age spectra at ca. 800 Ma of the primary K-feldspar may be the result of argon loss associated with the metasomatic event, rather than accelerated exhumation at this time. Elsewhere in the

Pecos Valley, primary K-feldspars (especially from pegmatites) yield spectra similar to secondary K-feldspars. This latter result would be at odds with our conjecture that the region was already exhumed and cooled to ~150 °C by 800 Ma. Because of certain conflicting data, we are further investigating argon loss from K-feldspars that is driven by recrystallization and/or reprepicitation during metasomatism rather than simply modeling argon transport via a thermally activated diffusion process.

We have dated several additional occurrences of metasomatic K-feldspars in the Pecos Valley and find a variety of Proterozoic and early Phanerozoic ages. Aside from the ca. 800 and 350–400 Ma events there is a well-established secondary K-feldspar growth prior to 1100 Ma, and lesser established events at ca. 600 and 700 Ma. We suggest that the multiple events between ca. 800 and 600 Ma are related to the long and punctuated breakup history of the Rodinia supercontinent. We suggest fluid flow along active rift faults or deep circulation of brines in rift basins may be the primary metasomatic process. As will be further discussed at Stop 2-2, being able to directly date K-feldspar can allow us to bracket the timing of brecciation. Separating the signal recorded by the argon systematics in K-feldspar that is related to the regional exhumation history from that related to metasomatism allows us to strongly argue for a Proterozoic brecciation event.

Driving instructions: Leave Windy Bridge, and head north ~1 mi (1.6 km) to Brush Ranch Camp. Turn right on the drive immediately past the tennis courts. Park in the clearing near Pecos River.

Stop 2-2: Brush Ranch Camp: Brecciation/Metasomatism

The drainages east of Brush Ranch Camp incise the largest area of exposed Proterozoic basement in the Pecos Valley. The Windy Bridge Tonalite transitions eastward to the Pecos

Granodiorite along a non-obvious contact, and the Indian Creek Granite crops out further to the east and to the south toward Windy Bridge where tonalite reappears (Fig. 10). Lithologies exhibit variable degrees of deformation from non-foliated to highly sheared mylonitic textures. Intermittent zones of intensely fractured to brecciated basement and K-metasomatic alteration overprint sheared Windy Bridge tonalite. Breccia is exposed as subtle north-trending ridges up to tens of meters wide that apparently define the structural orientation of these zones.

K-metasomatism is most intense in brecciated zones where the primary foliation of fragmented country rock, and in some instances the breccia texture, is completely obscured by secondary K-feldspar–epidote–quartz replacement. Metasomatism is not restricted to brecciated zones and grades in and out of more and less altered rock eastward where the Pecos granodiorite is pervasively cross-cut by randomly oriented metasomatic K-feldspar–epidote veinlets. Fractures exhibit alteration selvages up to several cm wide that grade to a more disseminated style of plagioclase replacement in the host rock. Primary microcline in granodiorite has remained unaffected by K-metasomatic fluids relative to coexisting plagioclase that has undergone partial to complete replacement by K-feldspar ± epidote ± sericite (Fig. 12A).

$^{40}Ar/^{39}Ar$ dating of both primary microcline and secondary K-feldspar at this location reveals disparate results (Fig. 12B). Primary K-feldspar argon release patterns are like most primary samples from the region whereas the younger results from the secondary K-feldspar record the commonly observed, complex ca. 800–850 Ma spectra given by several analyses within the Pecos Valley. We have not fully investigated if the primary K-feldspars' argon concentration profiles have been disturbed by the fluids that formed the secondary K-feldspar. The epidote+K-feldspar mineral assemblage could record hydrothermal temperatures of

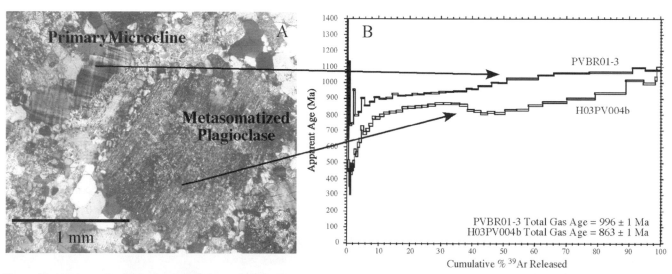

Figure 12. (A) Example of selective replacement of plagioclase without apparent alteration of primary microcline. (B) Age spectra of secondary and primary K-feldspars. Sample PVBR01-3 has an age spectrum that is similar to many primary K-feldspars from the region, whereas secondary K-feldspar from H03PV004 is significantly younger and more similar to many metasomatic K-feldspars from the Pecos Valley.

at least 200 °C and therefore the primary K-feldspar could be significantly degassed during this later event.

In addition to $^{40}Ar/^{39}Ar$ analyses of metasomatic K-feldspar, we have conducted limited U-Pb geochronology on hydrothermal epidote. A sheared, metasomatized tonalite sample records two distinct episodes of epidote growth at 1197 ± 130 and 720 ± 80 Ma, the later of which correlates well with the K-feldspar data. Both the spatial association of brecciation and metasomatism, and the textural overprint of metasomatism on breccias in the Brush Ranch area, convincingly demonstrate that metasomatism was coeval with, or possibly postdated, brittle deformation. Therefore the $^{40}Ar/^{39}Ar$ results of the secondary K-feldspar constrain the timing of at least one brecciation event to be early Neoproterozoic.

Driving instructions: Return south through Pecos on NM 63 to I-25 exit 307 (Pecos). From this exit, travel east 28 mi (45 km) on I-25N to exit 335 (Tecolote). Go under I-25, and take the frontage road east toward Las Vegas. After 4 mi (6.4 km) on the frontage road, park beside the large outcrop.

Stop 2-3: Monoclinal Exposure of Glorieta Sandstone and Thermochronology Evidence of Fault Reactivation

The eastern front of Laramide arches in the southern Rocky Mountains of Colorado and northern New Mexico is linear overall, trending N-S. This deformation front is dominated by NNW-trending thrust faults and folds that step right to the north, forming a sawtooth pattern of uplifted units. These structural steps range in size from 50 km between the en echelon Sangre de Cristo, Wet Mountain, and Front Range arches to 10 km along the margin of the northeastern Front Range. South of the Sangre de Cristo arch, the Laramide deformation front makes a major step to the west around the southern plunge of the Sangre de Cristo Mountains toward the Sandia Mountains outside of Albuquerque, where Laramide structures are extensively overprinted by Neogene Rio Grande Rift deformation. This field trip stop is on the southeasternmost part of the Sangre de Cristo Mountains where the Laramide front makes this arcuate step toward the west (Fig. 1). At this locality, the Permian Glorieta Sandstone is folded within the asymmetrical Creston Anticline, a major monoclinal structure with at least 3 km of throw that extends to the outskirts of Las Vegas, New Mexico.

Baltz and Myers (1999) compiled geologic mapping and stratigraphic evidence in the area, showing a complicated Phanerozoic history of Ancestral Rocky Mountain and Laramide shortening overprinted by post-Laramide extension. Within this context, the diverse arrays of Laramide faults and folds along the southeastern corner of the Sangre De Cristo arch are not surprising (Fig. 13). Fold axis trends and faults show similar trimodal distributions (Fig. 13B), with major NNW-SSE modes and lesser N-S and NNE-SSW modes, consistent with mechanisms of fault-related folding. Along the NNE-trending fault-related fold structures in the area (Fig. 13A), Baltz and Myers (1999) mapped smaller, oblique NW-striking faults with left-lateral separations offsetting the traces of the main faults and anticlines. Such oblique faults are rare along NNW-trending fold and fault systems. Analogous cross-cutting

A. Simplified Geologic Map of the Southeastern Sangre de Cristo Mountains

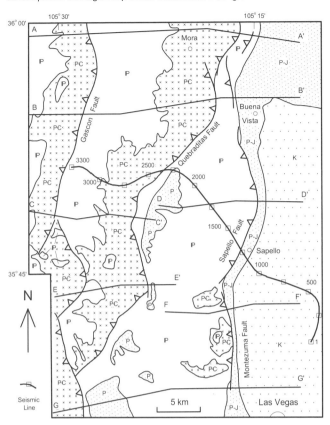

B. Rose Diagrams of Fault and Fold Trends from Baltz and Myers (1999)

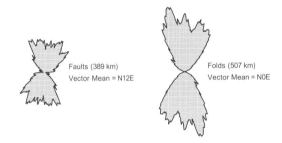

Figure 13. (A) Simplified geologic map of the eastern flank of the Sangre de Cristo arch near Mora, New Mexico after Baltz and Myers (1999). The heavy solid line shows the seismic acquisition line and the adjacent line segments show cross-sections in Figure 14. Abbreviations for units: PC—Proterozoic crystalline basement; IP—Pennsylvanian and older sedimentary rocks; (P)—Permian sedimentary rocks; P-J—Permian to Jurassic sedimentary rocks; (K)—Cretaceous sedimentary rocks. (B) Smoothed (10 degree window) rose diagrams of faults and folds mapped by Baltz and Myers (1999).

faults were mapped along parts of the NNE-trending East Kiabab monocline in Utah, which Tindall and Davis (1999) interpreted to be due to dextral-reverse slip on the underlying fault. Thus, we interpret the map pattern in the southeastern Sangre de Cristo arch as highly suggestive of ENE-shortening on NNW-striking thrust faults and oblique slip on NNE-striking faults, which probably have both dextral and thrust components to their slip. The lack of major strike-slip fault zones cutting the continuous Permian and Mesozoic strata that wrap around the southern plunge of the San-gre de Cristo arch suggests that major Laramide strike-slip faults, if they exist, must be farther west.

The structural geometries defined by Baltz and Myers (1999) suggest listric, largely dip-slip Laramide fault-propaga-tion folding and rotational fault-bend folding (Erslev, 1986) over listric faults that shallow at depth (Fig. 14). The fact that back-limb tilts increase from west to east in nearly every hanging wall (Fig. 14A) suggests that the fault curvature increases toward the surface, causing progressively more tilting of the hanging wall as the fault tip is approached.

This structural hypothesis was tested in 1999 by a seismic line shot as part of the CD-ROM project (Karlstrom et al., 2002; Magnani et al., 2004). Reprints of Magnani et al. (2004) that detail the seismic acquisition, processing and data will be provided to participants and discussed in detail on the field trip. In short, this seismic data showed a remarkable coherence between fault tra-jectories predicted from geometric modeling using 2DMove soft-ware and the position of seismic reflections in the basement that project from the surface location of the major faults (Fig. 14C).

Intriguingly enough, these faults appear to project into a low angle reflector that could be the basal detachment responsible for the formation of the Sangre de Cristo arch as a detachment fold (e.g., Erslev, 1993). Equally interesting is the occurrence

of a subparallel zone of basement reflections in the Las Vegas Basin. Because this reflective zone shows no involvement of the overlying Phanerozoic strata, movement on the zone was probably Proterozoic. If all of these west-dipping zones of reflec-tion originated together in the Proterozoic, then Laramide and Ancestral Rocky Mountain faults probably represent reactivated Precambrian faults. The Proterozoic movement on these faults may have had a strong dextral component considering that the Montezuma Fault parallels an aeromagnetic anomaly discontinu-ity remarkably similar to that along the Picuris-Pecos fault sys-tem (Cordell and Keller, 1984, Karlstrom and Daniel, 1993).

Thermochronology of the Montezuma Fault. Within the Proterozoic Supergroup sediments at the Grand Canyon, it is well documented that Laramide-age monoclines are cored by faults that reactivated faults that initially formed in the Mesopro-terozoic and Neoproterozoic (e.g., Timmons et al., 2001, 2005). Workers such as Marshak et al. (2000) have postulated that indeed many Phanerozoic structures have Proterozoic ancestry, but in most of the Rocky Mountains where no Proterozoic sedi-ments are preserved, there is often no direct way to determine if Phanerozoic faults have a Proterozoic ancestry.

M. Heizler and colleagues have begun using K-feldspar thermochronology across discrete faults systems to see if ther-mal history discontinuities can reveal fault movement histories. The Laramide Montezuma Fault located near Las Vegas, New Mexico, is an excellent situation for testing Proterozoic fault ancestry. Here, as shown in the cross section (Fig. 15A), there is

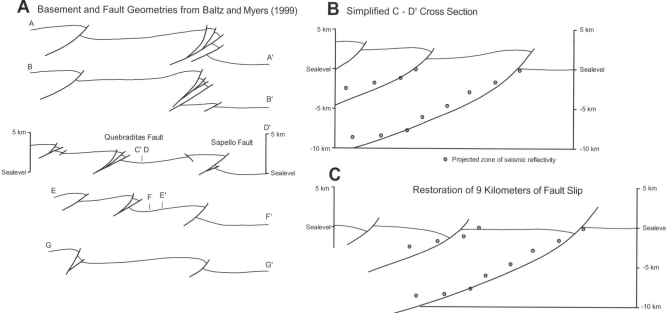

Figure 14. (A) Top of basement and fault geometries from cross sections of Baltz and Myers (1999). Section locations are in Figure 13. Com-posite cross section C–D′ is highlighted since it is closest to the CD-ROM seismic line. (B) Simplified section C–D′ modified to allow optimal restoration by fault-parallel flow and with all basement folding put into the hanging wall. Fault trajectories at depth projected parallel to the faults from the seismic interpretation of Magnani et al. (2004). (C) Restoration of simplified section C–D′ using fault-parallel flow and neglecting shal-low fault-propagation folding in the hanging wall.

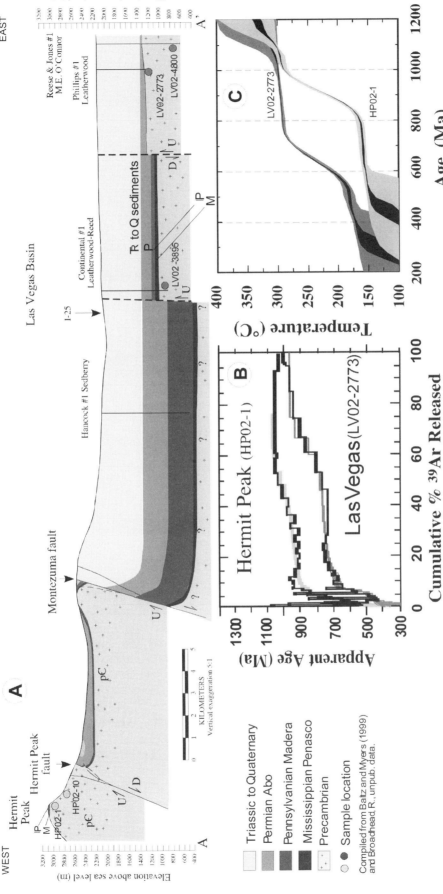

Figure 15. (A) Cross-section A–A' (see Fig. 4A) from Hermit Peak area across Las Vegas Basin. Significant structural relief (~2000 m) of the Mississippian nonconformity is related to Ancestral Rocky and Laramide faults on the Hermit Peak and Montezuma faults. (B) Representative age spectra for a sample at the summit of Hermit Peak (HP-02-1) and the subsurface under the Las Vegas Basin (LV02-2773). (C) MDD modeling results show that HP02-01 records accelerated cooling beginning at ca. 1100 Ma whereas LV02-2773 cooled later, beginning at ca. 750 Ma. This differential cooling records Laramide reactivation of the Proterozoic Montezuma Fault.

a combined ~2000 m of post-Mississippian throw on the Hermit Peak and Montezuma Faults. K-feldspar obtained from basement-penetrating drill holes on the east side of the Montezuma Fault yield distinctly different thermal histories compared to those given by samples on the west side of the fault (Fig. 4C). Two representative samples were singled out from similar structural position (relative to the Paleozoic nonconformity) on opposite sides of the Montezuma Fault (Fig. 15A). Sample HP02-1, from the summit of Hermit Peak, yields an age spectrum that has an age gradient that climbs steeply during the initial heating steps, but mainly records ages between ca. 900 and 1100 Ma (Fig. 15B). In contrast, sample LV02-2773, from the basement on the east side of the fault, is similar in form, but ~200 m.y. younger. Thermal histories derived from these samples show that HP02-1 underwent cooling from ~300 °C to 150 °C between 1000 and 800 Ma, whereas accelerated cooling did not occur on the east side of the Montezuma Fault until between 750 and 600 Ma (Fig. 15C).

There are a variety of geological scenarios that could explain the disparate thermal histories. For instance, these samples could have experienced differential reheating events or could have initiated at different crustal levels. From a regional perspective it is difficult to rationalize discrete heating events that would only affect the rocks on the east side of the Montezuma Fault and therefore we do not believe this to be the primary cause for age differences. We have explored the possibility that these samples were at different crustal levels in the Proterozoic and were brought to similar crustal positions during Ancestral Rocky Mountain faulting. We can reject this idea as being the sole cause for age discontinuity based on the inability of thermal history models to produce model age spectra that conform to the measured data. That is, model spectra and thermal history analyses cannot mimic a situation where the samples remained at different depths relative to each other throughout the Proterozoic and only were brought into juxtaposition during Phanerozoic faulting.

Our preferred explanation of the data distribution requires two periods of Proterozoic differential exhumation (Fig. 16). The thermal histories are recast schematically as exhumation paths and show that at ca. 1000 Ma, samples currently under the Las Vegas Basin and at Hermit Peak were at similar temperatures. At this point, the Hermit Peak samples underwent exhumation that caused cooling relative to samples to the east. This difference is interpreted to record initiation of the Montezuma Fault as a west-side-up thrust related to Grenville contraction. During exhumation of the western block the eastern block remained deep and hot. The cooling observed in the Las Vegas Basin samples that initiated at ca. 750 Ma is interpreted to record reactivation of the Montezuma Fault as an east-side-up extensional fault that brought the samples to nearly their present structural position. This late Neoproterozoic exhumation may have been caused by extension during rifting of the Rodinia supercontinent. The K-feldspar models poorly define the post 600 Ma thermal histories for these samples and there is likely some differential exhumation during the Ancestral Rocky Mountain and Laramide orogenies. For instance, sample LV02-2773 is from the Sierra Grande Uplift

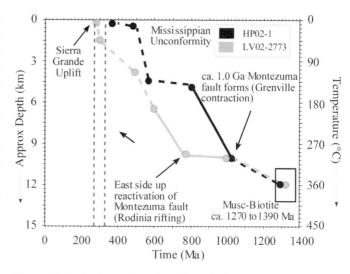

Figure 16. Geological exhumation history inferred from cooling histories derived from HP02-1 and LV02-2773. In order to accommodate the disparate cooling histories, we invoke disparate exhumation histories for these samples. We suggest that the Montezuma Fault was reactivated or initiated during west-side-up, Grenville-aged movement that was inverted during east-side-up, extensional tectonism associated with the rifting of Rodinia. Parts of the exhumation path that are dashed are inferred, whereas solid lines are constrained by the K-feldspar data.

that was active relative to Hermit Peak during Ancestral Rocky Mountain orogenesis (e.g., Baltz and Myers, 1999).

End of field trip—you can retrace your path toward Santa Fe or continue toward Las Vegas, New Mexico, on the frontage road.

ACKNOWLEDGMENTS

We would like to thank the many participants of our field trips to the area for sharing their insights, including Chuck Chapin, John Geissman, Laurel Goodwin, Karl Karlstrom, Shari Kelley, Thomas Neely, Geoff Rawling, Adam Read, Tim Wawrzyniec, and Dana Ulmer-Scholle, who also provided essential assistance with the petrographic identification of carbonate units. This work was partially funded by grants to Seth Fankhauser from the New Mexico Bureau of Geology and Mineral Resources, the American Association of Petroleum Geologists, the Rocky Mountain Association of Geologists, the Colorado Scientific Society, Edward Warner and ChevronTexaco.

REFERENCES CITED

Baltz, E.H., 1967, Stratigraphy and regional tectonic implications of part of Upper Cretaceous and Tertiary rocks, east-central San Juan Basin, New Mexico: U.S. Geological Survey Professional Paper 552, 101 p.

Baltz, E.H., 1999, Speculations and implications for regional interpretation of Ancestral Rocky Mountain paleotectonics, in Baltz, E.H., and Myers, D.A., eds., 1999, Stratigraphic framework of upper Paleozoic rocks, southeastern Sangre de Cristo Mountains, New Mexico, with a section on speculations and implications for regional interpretations of Ancestral Rocky Mountains paleotectonics: New Mexico Bureau of Mines and Mineral Resources Memoir 48, p. 140–170.

Baltz, E.H., and Myers, D.A., 1999, Stratigraphic framework of upper Paleozoic rocks, southeastern Sangre de Cristo Mountains, New Mexico, with a section on speculations and implications for regional interpretations of Ancestral Rocky Mountains paleotectonics: New Mexico Bureau of Mines and Mineral Resources Memoir 48, 269 p.

Bauer, P.W., 1988, Precambrian geology of the Picuris Range, north-central New Mexico: New Mexico Bureau of Mines and Mineral Resources Open-File Report OF-325, 280 p.

Bauer, P.W., and Ralser, S., 1995, The Picuris-Pecos fault—repeatedly reactivated, from Proterozoic(?) to Neogene: New Mexico Geological Society, 46th Field Conference, Guidebook, p. 111–115.

Bauer, P.W., Johnson, P.S., and Kelson, K.I., 1999, Geology and hydrogeology of the southern Taos Valley, Taos County, New Mexico, Final Technical Report: Santa Fe, New Mexico Office of the State Engineer, 56 p.

Beck, W.C., and Chapin, C.E., 1994, Structural and tectonic evolution of the Joyita Hills, central New Mexico: implications of basement control on Rio Grande rift, in Keller, G., and Cather, S.M., eds., Basins of the Rio Grande rift: structure, stratigraphy and tectonic setting: Geological Society of America Special Paper 291, p. 187–205.

Booth, F.O., III, 1976, Geology of the Galisteo Creek area, Lamy to Cañoncito, Santa Fe County, New Mexico [M.S. Thesis]: Golden, Colorado School of Mines, 122 p.

Bowring, S.A., and Condie, K.C., 1982, U/Pb zircon ages from northern and central New Mexico: Geological Society of America Abstracts with Programs, v. 14, p. 304.

Budding, A.J., 1972, Geology of the Glorieta quadrangle, New Mexico: New Mexico Bureau of Mines and Mineral Resources Geologic Map 24, scale 1:24,000.

Cather, S.M., 1999, Implications of Jurassic, Cretaceous and Proterozoic piercing lines for Laramide oblique-slip faulting in New Mexico and rotation of the Colorado Plateau: Geological Society of America Bulletin, v. 111, p. 849–868, doi: 10.1130/0016-7606(1999)1112.3.CO;2.

Cather, S.M., 2000, Preliminary analysis of the Fresnal fault system—a major dextral normal fault of Late Pennsylvanian age in south-central New Mexico: American Association of Petroleum Geologists Bulletin, v. 84, no. 8, p. 1237.

Cather, S.M., 2004, The Laramide orogeny in central and northern New Mexico and southern Colorado, in Mack, G.H., and Giles, K.A., eds., The Geology of New Mexico, A Geologic History: New Mexico Geological Society Special Publication 11, p. 203–248.

Cather, S.M., and Harrison, R.W., 2002, Tectonic implications of isopach patterns of lower Paleozoic units in southern New Mexico: New Mexico Geological Society Guidebook, v. 53, p. 85–101.

Cather, S.M., and Lucas, S.G., 2004a, Comparative stratigraphy of the Dakota Sandstone across the Picuris-Pecos fault system south of Lamy, New Mexico: Definitive evidence of Laramide strike-slip: New Mexico Geology, v. 26, p. 69.

Cather, S.M., and Lucas, S.G., 2004b, Comparative stratigraphy of the Dakota Sandstone across the Picuris-Pecos fault system south of Lamy, New Mexico: Definitive evidence of Laramide strike-slip: Geological Society of America Abstracts with Programs, v. 36, no. 5 (in press).

Chapin, C.E., 1983, An overview of Laramide wrench faulting in the southern Rocky Mountains with emphasis on petroleum exploration, in Lowell, J.D., ed., Rocky Mountain foreland basins and uplifts: Denver, Rocky Mountain Association of Geologists, p. 169–179.

Chapin, C.E., and Cather, S.M., 1981, Eocene tectonics and sedimentation in the Colorado Plateau–Rocky Mountain area, in Dickinson, W.R., and Payne, M.D., eds., Relations of tectonics to ore deposits in the southern Cordillera: Arizona Geological Society Digest, v. 14, p. 173–198.

Compton, R.R., 1966, Analyses of Pliocene-Pleistocene deformation and stresses in northern Santa Lucia Range, California: Geological Society of America Bulletin, v. 77, p. 1361–1380.

Condie, K.C., 1982, Plate-tectonics model for continental accretion in the southwestern United States: Geology, v. 10, p. 37–42.

Cordell, L., and Keller, G.R., 1984, Regional structural trends inferred from gravity and aeromagnetic data in New Mexico–Colorado border region: New Mexico Geological Society, 35th Field Conference, Guidebook, p. 21–23.

Daniel, C.G., Karlstrom, K.E., Williams, M.L., and Pedrick, J.N., 1995, The reconstruction of a middle Proterozoic orogenic belt in north-central New

Mexico, U.S.A.: New Mexico Geological Society, 46th Field Conference, Guidebook, p. 193–200.

Deino, A., and Potts, R., 1992, Age-probability spectra for examination of single-crystal $^{40}Ar/^{39}Ar$ dating results: Examples from Olorgesailie, southern Kenya Rift: Quaternary International, v. 13-14, p. 47–53, doi: 10.1016/1040-6182(92)90009-Q.

Erslev, E.A., 1986, Basement balancing of Rocky Mountain foreland uplifts: Geology, v. 14, p. 259–262.

Erslev, E.A., 1993, Thrusts, back-thrusts, and detachment of Laramide foreland arches, in Schmidt, C.J., Chase, R., and Erslev, E.A., eds., Laramide basement deformation in the Rocky Mountain foreland of the western United States: Geological Society of America Special Paper 280, p. 339–358.

Erslev, E.A., 2001, Multistage, multidirectional Tertiary shortening and compression in north-central New Mexico: Geological Society of America Bulletin, v. 113, p. 63–74, doi: 10.1130/0016-7606(2001)1132.0.CO;2.

Erslev, E.A., and Rogers, J.L., 1993, Basement-cover geometry of Laramide fault-propagation folds, in Schmidt, C.J., Chase, R., and Erslev, E.A., eds., Laramide basement deformation in the Rocky Mountain foreland of the western United States: Geological Society of America Special Paper 280, p. 125–146.

Erslev, E.A., Rogers, J.L., and Harvey, M., 1988, The northeastern Front Range revisited: Horizontal compression and crustal wedging in a classic locality for vertical tectonics: Field Trip guide for the 1988 Geological Society of America Annual Meeting, p. 141–150.

Fankhauser, S.D., and Erslev, E.A., 2004, Unconformable and cross-cutting relationships indicate major Precambrian Faulting on the Picuris-Pecos fault system, southern Sangre de Cristo Mountains, New Mexico: New Mexico Geological Society, 55th Field Conference, Guidebook, p. 121–133.

Grambling, J.A., Williams, M.L., Smith, R.F., and Mawer, C.K., 1989, The role of crustal extension in the metamorphism of Proterozoic rocks in New Mexico, in Grambling, J.A., and Tewksbury, B.J., eds., Proterozoic geology of the southern Rocky Mountains: Geological Society of America Special Paper 235, p. 87–110.

Heizler, M.T., Ralser, S., and Karlstrom, K.E., 1997, Late Proterozoic (Grenville?) deformation in central New Mexico determined from single crystal muscovite $^{40}Ar/^{39}Ar$ age spectra: Precambrian Research, v. 84, no. 1-2, p. 1–15, doi: 10.1016/S0301-9268(96)00053-8.

Ilg, B., Bauer, P.W., Ralser, S., Rogers, J., and Kelley, S., 1997, Geology of the Glorieta 7.5-min. quadrangle, Santa Fe County, New Mexico: New Mexico Bureau of Mines and Mineral Resources, Open-File Geologic Map OF-GM 11, scale 1:12,000.

Karlstrom, K.E., and Daniel, C.G., 1993, Restoration of Laramide right-lateral strike-slip in northern New Mexico by using Proterozoic piercing points: Tectonic implications from the Proterozoic to the Cenozoic: Geology, v. 21, p. 1139–1142, doi: 10.1130/0091-7613(1993)0212.3.CO;2.

Karlstrom, K.E., Dallmeyer, R.D., and Grambling, J.A., 1997, $^{40}Ar/^{39}Ar$ evidence of 1.4 Ga regional metamorphism in New Mexico: implications for thermal evolution of lithosphere in the southwestern USA: Journal of Geology, v. 105, p. 205–223.

Karlstrom, K.E., Bowring, S.A., Chamberlain, K.R., Dueker, K.G., Eshete, T., Erslev, E.A., Farmer, G.L., Heizler, M., Humphreys, E.D., Johnson, R.A., Keller, G.R., Kelley, S.A., Levander, A., Magnani, M.B., Matzel, J.P., McCoy, A.M., Miller, K.C., Morozova, E.A., Pazzaglia, F.J., Prodehl, C., Rumpel, H.-M., Shaw, C.A., Sheehan, A.F., Shoshitaishvili, E., Smithson, S.B., Snelson, C.M., Stevens, L.M., Tyson, A.R., and Williams, M.L., 2002, Structure and evolution of the lithosphere beneath the Rocky Mountains: Initial results of from the CD-ROM experiment: GSA Today, v. 13, no. 3, p. 4–10, doi: 10.1130/1052-5173(2002)0122.0.CO;2.

Karlstrom, K.E., Amato, J.M., Williams, M.L., Heizler, M., Shaw, C.A., Read, A.S., and Bauer, P.W., 2004, Proterozoic tectonic evolution of the New Mexico region: a synthesis, in Mack, G.H., and Giles, K.A., The geology of New Mexico: A geologic history: New Mexico Geological Society, Special Publication 11, p. 1–34.

Kelley, S.A., and Chapin, C.E., 1995, Apatite fission-track thermochronology of Southern Rocky Mountain-Rio Grande rift-western High Plains provinces: New Mexico Geological Society, 46th Field Conference, Guidebook, p. 87–96.

Kellogg, K.S., 1999, Neogene basins of the northern Rio Grande rift; partitioning and asymmetry inherited from Laramide and older uplifts: Tectonophysics, v. 305, no. 1-3, p. 141–152, doi: 10.1016/S0040-1951(99)00013-X.

Lisenbee, A.L., 1999, Geology of the Galisteo 7.5-min. Quadrangle, Santa Fe County, New Mexico: New Mexico Bureau of Mines and Mineral Resources, Open-File Geologic Map OF-GM 30, scale 1:24,000.

Lisenbee, A.L., 2000, Geology of the Ojo Hedionda 7.5-min. quadrangle, Santa Fe County, New Mexico: New Mexico Bureau of Mines and Mineral Resources, Open-File Geologic Map OF-GM 39, scale 1:24,000.

Lovera, O.M., Richter, F.M., and Harrison, T.M., 1989, ^{40}Ar/^{39}Ar thermochronometry for slowly cooled samples having a distribution of diffusion domain sizes: Journal of Geophysical Research, B, Solid Earth and Planets, v. 94, no. 12, p. 17,917–17,935.

Magnani, M.B., Levander, A., Erslev, E.A., and Bolay-Koenig, N., 2004, Listric thrust faulting in the Laramide front of north-central New Mexico guided by Precambrian basement anisotropies, *in* Karlstrom, K.E., and Keller, G.R., eds., Lithospheric structure and evolution of the Rocky Mountains: American Geophysical Union Monograph (in press).

Marcoline, J.R., Heizler, M.T., Goodwin, L.B., Ralser, S., and Clarke, J., 1999, Thermal, structural and petrological evidence for 1400-Ma metamorphism and deformation in central New Mexico: Rocky Mountain Geology, v. 34, p. 93–119.

Marshak, S., Karlstrom, K., and Timmons, J.M., 2000, Inversion of Proterozoic extensional faults: An explanation for the pattern of Laramide and Ancestral Rockies intracratonic deformation, United States: Geology, v. 28, p. 735–738, doi: 10.1130/0091-7613(2000)0282.3.CO;2.

McDonald, D.W., and Nielsen, K.C., 2004, Structural and stratigraphic development of the Miranda graben constrains the uplift of the Picuris Mountains: New Mexico Geological Society, 55th Field Conference, Guidebook, v. 55, p. 219–229.

McDougall, I., and Harrison, T.M., 1999, Geochronology and thermochronology by the ^{40}Ar/^{39}Ar method: Oxford University Press, New York, New York, 2nd ed., 269 p.

Melis, E.A., 2001, Tectonic history of the Proterozoic basement of the southern Sangre de Cristo Mountains, New Mexico [M.S. thesis]: Socorro, New Mexico Institute of Mining and Technology, 131 p.

Melis, E.A., Harpel, C.J., Kelley, S.A., and Bauer, P.W., 2000, Latest Eocene felsic volcanic rocks from the southern Sangre de Cristo Mountains, New Mexico: New Mexico Geology, v. 22, p. 43.

Miller, J.P., Montgomery, A., and Sutherland, P.K., editors, 1963, Geology of part of the Sangre de Cristo Mountains, New Mexico: New Mexico Bureau of Mines and Mineral Resources Memoir 11, 106 p.

Moench, R.H., Grambling, J.A., and Robertson, J.M., 1988, Geology of the Pecos Wilderness, Santa Fe, San Miguel, Mora, Rio Arriba, and Taos Counties, New Mexico: U.S. Geological Survey Miscellaneous Field Studies Map MF-1921-B, scale 1:48,000.

Montgomery, A., 1953, Precambrian geology of the Picuris Range, north-central New Mexico: New Mexico Institute of Mining and Technology, State Bureau of Mines and Mineral Resources Bulletin 30, 89 p.

Montgomery, A., 1963, Precambrian rocks, *in* Miller, J.P., Montgomery, A., and Sutherland, P.K., eds., 1963, Geology of part of the Sangre de Cristo Mountains, New Mexico: New Mexico Bureau of Mines and Mineral Resources, Memoir 11, p. 7–21.

Mount, V.S., and Suppe, J., 1987, State of stress near the San Andreas fault; implications for wrench tectonics: Geology, v. 15, p. 1143–1146.

Read, A.S., Allen, B.D., Osburn, G.R., Ferguson, C.A., and Chamberlin, R.M., 1998 (*last revised: 14 Feb 2000*), Geology of the Sedillo 7.5-min. quadrangle, Bernalillo County, New Mexico: New Mexico Bureau of Mines and Mineral Resources, Open-file Geologic Map OF-GM 20, scale 1:24,000 (mapped at 1:12,000 and 1:24,000).

Read, A.S., and Rawling, G.C., 2002, Geology of the Pecos 7.5-min. quadrangle, San Miguel and Santa Fe Counties, New Mexico: New Mexico Bureau of Geology and Mineral Resources, Open-File Geologic Map OF-GM 52, scale 1:24,000.

Peters, L., and McIntosh, B., 1999, ^{40}Ar/^{39}Ar geochronology results from the Galisteo and Eagle Rock Dikes: New Mexico Geochronological Research Laboratory Internal Report 20, 7 p.

Quidelleur, X., Grove, M., Lovera, O.M., Harrison, T.M., Yin, A., and Ryerson, F.J., 1997, The thermal evolution and slip history of the Renbu Zedong Thrust, southeastern Tibet: Journal of Geophysical Research, v. 102, p. 2659–2679, doi: 10.1029/96JD02483.

Scholle, P.A., and Ulmer-Scholle, D.S., 2003, A Color Guide to the Petrography of Carbonate Rocks: Grains, textures, porosity, diagenesis: American Association of Petroleum Geologists Memoir 77, 474 p.

Shaw, C.A., Heizler, M.T., and Karlstrom, K.E., 2004, ^{40}Ar/^{39}Ar thermochronologic record of 1.45–1.35 Ga intracontinental tectonism in the southern Rocky Mountains: Interplay of conductive and advective heating with intracontinental deformation, *in* Karlstrom, K.E., and Keller, G.R., eds., Lithospheric structure and evolution of the Rocky Mountains: American Geophysical Union Monograph (in press).

Smith, G.A., 2004, Middle to late Cenozoic development of the Rio Grande rift and adjacent regions in northern New Mexico, *in* Mack, G.H., and Giles, K.A., eds., The geology of New Mexico, a geologic history: New Mexico Geological Society Special Publication 11, p. 331–358.

Stacey, J.S., Doe, B.R., Silver, L.T., and Zartman, R.E., 1976, Plumbotectonics IIA, Precambrian massive sulfide deposits: U.S. Geological Society Open-File report 76-476.

Stewart, J.H., Gehrels, G.E., Barth, A.P., Link, P.K., Christie-Blick, N., and Wrucke, C.T., 2001, Detrital zircon provenance of Mesoproterozoic to Cambrian arenites in the western United States and northwestern Mexico: Geological Society of America Bulletin, v. 113, p. 1343–1356, doi: 10.1130/0016-7606(2001)1132.0.CO;2.

Sutherland, P.K., 1963, Laramide orogeny, *in* Miller, J.P., Montgomery, A., and Sutherland, P.K., eds., Geology of part of the Sangre de Cristo Mountains, New Mexico: New Mexico Bureau of Mines and Mineral Resources Memoir 11, p. 47–49.

Tavarnelli, E., 1998, Tectonic evolution of the southern Salinian block, California, USA: Paleogene to recent shortening in a transform-fault bounded continental fragment, *in* Holdsworth, R.E., Strachan, R.A., and Dewey, J.F., eds., Continental transpressional and transtensional tectonics: Geological Society [London] Special Publication 135, p. 107–118.

Timmons, J.M., Karlstrom, K.E., Heizler, M.T., Bowring, S.A., Gehrels, G.E., and Crossey, L.J., 2005, Tectonic inferences from the ca. 1254–1100 Ma Unkar Group and Nankoweap Formation, Grand Canyon: Intracratonic deformation and basin formation during protracted Grenville orogenesis: Geological Society of America Bulletin (in press).

Timmons, J.M., Karlstrom, K.E., Dehler, C.M., Geissman, J.W., and Heizler, M.T., 2001, Proterozoic multistage (ca. 1.1 and 0.8 Ga) extension recorded in the Grand Canyon Supergroup and establishment of northwest- and north-trending tectonic grains in the Southwestern United States: Geological Society of America Bulletin, v. 113, p. 163–181, doi: 10.1130/0016-7606(2001)1132.0.CO;2.

Tindall, S.E., and Davis, G.H., 1999, Monocline development by oblique-slip fault-propagation folding; the East Kaibab Monocline, Colorado Plateau, Utah: Journal of Structural Geology, v. 21, p. 1303–1320, doi: 10.1016/S0191-8141(99)00089-9.

Ulmer, D.S., 1992, Multistage evaporite replacement within Upper Paleozoic carbonate units from northern New Mexico, Wyoming, and west Texas-New Mexico [unpublished Ph.D. thesis]: Dallas, Southern Methodist University, 204 p.

Ulmer, D.S., and Laury, R.L., 1984, Diagenesis of the Mississippian Arroyo Peñasco Group, north-central New Mexico: New Mexico Geological Society, 35th Field Conference, Guidebook, p. 91–100.

Wawrzyniec, T.F., Geissman, J.W., Melker, M.D., and Hubbard, M., 2002, Dextral shear along the eastern margin of the Colorado Plateau: a kinematic link between Laramide contraction and Rio Grande rifting (ca. 75–13 Ma): Journal of Geology, v. 110, no. 3, p. 305–324, doi: 10.1086/339534.

Woodward, L.A., Anderson, O.J., and Lucas, S.G., 1997, Mesozoic stratigraphic constraints on Laramide right slip on the east side of the Colorado Plateau: Geology, v. 25, p. 843–846, doi: 10.1130/0091-7613(1997)0252.3.CO;2.

Woodward, L.A., Anderson, O.J., and Lucas, S.G., 1999, Late Paleozoic right-slip faults in the Ancestral Rocky Mountains: New Mexico Geological Society, 50th Field Conference, Guidebook, p. 149–153.

Yin, A., and Ingersoll, R.V., 1997, A model for evolution of Laramide axial basins in the Southern Rocky Mountains, U.S.A: International Geology Review, v. 39, p. 1113–1123.

Yonkee, W.A., and Mitra, G., 1993, Comparison of basement deformation styles in parts of the Rocky Mountain foreland, Wyoming, and the Sevier orogenic belt, northern Utah, *in* Schmidt, C.J., Chase, R., and Erslev, E.A., eds., Laramide basement deformation in the Rocky Mountain foreland of the western United States: Geological Society of America Special Paper 280, p. 197–228.

Geological Society of America
Field Guide 5
2004

Structural implications of underground coal mining in the Mesaverde Group in the Somerset Coal Field, Delta and Gunnison Counties, Colorado

Christopher J. Carroll*
Colorado Geological Survey, 1313 Sherman St. No. 715, Denver, Colorado 80203, USA

Eric Robeck
Department of Geology, Brigham Young University, S-389 ESC, Provo, Utah 84602, USA

Greg Hunt
Bowie Resources Limited, P.O. Box 1488, Paonia, Colorado 81428, USA

Wendell Koontz
Arch Coal, Inc., Mountain Coal Company, P.O. Box 591, Somerset, Colorado 81434, USA

ABSTRACT

Paleogene and Neogene faults and fractures on the eastern edge of the Colorado Plateau are present in Mesaverde Group coal and sandstone beds. Recent observations of coal cleat orientation in relation to faults in coal mines have significant impacts for mine planning in the area. Faults, coal cleats, and natural fractures are interpreted to show a structural evolution of the Mesaverde Group through time.

This field trip will include a visit to two active underground coal mines, the Bowie Resources' Bowie No. 2 Mine, and Mountain Coal's West Elk Mine. Mine geologists will discuss structural styles including fault orientations and timing, cleat development, and rotation. Geologic encounters ranging from fault flooding, subsidence, mine fires, methane gas problems, and land use restrictions will also be discussed. Coal cleat development and open-mode fractures in adjacent sandstones will be observed on outcrops and compared to underground measurements in coal mines in the Somerset Coal Field, Colorado's most productive. Coal cleat orientations along a reverse fault in one mine will show rotation in relation to possible Neogene age displacement.

This two-day trip begins at the Convention Center in downtown Denver, Colorado. Participants will be transported in vans westbound on Interstate 70 to Glenwood Springs, then south on State Highway 82 to Carbondale, then southwest on State Highway 133 to Somerset, with a lunch stop in Redstone to observe 100-year-old coking coal beehive-shaped ovens. The first afternoon will include a stop at Paonia Reservoir Dam for introductory remarks on the regional fracture development of the Somerset Coal

*Chris.Carroll@state.co.us

Carroll, C.J., Robeck, E., Hunt, G., and Koontz, W., Structural implications of underground coal mining in the Mesaverde Group in the Somerset Coal Field, Delta and Gunnison Counties, Colorado, *in* Nelson, E.P. and Erslev, E.A., eds., Field trips in the southern Rocky Mountains, USA: Geological Society of America Field Guide 5, p. 41–58. For permission to copy, contact editing@geosociety.org. © 2004 Geological Society of America

Field and a mine tour at the West Elk Mine, which has encountered warm water flooding in large-scale faults associated with the West Elk Mountain uplift. We will head to Paonia for dinner and overnight. The second day will include a stop at the reclaimed Bowie No. 1 portal to observe fracture patterns on the west side of the coal field and an underground tour of the Bowie No. 2 Mine and possibly the new Bowie No. 3 Mine, where the mine geologist will discuss observations of coal cleat orientation changes in relation to faulting. Recent coalbed methane exploration in the southern Piceance Basin will also be addressed. Then we will drive a four-hour route back to Denver with a quick stop at the Muddy Creek Landslide, a 2.5 mi^2 (6.5 km^2) active earth flow complex.

Keywords: cleat, fractures, shear, faults, mining, Laramide.

INTRODUCTION

The purpose of this field trip is to compare fractures and faults and explore coal cleat development over time associated with both Neogene and Eocene tectonic events in the Cretaceous Mesaverde Group. Field trip participants will observe the variability in fracture patterns in surface outcrops of sandstone and coal beds and compare those to patterns of fractures underground. Prediction of fractures at depth can be verified by observing those occurring in coal mines 800 ft (244 m) to 1700 ft (518 m) beneath the surface. We will also see active coal mine operations at some of the state's largest underground coal mines during a time when Colorado's compliance coal is being produced at an all-time high. Colorado now produces over 38 million short tons (34.5 metric tons) of coal annually.

Previous Investigators

Willis T. Lee of the U.S. Geological Survey did much of the pioneering coal geology in the Somerset Coal Field. His early work and coal resource estimates marked the southern Piceance Basin as a noteworthy coal region. Many subsequent authors contributed to the knowledge base of the Somerset Coal Field. In 1995, the Texas Bureau of Economic Geology, in cooperation with the Gas Research Institute, published a comprehensive evaluation on the geologic assessment of natural gas from coal in four major western coal regions. One of these was for the Piceance Basin, and the publication is considered a landmark for fracture data, coal rank and gas composition, and coalbed methane production targets (Tyler et al., 1995).

Cretaceous Setting

During the Cretaceous, west-central Colorado was at ~42° north latitude, with a humid subtropical climate (Robinson Roberts and Kirschbaum, 1995). A large epicontinental seaway known as the Western Interior Seaway (Hettinger et al., 2000; Robinson Roberts and Kirschbaum, 1995; many previous authors) existed in what is now central North America. Shoreline sediments were deposited in this shallow seaway eastward from a tectonically active highland called the Sevier orogenic belt west of present-day Colorado. This region around the present-day Somerset Coal Field was an area of low relief, near-shore deposition. The resulting sediment supply and fluctuating sea levels created a complex depositional environment of shoreline and near-shore deposits. Up to 11,000 ft of sediment was deposited within the seaway during this time (Haun and Weimer, 1960).

Cretaceous coal-bearing rocks in the fluvial Mesaverde Group were deposited in a coastal plain. In response to tectonic activity in the Sevier orogenic belt, the Cretaceous shoreline prograded northeastward across Colorado (see Fig. 1). The basal unit, called the Rollins Sandstone, represents the youngest shoreface to retreat. Coal beds just above the thick near-shore sandstones were probably deposited as peat in freshwater swamp environments. Peat beds located higher in the section were deposited in swamps preserved between distributary channel systems.

At the end of the Cretaceous, the Western Interior Seaway withdrew from Colorado. Structural deformation commenced before the end of the Cretaceous as the onset of the Laramide Orogeny. Structural deformation during this time, and in subsequent post-Laramide uplifts, all combined to fold, fault, and fracture the Cretaceous and older rocks in the southern Piceance Basin.

FIELD TRIP DESCRIPTION

Day 1. Denver, Colorado, to Paonia, Colorado

Denver is often referred to as the "Mile High City," and it was founded on the banks of Cherry Creek, downstream from where gold was first discovered in 1858. Denver is situated near the axis of the Denver Basin, a sedimentary structural basin of Mesozoic and Cenozoic strata formed by Laramide tectonics in the early Cenozoic. The Front Range is a large Late Cretaceous–early Tertiary anticline that was tilted twice in the middle and late Cenozoic as part of regional deformation (Steven et al., 1997). Neogene deformation occurred in the middle and latest Miocene, Pliocene, and possibly Quaternary. The Front Range is the largest uplift in the southern Rocky Mountain Province. Although surface relief from Mount Evans to Denver is over 8000 ft, the maximum basement relief from Mount Evans to the Precambrian rocks below the Denver Basin is over 21,000 ft (see route map).

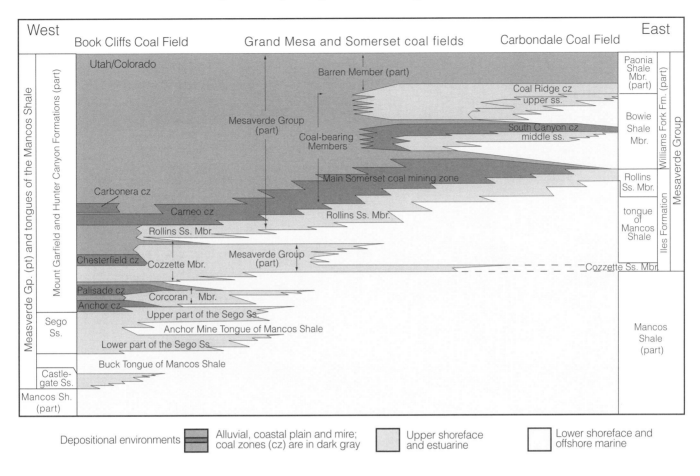

Figure 1. Stratigraphic correlations and facies relationships in the Mesaverde Group, southern part of the Piceance Basin, Colorado. Line of section drawn perpendicular to depositional strike. Modified after Hettinger et al. (2000). Cz on the figure stands for major coal zone.

The most significant tectonic event forming the geomorphic style of the Front Range was the Laramide Orogeny, which caused a large block of Proterozoic basement rock to be uplifted and shortened by reverse faults during the latest Cretaceous and Tertiary. Debate rages over the origin, but the most popular theory states that the subduction of the Farallon oceanic plate under western North America changed to a very shallow angle, affecting Colorado. This drove subduction-related processes eastward into the Rocky Mountain region and caused the uplifts and plutons there while the Cordilleran arc to the west was relatively inactive. Other theories consider collision with Wrangellia and the resulting compression to be responsible for the Laramide Orogeny (Sonnenberg and Bolyard, 1997; Steven et al., 1997). Along with uplift of the Front Range was the structural formation of the Denver Basin.

| Cumulative | | Description |
mi	(km)	
0.0	(0.0)	Leave the Colorado Convention Center at 8:00 a.m. Driver around to Colfax Ave. Turn right (west) and head 0.2 mi (0.32 km) to Kalamath St.
0.2	(0.3)	Turn left and head south on Kalamath 1.2 mi (1.9 km) to the 6th Avenue Freeway, and turn right.
1.4	(2.3)	Drive west on 6th Avenue for 12 mi (19.3 km). This part of the drive is on the west side of the Denver Basin, a Laramide-age (67–35 Ma) structural basin containing over 12,000 ft (3,657 m) of Paleozoic through Cenozoic strata. Cretaceous coal-bearing strata are located in the Laramie Formation. The Front Range looms ahead, topped by the "late Eocene erosion surface" (Epis and Chapin, 1975). Note the Continental Divide in the distance. Just before the I-70 interchange at 3-o'clock is South Table Mountain in Golden, a 63–64 Ma basaltic (shoshonite) flow. The Cretaceous-Tertiary boundary is located about two-thirds of the way up the hill.

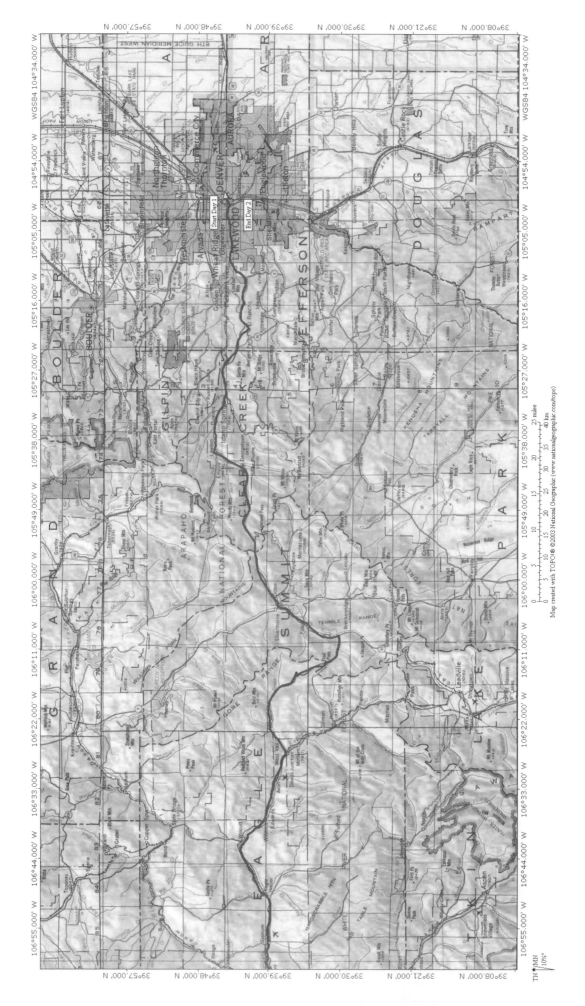

Route Map. Maps (*this and following page*) of the field trip area, outlining the entire route traveled.

Route Map (*continued*).

13.4 (21.6) Turn onto I-70 and head west into the mountains. Note the outcrops of Mesozoic strata at the Morrison Rd. exit, Exit 260. This marks the western edge of the Denver Basin. The coal-bearing formation here is the Laramie Formation, a Late Cretaceous non-marine swamp deposited behind the Fox Hills Sandstone prograding shoreface. Coal was mined from vertical beds in Golden during the 1890s. Jurassic strata of the Morrison Formation can be seen in the excellent road cut at Morrison Rd. The underlying Permian Fountain Formation forms the red rock outcrops at Red Rocks Park, one mile south of this location. This formation unconformably overlies the igneous Boulder Granodiorite and metamorphic basement complex with a 1.4 Ga hiatus in time. While driving westbound, observe the Precambrian exposures in the Colorado Mineral Belt from Golden to Silverthorne.

55.0 (88.5) Eisenhower Tunnel. At 11,000 ft (3,352 m), this is the high point for the field trip (beneath the Continental Divide). Welcome to the western slope of Colorado.

65.6 (105.6) Silverthorne. Here the footwall of the Williams Fork thrust fault exposes Cretaceous strata again, this time on the western side of the divide. Excellent exposures of the Dakota Sandstone and the Mancos Shale can be observed in a road cut on the right, just before the Silverthorne exit.

81.0 (130.3) Vail Pass. Phanerozoic strata are exposed here again, this time the Maroon Formation (this is the Pennsylvanian-Permian equivalent to the Fountain Formation on the eastern slope).

95.2 (153.2) Vail. Bedrock here is the Pennsylvanian Minturn Formation. The unit forms cliffs of shale and sandstone and is landslide prone in dip slope. Note that coal is completely absent in the Pennsylvanian in western North America.

101.2 (162.9) Dowd's Junction. Interstate 70 crosses the Eagle River here. This former train line carried coal and freight over Tennessee Pass until the train line's closure in 1998. All coal sales on the western slope must now travel through the Moffat Tunnel, ~22 million tons per year.

145.0 (233.3) Glenwood Canyon. The Colorado River cuts down into the lower Paleozoic section here. On the east side of the canyon is the Mississippian Leadville Limestone. The Cambrian Sawatch Quartzite and Ordovician Manitou Dolomite are exposed in the central part of the canyon. Uplift of the nearby White River Plateau on the north side of the canyon resulted in the Colorado River down-cutting into the Precambrian rocks.

157.0 (252.7) Drive to Glenwood Springs. Turn off at the main exit (116), and turn right. Go to the light, and turn right again, then drive to the bridge across the Colorado River and turn right over the bridge. Head north on this road, SH 82, to Carbondale, 20 miles. The Colorado Geological Survey and the U.S. Geological Survey studied this area extensively in the 1990s for its tectonic salt collapse features and subsequent hydrocompactive soils in the valley floor (Kirkham et al., 1996). Publications pertaining to this story are available on the field trip.

165.9 (267) Turn right at Carbondale onto SH-133. Carbondale was a former large coal mining town at the turn of the twentieth century. The Carbondale field produced 31.7 million short tons (28.7 metric tons) of bituminous coal and anthracite.

177.0 (284.9) Stop 1.

Stop 1. Redstone, Colorado

This small art colony was originally the home of John Osgood, founder of the Colorado Fuel and Iron (CF&I) corporation in 1892. He built the Redstone Castle and started the coking coal industry in Colorado at the Mid-Continent Mines northwest of Redstone. We will eat lunch and have a short walking tour of the beehives. Feel free to walk into town to shop, browse, or use the facilities.

Geology and Coal Mining History of Redstone, Colorado

Cretaceous and older strata here are sharply folded and faulted. Tilted beds of the coal-bearing Mesaverde Group reflect Late Cretaceous–Tertiary Laramide orogenic movements. The Elk and West Elk Mountain intrusive complexes of Tertiary-age sills, laccoliths, dikes, and associated structures completed the tectonic development in this part of the Piceance Basin (see Fig. 2). A high geothermal heat gradient increased the rank of coals in the Mesaverde Group to produce localized deposits of coking coal created by the numerous intrusions in the area. The coal beds consist of "premium grade" medium-volatile bituminous to "marginal grade" high-volatile bituminous coking coal (Goolsby et al., 1979). Coking coal is a carbonaceous solid produced from coal by thermal decomposition. This part of the Piceance Basin contains an estimated 500 million short tons (453 metric tons) of coking coal in reserves less than 3,000 ft (914 m) deep.

Coal was mined near Redstone between 1899 and 1985. The area was once Colorado's main source for coking coal. This type of metallurgical coal byproduct was used to supply the railroads with boiler fuel and the mining industry with smelter fuel. Slag from the railroads was coked and sold for metallurgical purposes

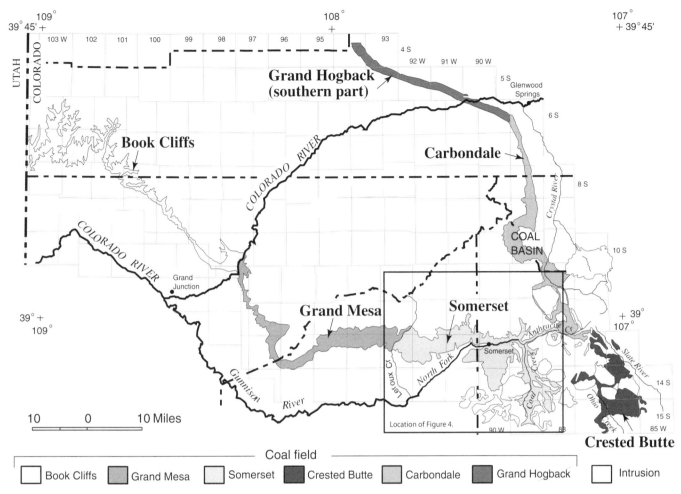

Figure 2. Location of coal fields in the southern part of the Piceance Basin, Colorado. Coal fields are defined by outcrop of the Mesaverde Group. Box marks the Somerset Coal Field study area. Modified after Hettinger et al. (2000).

in steel mills in Pueblo. The coking ovens, called beehives, were aligned adjacent to each other for maximum efficiency. The beehives at Redstone received coke from the Coal Basin coal area, just 4.5 mi (7.24 km) west of Redstone. By 1900, a rail line, the Crystal River Railroad, was in place and delivering coking coal to iron processors in Pueblo.

The coking coal mining industry slowed by 1930, but the Mid-Continent Coal and Coke Company revived mining in Coal Basin in the 1970s and 1980s with five mines. Coal Basin has the highest recorded heat values for coal in the state (over 14,000 Btu/lb as received). The mining challenges in this area were unique in that overburden depths exceeded 2500 ft (762 m), and coal beds dip relatively steeply (nearly 20°). High methane contents were a dangerous byproduct of mining, and mine elevations were over 10,000 ft (3048 m), the highest elevation for coal mines in the nation. The last mine to operate was the Dutch Creek No. 2 Mine, which closed in 1989. Over 20 million tons of coal and coke were mined from Coal Basin (see Fig. 3).

Return to SH-133 and continue south toward Paonia.

Cumulative		
mi	*(km)*	*Description*
184.1	(296.3)	Bypass the turnoff to Marble, Colorado. This is the home of the Yule Marble Quarry, where marble for the Lincoln Memorial and the Tomb of the Unknown Soldier in Washington, D.C., were quarried. Parts of the Colorado State Capitol also contain Yule Marble. Yule Marble is also famous for its mechanical properties and has been used in many rock deformation experiments (E. Nelson, 2004, personal commun.).
188.6	(303.5)	McClure Pass. Here we see the first exposure of the Mesaverde Group strata in the steep, rock-fallen cliff face along the highway.
201.1	(323.6)	Muddy Creek Landslide. This very active slide moved catastrophically in 1985–1986. The Bureau of Reclamation, Colorado Department of Transportation, and the

Figure 3. Smoke pours out of the boiler house at the CF&I mine at Coal Basin. Building was constructed of corrugated tin and several smokestacks extend from the roof. Circa 1900. Photo courtesy of the Colorado Historical Society (permission granted July 2004).

Colorado Geological Survey have had a continuing monitoring project since then. It moves ~10 ft (3 m) per year horizontally into Muddy Creek.

203.0   (326.7)   Stop 2.

Stop 2. Paonia Reservoir and Dam

This stop will present an introduction to fracture styles in the upper Mesaverde Group in the rock outcrops above Paonia Dam (see Fig. 4). With the current drought conditions, this reservoir has been low for ~5 yr. This location marks the eastern edge of the Somerset Coal Field. Looking west, the North Fork Valley can be seen. It is marked by incision of the North Fork of the Gunnison River into soft Cretaceous and Paleocene strata. The Mesaverde Group dips gently north throughout the valley. A recent study by the Colorado Geological Survey estimated 5.1 billion short tons (4.6 billion metric tons) of coal available in the Somerset Coal Field (Schultz et al., 2000). To the south is the Kebler Pass region to Crested Butte, home of Colorado's only resource of anthracite.

Cretaceous Mesaverde Group rocks of the Somerset Coal Field contain three distinct fracture types: open-mode fractures (including fractures in sandstone and coal cleats), shear-mode (faults and shear) fractures, and surface joints (Carroll, 2003). Coal cleat strikes range from N58E to N75E with vertical dip and well developed butt cleats consistently across the study area. Mine plans are oriented to extract coal within 15° of the main anisotropic coal cleat direction. Two primary, systematic sets of open-mode fractures called Set 1 and Set 2 have average orientations of N76E, 82 SE and N18W, 84 SW, respectively, throughout the coal field. The northeast-striking fractures are most common and crosscut the northwest striking fractures. All fractures are generally non-mineralized, orthogonal to bedding, and regularly spaced with small apertures. Shear fractures observed underground strike N45W to N78W and dip between 60° and 76° southwest. The shear fractures are associated with tectonic faults in the mines. Major faults in the Somerset region strike northeast on the west side near Bowie No. 2 Mine, and northwest on the east side near West Elk Mine. All regional fracture systems are considered post-Laramide (see Table 1). Some fractures around the Iron Point intrusive stock may be younger, but not all workers agree with this.

Continue on SH-133 westbound below the dam.

Cumulative		
mi	(km)	Description
209.5	(337.2)	West Elk Mine, operated by Arch Coal since 1998. Turn left and head up to the mine office.
210.5	(338.8)	Stop 3.

Stop 3. West Elk Mine Underground Tour

Wendell Koontz, mine geologist for the mine, will lead this tour.

Mountain Coal Company operates the West Elk Mine for Arch Coal; the mine currently produces over 6 million short tons (5.4 million metric tons) of coal annually. West Elk is the only active coal mine situated on the south side of the North Fork Valley (see Fig. 5). Elevated geothermal gradients from a deeper-seated igneous cupola related to the nearby West Elk Mountains make the mine warmer than usual. This massive Tertiary intrusive complex affects coal rank within its influence, including anthracite higher in the surrounding mountains. Large faults and high groundwater pressures and temperatures affect West Elk Mine. Coal is mined from the B and E Seams of the Mesaverde Formation.

West Elk has a variety of geologic features of interest, including tectonic normal and strike-slip faults, compactional faults, clastic dikes, and paleo-sand channels. The mine tour will focus on structural deformation features exposed in the mine

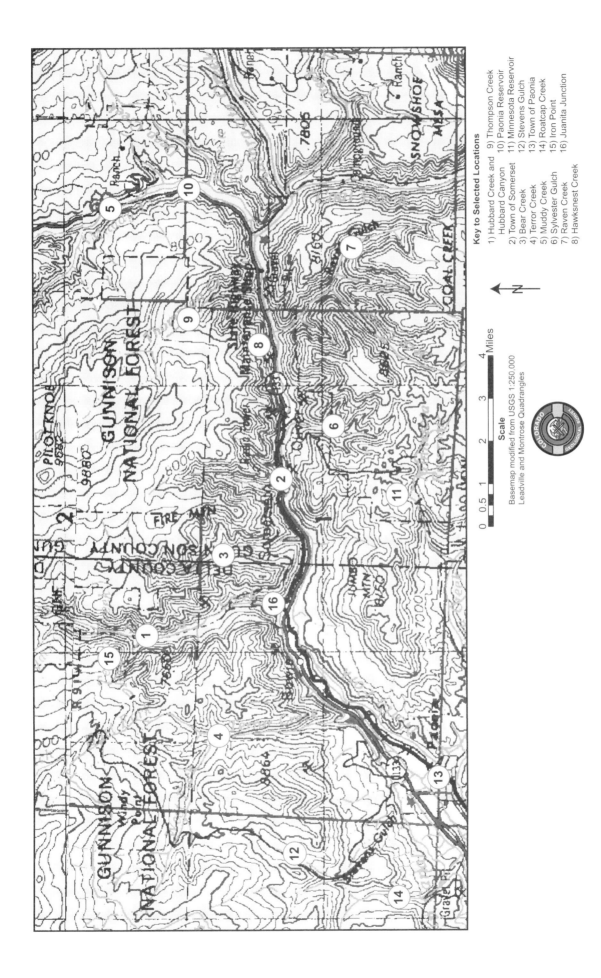

Key to Selected Locations

1) Hubbard Creek and Hubbard Canyon
2) Town of Somerset
3) Bear Creek
4) Terror Creek
5) Muddy Creek
6) Sylvester Gulch
7) Raven Creek
8) Hawksnest Creek
9) Thompson Creek
10) Paonia Reservoir
11) Minnesota Reservoir
12) Stevens Gulch
13) Town of Paonia
14) Roatcap Creek
15) Iron Point
16) Juanita Junction

Scale

0 0.5 1 2 3 4 Miles

Basemap modified from USGS 1:250,000
Leadville and Montrose Quadrangles

Figure 4. Location map of selected key areas in the Somerset Coal Field.

TABLE 1. AVERAGE ORIENTATIONS OF ALL FRACTURE STATIONS IN THE MESAVERDE GROUP, SOMERSET COAL FIELD

Location	Set 1 orientation (Strike, Dip)	Set 2 orientation (Strike, Dip)	Shear (Strike, Dip)	Aperture	Spacing	Coal Face Cleat (Strike, Dip)	Coal Butt Cleat (Strike, Dip)	Geologic Formation
West side								
Bowie Mine #1 Portal (n = 65)	N 71 E, 80 SE	N 24 W, 81 SW		<0.1 in.	1.2 ft	N 64 E, 87 NW	N 22 W, 88 NE	Bowie Shale
Hubbard Creek (n = 42)	N 84 E, 90	N 5 W, 85 SW		0–25 in.	1–2 ft	none	none	Rollins SS
Iron Pt (n = 24)	N 78 W, 71 NE			.01 in.	1–2 in.	none	none	Igneous Stock
East side								
Coal Gulch near Sanborn Creek (n = 68)	N 80 E, 84 NW	N 32 W, 80 SW		<0.1 in.	0.5–4.0 in.	N 68 E, 85 NW	N 27 W, 87 NE	Paonia Shale, E seam
West Elk Mine Shaft #3 (n = 145)	N 53 E, 89 SE	N 26 W, 77 SW		<0.1 in.	1–3 in.	none	none	Kmv Barren Inverval
Paonia Reservoir (n = 63)	N 69 E, 80 SE	N 22 W, 82 SW	N 47 E, 66 SE	<0.1 in.	1.0 ft	none	none	Kmv Barren Interval

(see Fig. 6) including the normal, strike-slip, and compactional faults. Depending on the time available and the interests of the group, additional topics may include roof support, pillar design, and mining techniques. We will have a short safety discussion and certification class for the underground tour.

After the mine tour, head back down to SH-133. Look for steam from a smoldering coal fire from the abandoned Oliver No. 3 Mine at 3-o'clock. Turn left on SH-133.

Cumulative mi	(km)	Description
212.0	(341.2)	Somerset, Colorado, home of Oxbow Mining's Elk Creek Mine. This mine opened in April 2003 and produced nearly 4.6 million short tons (4.17 metric tons) of coal in 2003 (Carroll, 2004). Elk Creek is capable of producing over 700,000 short tons (635,000 metric tons) of coal monthly with its new longwall mining equipment.
214.0	(344.4)	Bowie Resources' coal operations. This will be the focus of Day 2.
223.0	(358.9)	Paonia, Colorado. This is the last stop for today. Stay at the Redwood Arms Motel. Dinner will be in town at a Mexican restaurant, Fiesta Vallarta. There are also several other options within walking distance. End of Day 1.

Day 2. Paonia, Colorado, to Denver, Colorado

Start with breakfast at Butch's Café next to the Redwood Arms Motel. It is next to the Gunnison Energy field office; Gunnison Energy is the petroleum division of Oxbow Mining's operations in the North Fork Valley. Check out and be on the road by 8:00 a.m.

Cumulative mi	(km)	Description
0.0	(0.0)	Paonia. Drive east on SH-133 toward Somerset.
0.3	(0.48)	Lump and stoke coal sales in Paonia.
1.6	(2.57)	Turn left at Steven's Gulch Rd. Drive up to the Bowie No. 1 Mine reclamation site.
3.5	(5.6)	Stop 1.

Stop 1. Bowie Mine No. 1 Portal

Observe fractures in coal cleat and adjacent rocks around the portal. There are excellent exposures here of coal cleat, and sandstone fractures are exposed in the artificial cut.

Cumulative mi	(km)	Description
5.5	(8.8)	Drive back down Steven's Gulch to SH-133. Turn left and head east to the North Fork Valley.
6.5	(10.5)	Orchard Valley Farms. Fruit and wine sales in the summer and fall.

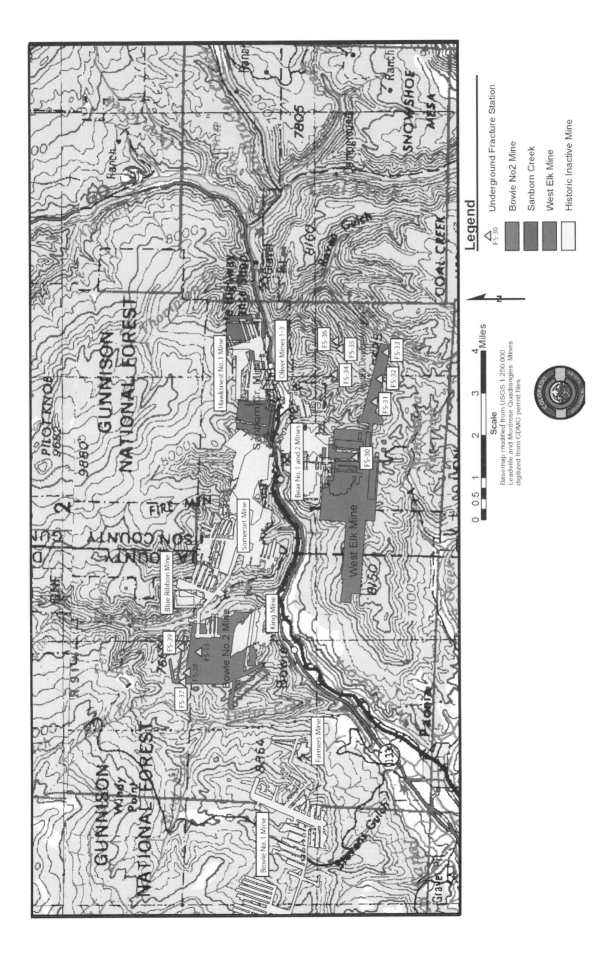

Figure 5. Location map of the Somerset Coal Field showing areas of underground mining. Fracture stations are noted as such (i.e., FS-37).

7.4 (11.9) Pass the Terror Creek Winery on the left. This small western slope winery takes advantage of mild climate, good sun on the slopes of Orchard Mesa, a south-facing Pleistocene terrace gravel surface 500 ft (152 m) above the North Fork of the Gunnison River. The basalt clasts making up the terrace cobbles are derived from debris flow fans from Grand Mesa to the north.

9.7 (15.6) Turn left onto Bowie Rd. Note the conveyors and trains ahead at 1-o'clock.

10.7 (17.2) Turn left at the mine entrance. Drive up to the small house on the right and park in the upper parking lot. This is the geotechnical office for Bowie Resources. Note the new wash plant now in service for the Bowie No. 3 Mine.

11.6 (18.7) Stop 2.

Figure 6. Photograph of main marker parting, B seam, West Elk Mine.

Stop 2. Bowie Mine No. 2 Tour

Greg Hunt, Bowie Mine geologist, and Eric Robeck, a graduate student from Brigham Young University, will discuss the geology and coal cleat before heading up to the mine (see Fig. 7).

Cumulative		
mi	(km)	Description
12.2	(19.6)	Drive up to the main office building at the top of the hill. Park in the upper lot and head into the building for safety training before the tour.

Underground stop 1: Fault crossing in the DU seam. Five major faults have been found underground in the Bowie No. 2 Mine (see Fig. 8). Each fault falls into one of two fault systems: an earlier northwest striking system of normal faults with offsets up to 49 ft (15 m) dipping either northeast or southwest or a later strike-slip system with a series of transtensional normal faults on the footwall.

The Mains fault was the earliest fault to form and is the best example of the earliest faulting episode. The numerous soft-sediment deformation features observed here suggest that the fault formed shortly after or during deposition of the coal-bearing sequence at Bowie. The fault strikes on average 135° and dips 57° to the northeast. Throw on the Mains fault is variable and difficult to interpret. In most places where the fault has been encountered, an upper and a lower fault plane exist. The fault block in between is up to 50 ft (15 m) wide and exhibits severe normal drag with beds dipping up to 45°. Fault reactivation selectively favored both fault planes and is complicated by the steeply dipping fault block.

Deposition of the coal-bearing strata at Bowie probably occurred between 74 and 73 Ma (Franczyk et al., 1991), nearly coincident with the initiation of the Laramide Orogeny at ca. 72 Ma (Scott et al., 1996), represented by the activation of several structural features flanking the basin. However, the maximum far-field stress in the early Laramide was probably oriented roughly east-west (as it was throughout most of the Laramide), which is not conducive to the formation of a northwest-striking normal fault. Two mechanisms may be used to explain the Mains fault orientation. (1) Depositional trends near Bowie prograde to the northeast (Franczyk et al., 1991). Perhaps these early faults initiated as unloading (growth) faults on an unstable delta front, although the relative constancy of dip (or absence of listricity) in the vertical section and the lack of observable B-D seam interburden thickness variations seem to suggest otherwise. (2) Precambrian basement rock underneath the Piceance Basin has a strongly-defined northwesterly grain that has been implicated in numerous thrusts in the southern part of the basin. A transition from thin-skinned (Sevier) to thick-skinned thrusting in the early Laramide may have reactivated Precambrian basement blocks with little regard to the principal stress orientation. However, this argument fails to explain how a normal fault formed in an apparently active compressional regime.

Underground stop 2: Cleat study traverse across the Mains fault. The patterns of cleat orientation at Bowie show significant deviations from the regional norm in areas critical to mine engineering and safety, especially surrounding faults and shear zones, where roof and rib quality is often compromised. These deviations are consistent enough to be predictable and, in ideal cases, may even be *predictive of* geological phenomena such as faults.

Faults have the potential to significantly perturb far-field stresses, causing dramatic rotations of face cleat forming near the fault. Two conditions must be satisfied for a noticeable rotation to occur. First, the fault must form before or during face cleat formation. Second, the fault orientation must be oblique to the major stress axis controlling cleat strike; if the fault is either parallel or perpendicular to σ_1, little shear stress will be resolved onto the fault plane, and the potential for significant stress perturbation and cleat rotation is reduced. These two conditions are rarely satisfied for a particular fault. The amount of time between

Bowie Generalized Stratigraphic Column

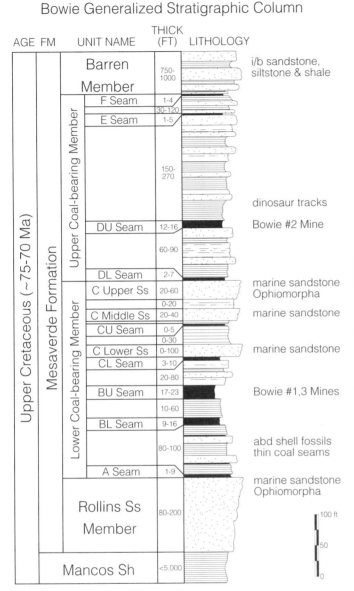

Figure 7. Generalized stratigraphic column of the Mesaverde Group coal beds at the Bowie Mines. Courtesy of Bowie Resources.

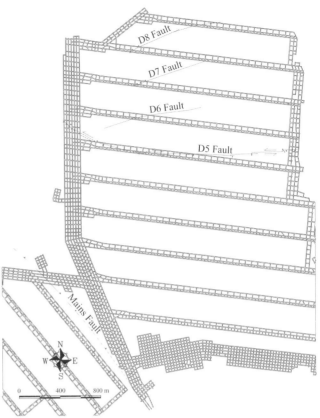

Figure 8. Plan view of the Bowie No. 2 Mine faults.

deposition and cleat formation is usually a small fraction of the total formation age, especially in eastern United States coals. Also, peat swamps form primarily in tectonically quiescent margins where penecontemporaneous faulting is uncommon. Even in areas where faults predate cleat formation, they may do so by such a small period of time that the far-field stresses remain relatively constant during both fault and cleat formation.

Because of the general regularity of face cleat orientations in specific coal basins around the world, it is possible to recognize and relatively date those faults that predate cleats on the basis of abrupt changes in face cleat orientation. Although recognized examples of this phenomenon are rare, they have been documented in the Bowen basin of Australia (Pattison et al., 1996), the San Juan basin of Colorado and New Mexico (Laubach et al., 1991; Tremain et al., 1991), the Beatrice Mine in Virginia (McCulloch et al., 1974), the Wasatch Plateau of Utah (Hucka, 1991), the Crested Butte coalfield southeast of Somerset (Lee, 1912), and most recently at Bowie (Robeck et al., 2004).

Of all the faults in the Bowie No. 2 Mine, the Mains fault appears to be the only one that satisfies the two criteria listed above for fault-induced cleat rotation. First, soft-sediment structures confirm that faulting initiated soon after deposition and was perhaps contemporaneous with it. Second, the Mains fault is the only fault with a fault style (reverse) and orientation not consistent with having formed from the ENE maximum horizontal compressive stress that formed the cleat and the D5–D9 fault system.

In the West Submains fault crossing area, ten face cleat measurements were taken at 60 stations in all entries up to 1,600 ft (500 m) away from the fault on the footwall and 900 ft (270 m) away from the fault on the hanging wall. A map of cleat rose diagrams is shown in Figure 9. The face cleat strikes are uniform and consistent with regional trends up to ~262 ft (80 m) away from the fault on both the footwall and hanging wall. Cleat strikes rotate, on average, northward into the fault plane, becoming abruptly fault-parallel within ~30 ft (10 m) of the fault (Fig. 9). The fault-parallel cleats near the fault are invariably slickensided. Face cleat dips become more random and on average shallower near the fault, especially on the hanging wall.

In addition to a general rotation, cleat strikes within the fault-perturbed stress field (<80 m from the fault) become more chaotic (i.e., their distribution becomes more random). In Figure 10, 95% confidence cones, measured in degrees from the mean pole-to-plane, were plotted from the ten measurements taken at each station. The 95% confidence cone width is never more than 6° at a distance of over 80 m from the fault. Near the fault, however,

it is widely variable, reaching as much as 19.4° on the hanging wall. Note that, although overall scatter increases, many stations near the fault have very well defined trends with some of the lowest scatter measured. This suggests, in conjunction with the wide scatter in cleat strike, that these cleat domains may have formed during and between paleoslip events on the Mains fault, with areas of wide scatter occurring where zones overlap. It is also obvious from Figures 8 and 9 that the hanging wall cleats were more affected than the footwall cleats, although the width of the fault-perturbed field is almost the same for both.

The common tendency for face cleats to form perpendicular to the fold axis, and butt cleats parallel to it (McCulloch, et al., 1974), is not supported in the Bowie area, where face cleats strike as much as 50° from the axis of folding. However, face

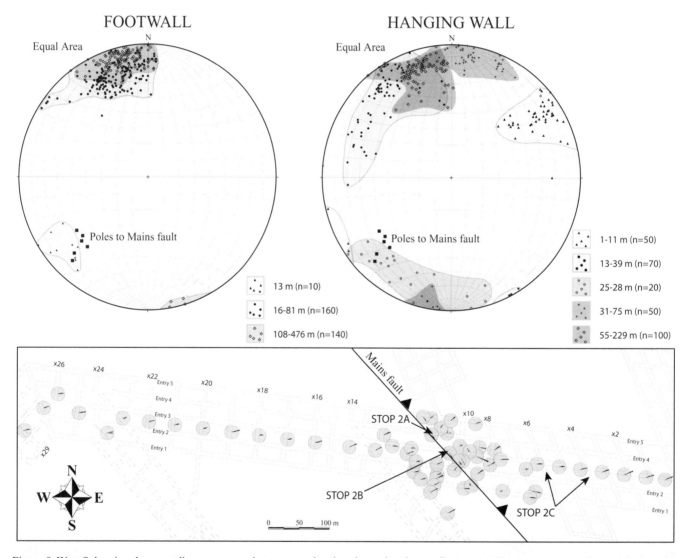

Figure 9. West Submains cleat rose diagram map and stereonets, showing cleat azimuth rose diagrams at 60 measurement stations on the hanging wall and footwall. Cleat trends vary from regional at distances >80 m from the fault, as shown in the stereonets above. Rose diagram petals are 5°. Shaded circles around rose diagrams are drawn at 50% of data; i.e., a rose diagram petal containing 5 of 10 measurements will plot to this circle. The three DU seam substops are shown.

cleats form nearly perpendicular to the synclinal axis of the Piceance basin and to the southern extension of the Grand Hogback monocline east of it (Tweto et al., 1976). This seems to illustrate the fact that although cleat formation and basinal folding are probably products of the same compressive event (the Laramide Orogeny), folding is a more complex process and is also related to both the distance from the center of the basin and the magnitude of folding. This is suggested by the fact that face cleats in the west-central Piceance basin do form nearly perpendicular to the axis of folding, which is there subparallel to the synclinal axis (Grout and Verbeek, 1992; Tyler et al., 1995).

Fault-rotated cleat domains could be of potential value to mine planning. Many faults are exposed poorly or not at all, and the detection of faults is often done by contouring widely-spaced drill hole data. Costly fault encounters may be avoided, or at least mitigated, by implementing systematic cleat mapping during mine advancement. This requires the establishment of a mine-specific face cleat baseline and is subject to the variables mentioned above; namely, fault timing and paleostress orientations. Where faults are observed to pre-date cleats may be an inexpensive and relatively effective method of fault detection.

Underground stop 3: Exfoliation cleat observations. We will next walk to crosscuts 4 and 6 between entries 2 and 3 to observe an unusual cleat style that we have called exfoliation cleat (see Fig. 11). Exfoliation cleat is characterized by concentric fractures in coal formed between regular preexisting face cleats. Depending on the degree of curvature in the dimension parallel to the face cleats, exfoliation cleats may resemble either

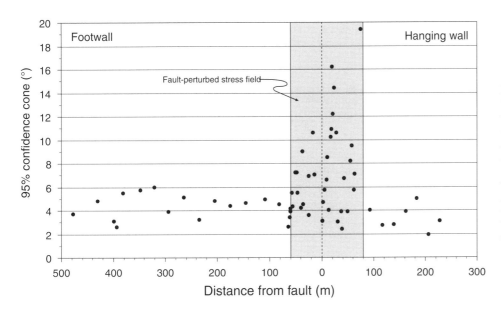

Figure 10. Over 95% confidence cones for each cleat station (n = 60) were determined using the program Stereonet (for Windows, version 1.2.0) by Richard Allmendinger (2002). Cleat orientations are very consistent at distances >60 m from the fault; however, <60 m from the fault, scatter increases abruptly. Although the zone of affected cleats is only slightly wider on the hanging wall than on the footwall, the hanging wall is much more affected.

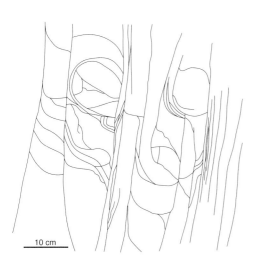

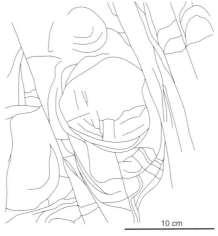

A. West Submains Xcut 4 Entry 2-3 W rib   B. West Submains Xcut 6 Entry 2-3 E rib

Figure 11. Cross sections of two exfoliation cleat zones. Exfoliation cleats form concentric shells with the long axis parallel to the primary stress direction. They are often concentrated, as they are here, between specific widely-spaced face cleats, with little exfoliation occurring between adjacent cleats. Where exfoliation occurred, it prevented the development of widespread vertical face cleats; note how face cleat frequency increases in non-exfoliated zones.

eggs (moderate curvature) or logs (no curvature). An extensive literature search turned up only one reference to this cleat style, that of Schopf (1959), who referred to these cleats as faceted cleats, eye coal, or fish coal. The authors have not observed examples of this cleat style anywhere but at Bowie, where it forms only in halos around fault and shear zones.

Exfoliation cleats seem to form from σ_1-parallel, uniaxial, and pore fluid–assisted tensile failure between early, widely spaced, 3–8 in (~8–20 cm) face cleats. Where exfoliation did not occur, face cleats are more closely spaced, 0.4–2 in (~1–5 cm). Exfoliation cleats are often grouped in horizontal bands, suggesting a partial dependence on compositional controls. It is unclear why exfoliation cleats form only around faults, but this observation suggests that fault-induced stress perturbation may be a causative mechanism, helping to explain the general rarity of this cleat style.

Exfoliation cleats formed early in the process of coal maturation at Bowie, before the coal reached its current rank. Because exfoliation cleats form between earlier cleats, the width of exfoliation cleat "eggs" is a good proxy for the average face cleat spacing at the time of formation. A histogram of 144 exfoliation cleat widths measured in the West Submains area is plotted along with the average Bowie face cleat spacing as measured by Tremain and Tyler (1997) (see Fig. 12). Exfoliation cleat width is, on average, much greater than face cleat spacings. The pronounced upper "tail" may suggest that exfoliation may have occurred since the onset of brittle failure at the peat-lignite transition, when there was no upper limit on cleat spacing.

Face cleat spacing has been correlated to coal maturity, measured as R_0 (vitrinite reflectance), by Law (1993) from data gathered from 39 coal mines across the United States. The results from the histogram in Figure 12 are plotted on the vitrinite reflectance versus cleat spacing plot of Law (1993) in Figure 13. Even an approximate fit to the curve suggests an early origin for exfoliation cleats, perhaps since the onset of brittle failure at the peat-lignite transition. In addition, modeling by Scott (1996) suggests that near Bowie the transition from peat to lignite occurred by ca. 56 Ma, or 15–20 m.y. after deposition, at the height of the Laramide Orogeny.

Underground stop 4: Fault crossing in BU seam. At the time of this publication, mining has advanced to within 200 ft (60 m) of the Mains fault in the BU seam. Based on recent drill holes, we expect to tunnel ~50 ft (15 m) through the fault block in beds dipping 35°–40°. Depending on what is encountered, a couple of additional stops will be planned with the purpose of describing the effects of complex faulting on mine planning and roof and rib quality.

Underground to surface stop 5: Mains fault perspective view from ridge. The saying that sometimes "you can't see the forest for the trees" is never truer than in the case of the Mains fault. The complex folding and faulting observed underground tend to blind us to the fact that on a larger scale, the Mains fault is a normal fault. Even where the *sum* of throws on the Mains fault planes is reverse (as in the DU seam West Submains), *regional* throw, measured by extrapolating undeformed hanging wall and footwall strata into the fault, is consistently ~29 ft (9 m) normal.

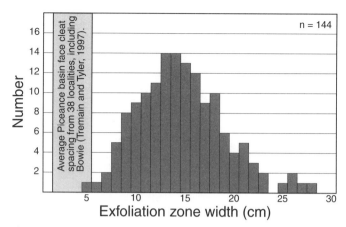

Figure 12. Histogram showing exfoliation zone widths, Bowie No. 2 Mine. The pronounced upper "tail" may suggest that exfoliation may have occurred since the onset of brittle failure at the peat-lignite transition, when there was no upper limit on cleat spacing. Widths are from the outermost concentric ring of exfoliated "eggs" or "logs." Measurements were taken from both the hanging wall and footwall of the Mains fault at distances of up to 100 m.

From the ridge, we look ~0.6 mi (1 km) to where the Mains fault offsets a ~60-ft-thick (20 m) sandstone channel above the mine. Throw on the channel has been surveyed at ~29 ft (9 m) down to the northeast. The central fault block has been eroded away, showing only the large-scale offset. From this vantage point, it is easy to see how a steeply dipping fault block could cause offset on both fault *planes* to be reverse but offset on the fault *zone* to be normal. Before this relationship was realized, fault offsets were taken at face value, and mine planners were thrown a few surprises when attempting to cross the fault.

This ridge also provides a good perspective view over the Bowie mine facilities and across the North Fork Valley. The lineament created by the Mains fault can be traced across the ridge south of the Gunnison River, where it was briefly encountered in the West Elk Mine.

After the mine tour and lunch, drive down the mountain and back to SH-133.

Cumulative mi	(km)	Description
14.5	(23.3)	Turn left at the Bowie Resources sign at the bottom of the hill. This is the old frontage road for Bowie.
14.8	(23.8)	Old King Mine Portal in the cliffs of the B seam outcrop at 9-o'clock.
15.3	(24.6)	Hubbard Creek Road. Fracture data in outcrops of the creek are similar to those observed underground at Bowie, but more numerous.
15.4	(24.8)	Cross railroad tracks. These tracks service the West Elk and Elk Creek mines up valley.

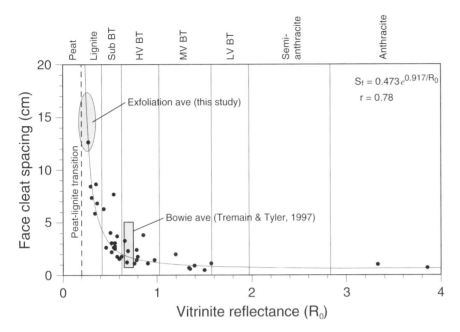

$$S_f = 0.473\,e^{0.917/R_0}$$
$$r = 0.78$$

Figure 13. Chart showing the timing of exfoliation cleats, Bowie Mine No. 2. Curve and data points for 39 locations across the U.S. are from Law (1993). Average exfoliation cleat widths are compared to average face cleat widths for the Bowie area collected by Tremain and Tyler (1997). While no time scale is implied for the horizontal axis, vitrinite reflectance may be a good proxy for time where coal undergoes progressive burial and maturation. The curve becomes asymptotic at the peat-lignite transition. This is not an artifact of regression, but reflects the fact that cleats do not form in peat.

Sixteen million short tons (14.5 million metric tons) of coal are shipped from Somerset annually (Carroll, 2004). Clients include coal-fired power plant operators in Kentucky, Illinois, Mississippi, and Tennessee. Cliffs here exhibit valley fractures parallel to the North Fork Valley. There is clinker in much of the coal zone above to the left.

Cumulative		
mi	(km)	Description
16.0	(25.7)	Cross railroad tracks several times.
17.3	(27.8)	At the junction of Bowie Rd and SH-133, turn left and drive toward Denver.
19.0	(30.5)	Town of Somerset and the Elk Creek Mine.
20.5	(33)	Entrance to West Elk Mine. Fractures in Coal Gulch at 9-o'clock measure N 80 E, 84 NW (Set 1), and N 32 W, 80 SW (Set 2) for sandstones in the Barren Interval of the Mesaverde Group.
21.6	(34.7)	Landslide occurred in 1996 during highway realignment.
22.3	(35.9)	The underground Hawksnest mines were located beneath the highway here.
25.4	(40.9)	Paonia Dam and Reservoir. Regular fractures here are in the Upper Mesaverde Group.
29.4	(47.3)	South Slide at Muddy Creek. Note rock buttress built by the Colorado Department of Transportation (CDOT) in 1986.
30.5	(49.1)	Stop 3.

Stop 3. Middle Landslide at Muddy Creek Earth Flow Complex
Pull off at road turnout and observe movement from 1985–1986 and subsequent movement since then.

38.0	(61.2)	Highway Landslide. Movement is slow, so patching by CDOT is the preferred mitigation pathway. Note old alignment while crossing.
41.9	(67.4)	Top of McClure Pass, elevation 8755 ft.
43.7	(70.3)	Rock fall of the Tertiary Wasatch Formation.
47.3	(76.1)	Mesaverde Group at road level. No coal mining in this area. Look for bears and a large fault here. As we cross over the fault, Mesozoic strata are quickly replaced at road level with Paleozoic strata.
48.3	(77.7)	Hays Creek Falls. Maroon Formation.
50.2	(80.8)	Redstone. Note Redstone Castle at 3-o'clock, just before town.
66.0	(106.2)	Carbondale.
67.4	(108.5)	Junction of SH-133 and the road to Aspen (right). Turn left onto SH-82 and head back to Glenwood Springs.
80.0	(128.7)	Glenwood Springs. If time permits, we can have dinner here or in Vail, Silverthorne, or Idaho Springs.
94.0	(151.3)	Glenwood Canyon. It took $580 million and 12 yr to construct the highway through this canyon in the 1980s and 90s.
96.5	(155.3)	Cross the Colorado River.
97.5	(156.9)	Dotsero crater. This is Colorado's youngest volcanic feature. The unconsolidated lapilli tuff basalt flow last erupted 4150 yr ago (Streufert et al., 1997).
230.0	(370.1)	Colorado Convention Center, Denver, Colorado. Two-day total mileage: 453 miles (729 km).

REFERENCES CITED

Allmendinger, R., 2002, Stereonet (for Windows), version 1.2.0: http://www.geo.cornell.edu/geology/faculty/RWA/RWA.html.

Carroll, C.J., 2003, Fractures in the Mesaverde Group at Somerset Coal Field, Delta and Gunnison Counties, Colorado, *in* Peterson, K.M., Olson, T.M., and Anderson, D.S., eds., Piceance Basin 2003 Guidebook, Rocky Mountain Association of Geologists, Denver, Colorado, CD-ROM.

Carroll, C.J., 2004, 2003 Summary of coal resources in Colorado: Denver, Colorado Geological Survey Special Publication 54, 25 p.

Epis, R., and Chapin, C., 1975, Geomorphic and tectonic implications of the post-Laramide, late Eocene erosion surface in the Southern Rocky Mountains, *in* Curtis, B.F., ed., Cenozoic history of the Southern Rocky Mountains: Geological Society of America Memoir 144, p. 45–74.

Franczyk, K.J., Fouch, T.D., Johnson, R.C., Molenaar, C.M., and Cobban, W.A., 1991, Cretaceous and Tertiary paleogeographic reconstructions for the Uinta-Piceance basin study area, Colorado and Utah: U.S. Geological Survey Bulletin 1787-Q, 37 p.

Goolsby, S.M., Reade, N.S., and Murray, D.K., 1979, Evaluation of coking coals in Colorado: Colorado Geological Survey Resource Series 7, 72 p., 3 plates.

Grout, M.A., and Verbeek, E.R., 1992, Fracture history of the Divide Creek and Wolf Creek anticlines and its relation to Laramide basin-margin tectonism, southern Piceance basin, northwestern Colorado: U.S. Geological Survey Bulletin 1787-Z, 32 p.

Haun, J.D., and Weimer, R.J., 1960, Cretaceous stratigraphy of Colorado, *in* Weimer, R.J., and Haun, J.D., eds., Guide to the Geology of Colorado: Geological Society of America, Rocky Mountain Association of Geologists, and Colorado Scientific Society Guidebook, p. 58–65.

Hettinger, R.D., Roberts, L.N.R., and Gognat, T.A., 2000, Investigations of the distribution and resources of coal in the southern part of the Piceance Basin, Colorado, Chapter O, *in* Kirschbaum, M.A., Roberts, L.N.R., and Biewick, L.R.H., eds., Geologic assessment of coal in the Colorado Plateau: Arizona, Colorado, New Mexico, and Utah: U.S. Geological Survey Professional Paper 1625-B, http://greenwood.cr.usgs.gov/energy/coal/PP1625B/Reports/Chapters/Chapter_O.pdf

Hucka, B.P., 1991, Analysis and regional implications of cleat and joint systems in selected coal seams, Carbon, Emery, Sanpete, Sevier, and Summit counties, Utah: Utah Geological Survey Special Study 74, 47 p.

Kirkham, R.M., Streufert, R.K., Hemborg, H.T., and Stelling, P.L., 1996, Geologic map of the Cattle Creek Quadrangle, Garfield County: Denver, Colorado Geological Survey Open-File Report 96-1, scale 1:24,000, one plate.

Laubach, S.E., Tremain, C.H., and Ayers, W.B. Jr., 1991, Coal fracture studies: guides for coalbed methane exploration and development: Journal of Coal Quality, v. 10, p. 81–88.

Law, B.E., 1993, The relationship between coal rank and cleat spacing: Implications for the prediction of permeability in coal: Proceedings of the 1993 International Coalbed Methane Symposium, University of Alabama/Tuscaloosa, p. 435–441.

Lee, W.T., 1912, Coal Fields of Grand Mesa and the West Elk Mountains, Colorado: U.S. Geological Survey Bulletin 510, 237 p.

McCulloch, C.M., Deul, M., and Jeran, P.W., 1974, Cleat in bituminous coalbeds: U.S. Bureau of Mines Report of Investigations 7910, 25 p.

Pattison, C.I., Fielding, C.R., McWatters, R.H., and Hamilton, L.H., 1996, Nature and origin of fractures in Permian coals from the Bowen Basin, Queensland, Australia, *in* Gayer, R., and Harris, I., eds., Coalbed Methane and Coal Geology: Geological Society [London] Special Publication 109, p. 133–150.

Robeck, E.D., Harris, R.A., and Hunt, G.L., 2004, The effects of fault-induced stress anisotropy on fracturing, folding and sill emplacement: A study of the Bowie No. 2 coal mine, southern Piceance basin, western Colorado: Geological Society of America Abstracts with Programs, v. 36, no. 4, p. 34.

Robinson Roberts, L.N., and Kirschbaum, M.A., 1995, Paleogeography of the Late Cretaceous of the Western Interior of Middle North America—Coal distribution and sediment accumulation: U.S. Geological Survey Professional Paper 1561, 115 p.

Schopf, J.M., 1959, Field description and sampling of coal beds: Washington, D.C., U.S. Geological Survey Bulletin 1111-B, p. 25–70.

Schultz, J.E., Eakins, W., Scott, D.C., and Teeters, D.D., 2000, Availability of coal resources in Colorado: Somerset Coal Field, west-central Colorado: Colorado Geological Survey Resource Series, v. 38, p. 84.

Scott, A.R., 1996, Coalification, cleat development, coal gas composition and origins, and gas content of Williams Fork coals in the Piceance basin, Colorado, *in* Tyler, R., et al., eds., Geologic and hydrologic controls critical to coalbed methane producibility and resource assessment: Williams Fork Formation, Piceance Basin, northwest Colorado: Gas Research Institute Topical Report GRI-95/0532, p. 220–251.

Scott, A.R., Tyler, R., and Comer, J.B., 1996, Evaluating changes in maximum horizontal stress in the Piceance Basin, Colorado, on the basis of timing of cleat development: Geological Society of America Abstracts with Programs, v. 28, no. 1, p. 63.

Sonnenberg, S.A., and Bolyard, D.W., 1997, Tectonic history of the Front Range in Colorado, *in* Bolyard, D.W., and Sonnenberg, S.A., eds., Geologic History of the Colorado Front Range: Denver, Rocky Mountain Association of Geologists, RMS-AAPG Field Trip 7, p. 1–7.

Steven, T.A., Evanoff, E., and Yuhas, R.H., 1997, Middle and late Cenozoic tectonic and geomorphic development of the Front Range of Colorado, *in* Bolyard, D.W., and Sonnenberg, S.A, eds., Geologic history of the Colorado Front Range: Denver, Rocky Mountain Association of Geologists, RMS-AAPG Field Trip 7, p. 115–134.

Streufert, R.K., Kirkham, R.M., Schroeder, T.J., II, and Widmann, B.L., 1997, Geologic map of the Dotsero Quadrangle, Eagle and Garfield Counties, Colorado: Denver, Colorado Geological Survey Open-File Report 97-2, scale 1:24,000, one plate.

Tremain, C.M., and Tyler, R., 1997, Cleat, fracture, and stress patterns in the Piceance Basin, Colorado: controls on coalbed methane producibility, *in* Hoak, T.E., Klawitter, A.L, and Blomquist, P.K., eds., Fractured reservoirs: characterization and modeling guidebook: Denver, Colorado, Rocky Mountain Association of Geologist, p. 103-114.

Tremain, C.M., Laubach, S.E., and Whitehead, N.H., 1991, Coal fracture (cleat) patterns in Upper Cretaceous Fruitland Formation, San Juan Basin, Colorado and New Mexico: Implications for coalbed methane exploration and development, *in* Schwochow, S., ed., Coalbed methane of western North America: Denver, Rocky Mountain Association of Geologists, Field Conference Guidebook, p. 49–59.

Tweto, O., Steven, T.A., Hail, W.J. Jr., and Moench, R.H., 1976, Preliminary geologic map of the Montrose 1° × 2° quadrangle, southwestern Colorado: U.S. Geological Survey Miscellaneous Field Studies Map MF-761.

Tyler, R., Kaiser, W.R., Scott, A.R., Hamilton, D.S., and Ambrose, W.A., 1995, Geologic and hydrologic assessment of natural gas from coal: Greater Green River, Piceance, Powder River, and Raton Basins, Western United States: Gas Research Institute Report of Investigations 228, 219 p.

Printed in the USA

Geological Society of America
Field Guide 5
2004

West Bijou Site Cretaceous-Tertiary boundary, Denver Basin, Colorado

Richard S. Barclay
Kirk R. Johnson
Denver Museum of Nature & Science, Department of Earth Sciences, 2001 Colorado Blvd., Denver, Colorado 80205, USA

ABSTRACT

The Cretaceous-Tertiary (K-T) boundary section at the West Bijou Site is remarkable because many of the methods used to constrain the position of a terrestrial K-T boundary have been successfully applied to a local section. These include palynology, magnetostratigraphy, shocked quartz and iridium analysis, vertebrate paleontology, geochronology, and paleobotany. The West Bijou Site K-T boundary records the extinction of the *Wodehouseia spinata* Assemblage Zone palynoflora (21%), followed immediately by the presence of a fern-spore abundance anomaly (74%) and the subsequent appearance of the P1 palynoflora. This palynological extinction is coincident with the presence of shock-metamorphosed quartz grains (5+ planes of parallel lamellae) and an iridium spike of 619 ± 32 parts per trillion within the 3-cm-thick boundary claystone. The boundary lies within a reversely magnetized interval, recognized as subchron C29r, substantiated by a radiometrically dated tuff 4.5 m below the boundary with an age of 65.73 ± 0.13 Ma. Dinosaur remains attributable to the late Maastrichtian *Triceratops* Zone were discovered 4 m below the boundary clay, and a partial jaw of a diagnostic Pu1 mammal was discovered 12 m above. Fossil plants are most abundant in the Paleocene and document a low diversity ecosystem recognizable as the southernmost extension of the FUI disaster recovery flora that radiated in North America following the K-T boundary cataclysm.

Keywords: K-T boundary, extinction, iridium, geochronology, Denver Basin, Colorado.

INTRODUCTION

West Bijou Site Field Trip

The purpose of this field trip is to visit the recently discovered West Bijou Site Cretaceous-Tertiary (K-T) boundary located in the eastern part of the Denver Basin, Colorado (Fig. 1). This K-T boundary section was discovered in August 2000 on the west slope of the West Bijou Creek valley (Barclay, 2002). The K-T boundary at this site has been constrained using multiple lines of evidence, combining the efforts of many scientists working

on the Denver Basin Project at the Denver Museum of Nature & Science and the U.S. Geological Survey.

The drive from Denver to the West Bijou Site K-T boundary section takes you east on I-70 across the high plains of eastern Colorado. Layered beneath you are 3700 m of sedimentary rocks that record the erosion of the Ancestral Rocky Mountains, preserve the time when Colorado was covered by a marine seaway, and then document the uplift and erosion of the present Rocky Mountains. Take the Strasburg exit south, the frontage road east for a short distance, and then turn south and continue on this road for 19 km until you reach a "dead end" sign at Arapahoe County

Barclay, R.S., and Johnson, K.R., West Bijou Site Cretaceous-Tertiary boundary, Denver Basin, Colorado, *in* Nelson, E.P. and Erslev, E.A., eds., Field Trips in the Southern Rocky Mountains, USA: Geological Society of America Field Guide 5, p. 59–68. For permission to copy, contact editing@geosociety.org. © 2004 Geological Society of America

Road 46. Turn left and drive for 1 mi until the road turns to the left with a red gate on the right. This is the gated entrance to the West Bijou Site K-T boundary, located on private property managed by the Plains Conservation Center. Permission to access the property must be obtained prior to entrance.

The Plains Conservation Center has been in operation for over 50 years. It presently is operating two sites, the original East Hampden Site and the newly acquired West Bijou Site that contains the K-T boundary section. Its principal goals are to educate the public about the valuable short grass prairie ecosystems that are rapidly disappearing along the Front Range of Colorado, as well as to actively participate in the preservation and study of

these ecosystems. The Plains Conservation Center is a semi-private organization owned and operated by the West Arapahoe Conservation District. You may contact the Plains Conservation Center for more information about their activities at Plains Conservation Center, 21901 E. Hampden Ave., Aurora, Colorado 80013, USA; 303-693-3621; e-mail: info@plainscenter.org; Web site: http://www.plainsconservationcenter.org/.

K-T Boundary Search

Roland Brown located the first North American Cretaceous-Tertiary boundary section on South Table Mountain in Golden,

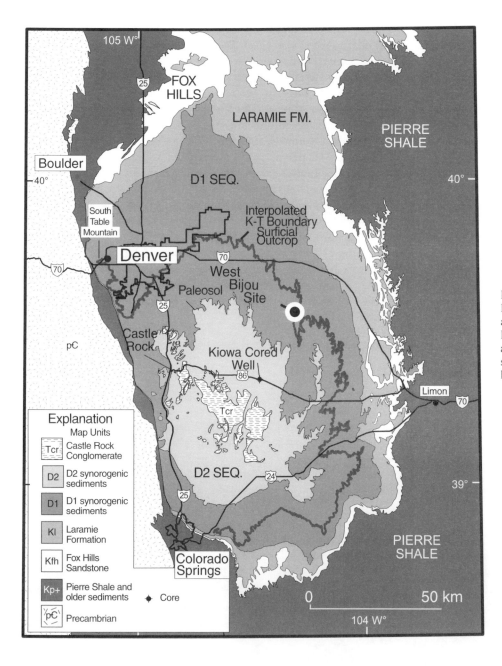

Figure 1. Geological map of Denver Basin synorogenic sedimentary rocks and position of the West Bijou Site. Map created by Robert G. Raynolds and Adrian Kropp using ArcInfo GIS software database of subsurface electric log data interpolated to surface.

Colorado (Brown, 1943). He found Paleocene mammals 24 m above *Triceratops* dinosaur remains and placed the boundary level just below the mammal-bearing horizon. The discovery of iridium at the K-T boundary (Alvarez et al., 1980) and the resulting extraterrestrial impact hypothesis for K-T extinctions demanded greater stratigraphic resolution and presented new tools to achieve that resolution. In the 1980s, precisely located K-T boundary layers were described in the Raton, Powder River, and Williston Basins, but no one was able to locate the precise K-T boundary in the Denver Basin. Roland Brown's K-T boundary at South Table Mountain (Brown, 1943) occurs in the coarse-grained facies typical of the proximal synorogenic deposits. As such, it did not provide appropriate strata necessary for fine-scale palynology, magnetostratigraphy, and geochemistry (Kauffman et al., 1990; Benson, 1998).

The Denver Basin Project began in 1997 with the goals of creating a chronostratigraphic framework for the Denver Basin synorogenic sediments and locating an exposure of the K-T boundary that contained evidence of extraterrestrial impact (Raynolds and Johnson, 2003). The work leading up to this project began in 1991, when scientists at the Denver Museum of Nature & Science started collecting plant and vertebrate fossils in the Denver metro area. Due to the paucity of continuous outcrop in the Denver Basin, the relationship of these localities to each other was not understood. In 1999, Kirk Johnson and Bob Raynolds coordinated a project to drill a 688-m-deep cored well in the town of Kiowa, Colorado (Raynolds and Johnson, 2002). The Kiowa Core served as a continuous subsurface stratigraphic record in the middle portion of the Denver Basin, useful for correlating the poorly constrained surface outcrops and fossil localities. The core was systematically analyzed using palynology and magnetostratigraphy to aid in correlation of surface sites and also to determine the position of the K-T boundary (Nichols and Fleming, 2002; Hicks et al., 2003). Once these samples were analyzed and the K-T boundary level was identified at a depth of 302 m, the Kiowa Core was used to calibrate a three-dimensional model of the basin's stratigraphy based on electric well logs (Raynolds, 2002). The resulting model was used to constrain the age and stratigraphic position of the numerous surface localities. The search for the boundary focused on the eastern portion of the Denver Basin because the distal portions of the synorogenic sequence contained the fine-grained rocks necessary to preserve the detail of events surrounding the K-T boundary extinction.

As a first step in constraining surface localities, the K-T boundary level in the Kiowa Core was projected onto a digital elevation model of the surface using ArcInfo GIS software and mapped relative to existing geological mapping and previously drawn K-T boundary lines. This interpolation used data from the recently drilled Kiowa Core, the Castle Pines Core drilled on the western side of the basin (Robson and Banta, 1993), and known surface exposures of the K-T boundary interval in the towns of Golden and Colorado Springs, Colorado.

Using the interpolated K-T boundary GIS map, potential outcrop areas were scouted by small aircraft and later spot-sampled

for pollen analysis. The West Bijou Site was chosen for intensive study because it contained a 50 m exposure of fine-grained fossiliferous strata with a Cretaceous pollen sample at the base and fossil leaves of early Paleocene aspect at the top. Analysis of systematically collected pollen samples through the 50 m of exposure subsequently allowed us to narrow the K-T boundary to the top of a 3 cm claystone in the middle of the hillside.

GEOLOGICAL SETTING

Denver Basin

The Denver Basin is a Rocky Mountain foreland basin that contains Late Cretaceous and Paleogene synorogenic sediments derived from the erosion of the Front Range mountains during the Laramide Orogeny. This asymmetrical geological structure stretches from Boulder in the north, down to Colorado Springs in the south, and eastward to Limon. The beds dip steeply on the western margin of the basin and become increasingly flat-lying toward the east.

Uplift of the Front Range occurred in two phases. The first commenced during the latest Cretaceous, ca. 68 Ma. Sedimentation continued in the Denver Basin for ~4 m.y., producing an unconformity-bounded package that Raynolds (2002) has termed the D1 sequence. The D1 sequence is predominantly arkosic near Colorado Springs, predominantly andesitic near Denver, and is a mixture of arkosic and andesitic in the center and eastern portions of the basin. A period of ~8 m.y. of nondeposition followed. Sediment accumulation began again in the last million years of the Paleocene, producing the D2 sequence, perhaps in response to renewed uplift along the front. This D2 sequence is an arkosic megafan derived from the Pikes Peak area west of Colorado Springs (Raynolds, 2002).

West Bijou Site

The West Bijou Site K-T boundary occurs in the D1 sequence in rocks previously mapped both as the Dawson Arkose and the Denver Formation (Dane and Pierce, 1936; Reichert, 1956). Raynolds (1997, 2002) places the Denver Formation and Dawson Arkose into his D1 sequence, and we put the exposures at the West Bijou Site in the middle of the 394-m-thick D1 sequence (thickness from the Kiowa Core). The beds at the site are effectively flat-lying, as the regional dip in this area is less than one-half degree to the west. In addition, the West Bijou Creek escarpment where the K-T boundary is located is roughly parallel to strike (Barclay et al., 2003).

Figure 2 is a detailed map of the field area created using differential GPS and contains the location of the three stops discussed in this field guide. The panel diagram in Figure 3 correlates the measured sections and positions of the sampling localities. A field datum was established for the field area surrounding the K-T boundary exposures to constrain the stratigraphic sections measured in the five gullies studied (Figs. 2 and 3). The

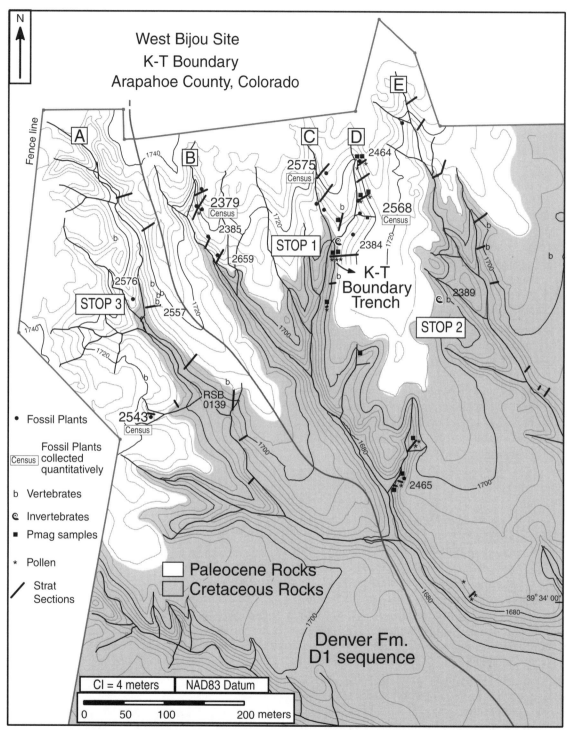

Figure 2. Topographic contour map (scaled in meters) of the field area on the Plains Conservation Center. Map was created using a Trimble® XRS differentially corrected GPS unit. The average horizontal and vertical precision was <1 m, providing the necessary detail to accurately plot the dense set of data. All Trimble® XRS field data was differentially corrected from a base station in Denver. At least 60 readings were taken per position, with less than 6% dilution of precision and a minimum of six satellites in view. Drainages were walked to define the topographic lows. Contour map created by Richard Barclay, Mark Gorman, Nicole Boyle, and Adrian Kropp. Research conducted on private property owned by the West Arapahoe Conservation District, managed by the Plains Conservation Center. See Figure 3 for lithologic descriptions of gullies A–E within the D1 sequence (Denver Formation).

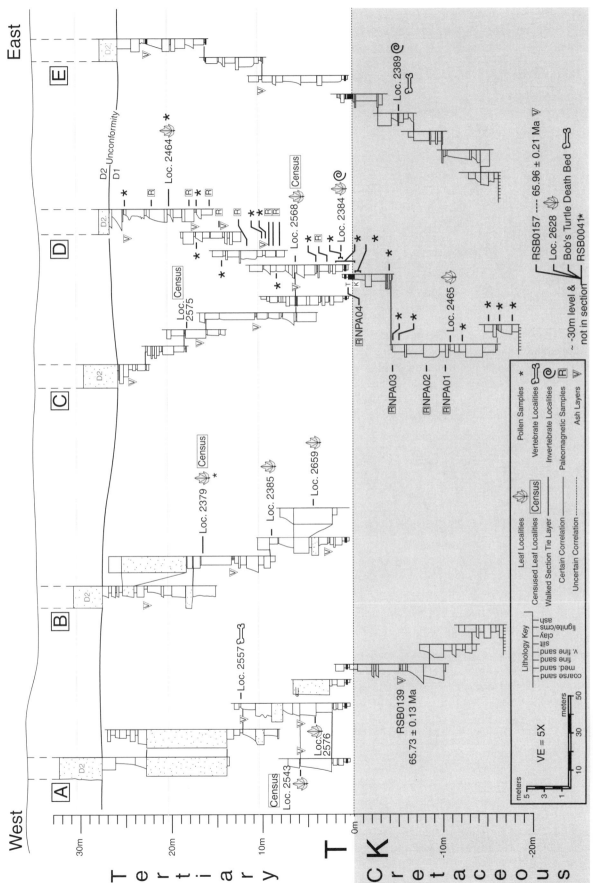

Figure 3. Stratigraphic correlation diagram within D1 sequence (Denver Formation) of the West Bijou Site K–T boundary section on the Plains Conservation Center. Positions of gullies A–E shown in Figure 2.

lowest in a diagnostic series of tuff beds, nicknamed the "Little Dude Ashes," forms the field datum. It has been trenched in each gully and correlated by walking along the characteristic series of tuff beds. It is present in the main research trench (Figs. 2–4).

The Paleocene D1 lithologies at the West Bijou Site are predominantly mudstone, siltstone, very fine-grained sandstone, and lignite. They preserve the depositional environments of floodplains, lakes, stream channels, and floodplain swamps, respectively. The Cretaceous strata contain fewer lignite beds, more sandstone units, and an absence of lacustrine siltstone units that are abundant in the Paleocene section and commonly preserve fossil leaves. The D1 sequence of strata is unconformably overlain by a thin lithified unit of very coarse sandstone of granitic provenance ascribed to the D2 sequence.

STOP 1. WEST BIJOU SITE K-T BOUNDARY TRENCH

Palynology

In the summer of 2000, a Cretaceous pollen sample was collected at the bottom of the 50 m of exposure at the West Bijou Site. Since fossil plants of probable Paleocene age were previously known from near the top of the exposures, we systematically collected pollen samples through the section in order to narrow the position of the K-T boundary. This initial effort constrained the boundary to a 3 m interval where we dug a deep section to expose fresh rock for detailed sampling. The palynological analysis conducted by Nichols and Fleming (2002) on these samples narrowed the K-T boundary to a 1 cm interval directly above a 3-cm-thick claystone bed found within a 90-cm-thick lignite. Palynomorphs attributable to the *Wodehousia spinata* Assemblage Zone are found in all samples below and within the 3 cm claystone. A 50% decrease of the species of palynomorphs is observed directly above this claystone. Nichols and Fleming (2002) note that this 50% local extirpation actually represents a 21% extinction because some of the taxa absent from the Paleocene of the West Bijou Site are present in other basins in the Western Interior above the K-T boundary layer (Nichols and Fleming, 2002). The boundary is overlain by a 4-cm-thick anomaly where fern spore abundance spikes to 74% from a Cretaceous background level of less than 5%. The abundance of fern spores then diminishes to 8% at the 7 cm level above the boundary claystone. The fern spore interval is replaced by a pollen assemblage attributable to palynofloral Zone P1 of the early Paleocene, as described by Nichols and Ott (1978).

Magnetostratigraphy

Jason Hicks of the Denver Museum of Nature & Science collected samples from twelve sites throughout the 50 m section at the West Bijou Site for paleomagnetic analysis (Hicks et al., 2003; Figs. 2 and 3 herein). All samples are of reversed magnetic polarity, placing the exposure in subchron C29r, based upon the association with the K-T palynological transition. This inter-

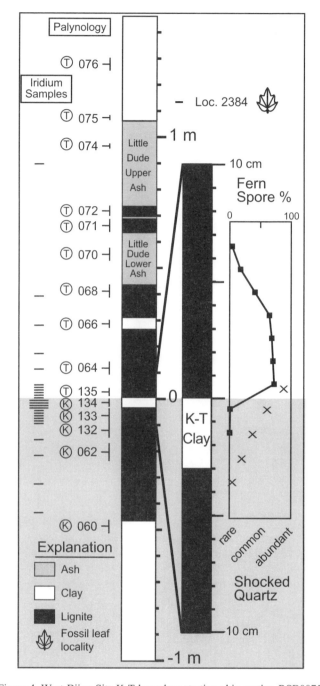

Figure 4. West Bijou Site K-T boundary stratigraphic section RSB0075. Iridium and pollen sample numbers are Denver Basin Project numbers (e.g., DBP-00-076). Iridium anomaly present in samples DBP-00-107, 108, 109, and 110 identified by double width bars on the left of the diagram.

pretation is supported by radiometric dates (see geochronology section). The reversed paleomagnetic signature of the rocks at the West Bijou Site provides the best means for determining the amount of time represented by the exposure. The duration of the Paleocene portion of subchron C29r is thought to be ~270,000 yr based upon the cyclostratigraphy of D'Hondt et al. (1996). In the

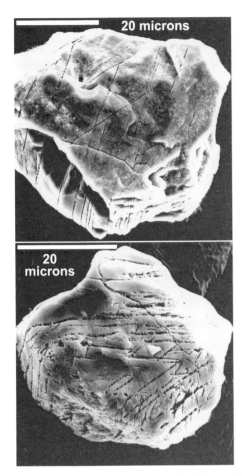

Figure 5. Electron-microscopic images of HF-etched shocked quartz grains from West Bijou Site K-T boundary claystone, section RSB0075. Lower grain shows 5+ sets of parallel shock lamellae.

Kiowa Core, the complete Paleocene section of subchron C29r is 35 m thick, but only 28 m (80%) of this section is preserved at the West Bijou Site. Based on this, we estimate that the Paleocene exposure at the West Bijou Site was deposited over ~216,000 yr. The majority of the fossil plant specimens are contained within the Paleocene section, so the recovery interval has been the focus of much of the paleobotanical work conducted at this boundary section.

Shocked Quartz and Iridium

The palynological extinction above the 3 cm boundary claystone directly coincides with both shocked quartz and iridium anomalies (Fig. 4). Five samples taken from a 5 cm interval across the K-T boundary claystone were analyzed by Bill Betterton of the U.S. Geological Survey for presence of shocked quartz grains. He found an increasing abundance of shocked quartz grains, with the highest concentration of grains occurring at the same stratigraphic position as the palynological K-T boundary (Barclay et al., 2003). Electron microscopic images show that these quartz grains record high levels of shock metamorphism, containing 5+ sets of parallel shock lamellae (Fig. 5). The HF-etching technique used by Betterton is described in Appendix F of Barclay (2002).

Frank Asaro at the Lawrence Berkeley National Laboratory analyzed samples from the West Bijou Site K-T boundary for presence of iridium with the Luis W. Alvarez Iridium Coincidence Spectrometer after neutron activation at the University of Missouri Columbia Research Reactor Center (Barclay et al., 2003). These samples were collected through a 3-m-thick section that included the palynological extinction and the shocked quartz grains (Fig. 4). He determined that there was an anomalously high Ir abundance of 619 ± 32 parts per trillion (ppt) within the 3 cm boundary claystone against a background level of <10 ppt. Iridium levels above 200 ppt in the interval 20 cm below the level of the maximum anomaly complicate the interpretation of this section.

Geochronology

The West Bijou Site contains multiple horizons of volcanic tuffs, both above and below the K-T boundary claystone. The majority of these tuffs have not been radiometrically dated but do contain sanidine and zircon crystals useful for analysis. Two tuffs were dated (4.5 m and ~30 m below the claystone) using the $^{40}Ar/^{39}Ar$ method on sanidine crystals (Hicks et al., 2003). The tuff 4.5 m below the K-T boundary was dated at 65.73 ± 0.13 Ma and is located within gully A, shown in Figures 2 and 3. This date supports the interpretation that the reversed polarity of the West Bijou paleomagnetic samples represents subchron C29r (Hicks et al., 2003).

The tuff ~30 m below the K-T boundary was dated at 65.96 ± 0.21 Ma, but it was not possible to directly measure its position in the stratigraphic section because it lies outside of the area mapped in Figure 2 and is ~4.5 km to the south of the main field area. Its stratigraphic position was determined solely upon its relative elevation compared to the boundary claystone. This tuff (RSB0041) is located in a small cutbank outcrop of West Bijou Creek at the coordinates 39°31′53″N, 104°17′38″W, NAD27. Both age determinations for these volcanic tuffs are consistent with all other West Bijou Site data and the recently recalibrated age of the K-T boundary as 65.51 ± 0.10 Ma (Hicks et al., 2002).

STOP 2. CRETACEOUS VERTEBRATE AND INVERTEBRATE FOSSIL SITE

Hadrosaurian dinosaur teeth and a ceratopsian dorsal vertebra were discovered in a thin, iron-rich sandstone located 4 m below the K-T boundary (DMNH loc. 2389; determinations made by Ken Carpenter of the Denver Museum of Nature & Science). The presence of these specimens provides evidence that the late Maastrichtian *Triceratops* Zone is present in the section below the palynological extinction horizon (Barclay et al., 2003).

This is the only locality at the West Bijou Site that has produced dinosaurian fossil material. The fauna from this site also includes turtles, crocodiles, champsosaurs, gar fish, lizards, gastropods, and bivalves.

Jaelyn Eberle of the University of Colorado at Boulder collected a partial jaw of the early Paleocene mammal *Protungulatum donnae* (DMNH loc. 2557, DMNH specimen 44371) 12 m above the K-T boundary (Eberle, 2003). This taxon is diagnostic of the basal portion of the Puercan North American Land Mammal "Age" (NALMA), dating the Paleocene strata at the West Bijou Site as Pu1.

STOP 3. PALEOCENE FOSSIL PLANT LOCALITY

West Bijou Site Paleobotanical Record

The Paleocene rocks at the West Bijou Site contain abundant fossil leaves, while the Cretaceous section yielded only two poor localities. This discrepancy is not well understood; however, the Cretaceous portion does not seem to contain the siltstone facies where the fossil leaves are abundantly preserved in the overlying Paleocene. The only age-diagnostic plant species found at the site is *Paranymphaea crassifolia* (Newberry) Berry, 4 m above the K-T boundary claystone (DMNH loc. 2567). *Paranymphaea crassifolia* is only known from the early Paleocene FUI megafloral zone of the northern Great Plains, which corresponds to the Puercan NALMA and an undetermined portion of the Torrejonian NALMA.

We collected over 2300 specimens from nine Paleocene localities, with 1548 of those specimens collected from four localities using a quantitative census method (Barclay et al., 2003). The sampled megaflora consists of 49 morphotypes and is taxonomically dominated by dicotyledonous angiosperms (74%), monocotyledonous angiosperms (10%), ferns and fern allies (11%), and conifers (5%). Diversity is low in the 28 m of Paleocene strata. The localities average 11.5 morphotypes at a rarefied richness level of 300 specimens per locality, with a range of 7.5–14.5 morphotypes per locality (Barclay et al., 2003). The localities are strongly dominated by a few morphotypes. The six most dominant taxa (Fig. 6) constitute 88% of the total specimens collected from the quantitatively collected localities. No statistically significant trend toward increasing diversity is observed during the first 28 m of the Paleocene at the West Bijou Site.

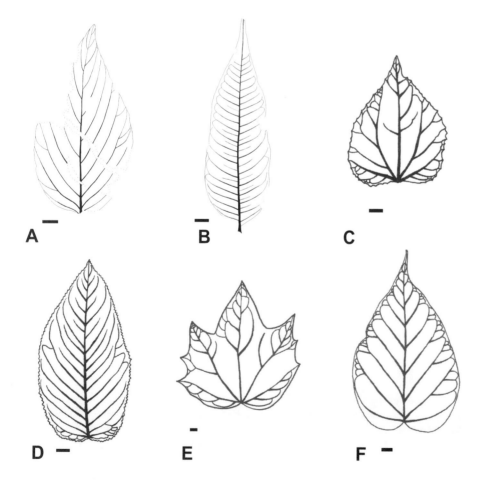

Figure 6. Line drawings of dominant Paleocene species at West Bijou Site. These six taxa dominate the flora, representing 88% of total specimens collected. (A) *Cornophyllum newberryi* (Hollick) McIver and Basinger (BC10 = 23.3%). (B) cf. *Averrhoites affinis* Hickey (BC09 = 21.6%). (C) *"Populus" nebrascensis* Newberry (BC37 = 15.6%). (D) *Dicotylophyllum anomalum* (Ward) Hickey (BC36 = 10.4%). (E) *"Cissites" panduratus* Knowlton (BC35 = 9.2%). (F) *Penosphyllum cordatum* (Ward) Hickey (BC27 = 7.9%). BC numbers refer to the Bijou Creek morphotype number and percentage of specimens from combined collections at all Paleocene sites on the West Bijou Site. Scale bar lengths: 1 cm.

The Paleocene West Bijou Site flora represents the southern-most example known of the FUI disaster-recovery flora, a flora that rose to dominance following the K-T boundary cataclysm in North America (Johnson, 2002; Barclay et al., 2003). This FUI flora spread across more than 400,000 km², stretching from Denver to southern Saskatchewan. The West Bijou Site contains the seven most abundant taxa present in the early Paleocene Fort Union Formation of the Williston Basin of North Dakota (700 km) and shares nine taxa in common with the early Paleocene Raven-scrag flora of southern Saskatchewan (1100 km). While the West Bijou Site shares many taxa in common with early Paleocene floras of the northern Great Plains, there is almost no overlap between coeval floras on the western margin of the Denver Basin, where high diversity floras exhibit rainforest characteristics in the early Paleocene (Johnson and Ellis, 2002; Johnson et al., 2003).

Paleoecology and Paleoclimate

The West Bijou Site is located in the distal portion of the Denver Basin. It was once a low relief floodplain dominated by broad-leafed angiosperms, with small, ephemeral lakes and occasional conifer-dominated swamps. A mean annual temperature estimate of 18.6 ± 2.6 °C (Fig. 7; estimate calculated using 34 dicot leaf morphotypes using the leaf margin analysis method of Wilf, 1997) is consistent with the interpretation of a warm climate for the Denver Basin in the early Paleocene (Johnson et al., 2003; Barclay et al., 2003). The presence of palm fronds in the localities suggests that the ecosystem did not experience sustained freezing (Sakai and Larcher, 1987). Mean annual precipitation was also elevated, with an average of 155 cm/yr (Fig. 8; using the leaf area analysis method of Wilf et al., 1998), consistent with the presence of many small, ephemeral lakes, and aquatic vertebrate taxa (Barclay et al., 2003). Versions of Figures 7 and 8 published in Barclay et al. (2003) contain significant typographical errors, which are corrected here.

REFERENCES CITED

Alvarez, L.W., Alvarez, W., Asaro, F., and Michel, H.V., 1980, Extraterrestrial cause for the Cretaceous-Tertiary extinctions: Science, v. 208, p. 1095–1108.

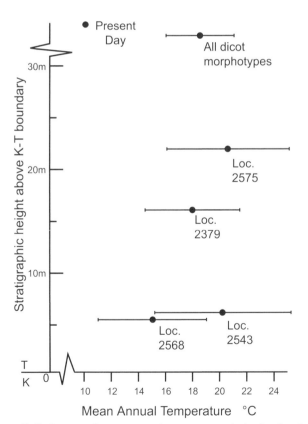

Figure 7. Estimates of mean annual temperature derived using leaf margin analysis on quantitatively collected Paleocene floras. The West Bijou Site value was calculated using 34 dicots from nine Paleocene localities. Present-day Denver mean annual temperature is from 51 mean values (modern data from High Plains Regional Climate Center, University of Nebraska).

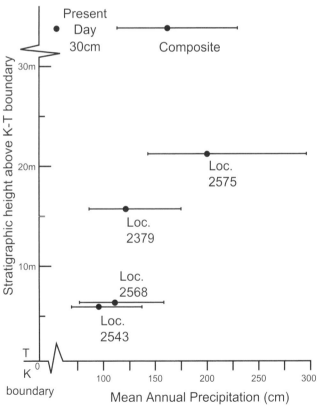

Figure 8. Estimates of mean annual precipitation derived using leaf margin analysis on quantitatively collected Paleocene floras. The West Bijou Site composite value was calculated using data from all four censused localities. Present-day mean annual precipitation for Denver is from 51 mean values (modern data from High Plains Regional Climate Center, University of Nebraska).

Barclay, R.S., 2002, The Cretaceous-Tertiary boundary and plant diversity in the earliest Paleocene, eastern Denver Basin, Colorado [M.S. thesis]: Gainesville, University of Florida, 195 p.

Barclay, R.S., Johnson, K.R., Betterton, W.J., and Dilcher, D.L., 2003, Stratigraphy and megaflora of a K-T boundary section in the eastern Denver Basin: Rocky Mountain Geology, v. 38, p. 45–71.

Benson, K.P., 1998, Floral diversity and paleoclimate of the Latest Cretaceous and Early Tertiary deposits, Denver Basin, Colorado, USA [B.A. Thesis]: Colorado Springs, Colorado College, 179 p.

Brown, R.W., 1943, Cretaceous-Tertiary boundary in the Denver Basin, Colorado: Geological Society of America Bulletin, v. 54, p. 65–86.

Dane, C.H., and Pierce, W.G., 1936, Dawson and Laramie formations in southeastern part of the Denver Basin, Colorado: AAPG Bulletin, v. 20, p. 1308–1328.

D'Hondt, S., Herbert, T.D., King, J., and Gibson, C., 1996, Planktic foraminifera, asteroids, and marine production: Death and recovery at the Cretaceous-Tertiary boundary, *in* Ryder G., Fastovsky D., and Gartner S., eds., The Cretaceous-Tertiary event and other catastrophes in Earth history: Geological Society of America Special Paper 307, p. 303–317.

Eberle, J.J., 2003, Puercan mammalian systematics and biostratigraphy in the Denver Formation, Denver Basin, Colorado: Rocky Mountain Geology, v. 37, p. 143–169.

Hicks, J.F., Johnson, K.R., Tauxe, L., Clark, D., and Obradovich, J.D., 2002, Magnetostratigraphy and geochronology of the Hell Creek and basal Fort Union Formations of southwestern North Dakota and a recalibration of the Cretaceous-Tertiary boundary, *in* Hartman, J.H., Johnson, K.R., and Nichols, D.J., eds., The Hell Creek Formation and the Cretaceous-Tertiary boundary in the northern Great Plains: An integrated continental record of the end of the Cretaceous: Geological Society of America Special Paper 361, p. 35–56.

Hicks, J.F., Johnson, K.R., Obradovich, J.D., Miggins, D.P., and Tauxe, L., 2003, Magnetostratigraphy of Upper Cretaceous (Maastrichtian) to lower Eocene strata of the Denver Basin, Colorado: Rocky Mountain Geology, v. 38, p. 1–27.

Johnson, K.R., 2002, Megaflora of the Hell Creek and lower Fort Union Formations in the western Dakotas: Vegetational response to climate change, the Cretaceous-Tertiary boundary event, and rapid marine transgression, *in* Hartman, J.H., Johnson, K.R., and Nichols, D.J., eds., The Hell Creek Formation and the Cretaceous-Tertiary boundary in the northern Great Plains: An integrated continental record of the end of the Cretaceous: Geological Society of America Special Paper 361, p. 329–390.

Johnson, K.R., and Ellis, B., 2002, A tropical rainforest in Colorado 1.4 million years after the Cretaceous-Tertiary boundary: Science, v. 296, p. 2379–2383, doi: 10.1126/SCIENCE.1072102.

Johnson, K.R., Reynolds, M.L., Benson, K.P., Werth, K.W., and Thomasson, J.R., 2003, Overview of the Late Cretaceous, early Paleocene, and early Eocene megaflora of the Denver Basin, Colorado: Rocky Mountain Geology, v. 38, p. 101–120.

Kauffman, E.G., Upchurch, G.R., and Nichols, D.J., 1990, The Cretaceous-Tertiary boundary interval at South Table Mountain, near Golden, Colorado, *in* Kauffman, E.G., and Walliser, O.H., eds., Extinction Events in Earth History: Berlin, Springer-Verlag, p. 365–389.

Nichols, D.J., and Fleming, R.F., 2002, Palynology of the Denver Basin: Rocky Mountain Geology, v. 37, p. 135–163.

Nichols, D.J., and Ott, H.L., 1978, Biostratigraphy and evolution of the *Momipites-Caryapollenites* lineage in the early Tertiary in the Wind River Basin, Wyoming: Palynology, v. 2, p. 93–112.

Raynolds, R.G., 1997, Synorogenic and post-orogenic strata in the central Front Range, Colorado, *in* Bolyard, D.W., and Sonnenberg, S.A., eds., Geologic history of the Colorado Front Range: Denver, Rocky Mountain Association of Geologists Field Trip Guide, p. 43–47.

Raynolds, R.G., 2002, Laramide synorogenic strata of the Denver Basin: Rocky Mountain Geology, v. 37, p. 111–134.

Raynolds, R.G., and Johnson, K.R., 2002, Drilling of the Kiowa Core, Elbert County, Colorado: Rocky Mountain Geology, v. 37, p. 105–109.

Raynolds, R.G., and Johnson, K.R., 2003, Synopsis of the stratigraphy and paleontology of the uppermost Cretaceous and lower Tertiary strata in the Denver Basin: Rocky Mountain Geology, v. 38, p. 171–181.

Reichert, S.O., 1956, Post-Laramide stratigraphic correlation in the Denver Basin, Colorado: Geological Society of America Bulletin, v. 67, p. 107–112.

Robson, S.G., and Banta, E.R., 1993, Data from core analyses, aquifer testing, and geophysical logging of Denver Basin bedrock aquifers at Castle Pines, Colorado: U.S. Geological Survey Open File Report 93-442, 59 p.

Sakai, A., and Larcher, W., 1987, Frost survival of plants: responses and adaptation to freezing stress: Berlin, Springer-Verlag, 321 p.

Wilf, P., 1997, When are leaves good thermometers? A new case for Leaf Margin Analysis: Paleobiology, v. 23, p. 373–390.

Printed in the USA

Geological Society of America
Field Guide 5
2004

Buried Paleo-Indian landscapes and sites on the High Plains of northwestern Kansas

Rolfe D. Mandel
Kansas Geological Survey, 1930 Constant Avenue, University of Kansas, Lawrence, Kansas 66047, USA

Jack L. Hofman
Department of Anthropology, University of Kansas, Lawrence, Kansas 66045, USA

Steven Holen
Department of Anthropology, Denver Museum of Nature & Science, Denver, Colorado 80205, USA

Jeanette M. Blackmar
Museum of Anthropology, University of Kansas, Lawrence, Kansas 66045, USA

ABSTRACT

Recent geoarchaeological research on the High Plains of northwestern Kansas has yielded new information regarding the location of buried landscapes that may harbor the material record of the earliest humans in the region. Soil-stratigraphic and geomorphic investigations in playas and in the valleys of small, intermittent streams (draws) located high in drainage networks indicate these geomorphic setting were zones of slow sedimentation and episodic soil development during the terminal Pleistocene and early Holocene. Deeply buried paleosols in the draws have yielded Clovis and possibly pre-Clovis cultural deposits, and a landscape/soil-stratigraphic model has been developed to systematically search for additional early sites in buried contexts. This field guide presents results of geoarchaeological investigations at five Paleo-Indian sites in the upper Beaver Creek drainage basin. Also, locations with buried Paleo-Indian landscapes but no recorded sites are described.

Keywords: geoarchaeology, High Plains, Paleo-Indian, pre-Clovis, draws, playas, paleosols.

INTRODUCTION

The emergence of archaeological geology, or geoarchaeology, in North America is strongly linked to Paleo-Indian studies in the Great Plains (Mandel, 2000a). These studies began in the mid to late 1920s with the discoveries at the Folsom site in New Mexico, but it was work at the Clovis site (New Mexico) during the 1930s that established a tradition of integrating geoscientific investigations with Paleo-Indian research (Holliday, 1997, p. 1).

This tradition has persisted into the twenty-first century, and geoarchaeology continues to play a significant role in analyzing early sites in the Great Plains. Geoscientific methods also have been used to develop predictive models for locating stratified late Wisconsinan and early Holocene cultural deposits in the region, a point that will be stressed during this field trip.

The open grasslands of the midcontinent have yielded some of the most important early sites in the Western Hemisphere (Holliday, 1997, p.1; Hofman and Graham, 1998; Stanford,

Mandel, R.D., Hofman, J.L., Holen, S., and Blackmar, J.M., Buried Paleo-Indian landscapes and sites on the High Plains of northwestern Kansas, *in* Nelson, E.P. and Erslev, E.A., eds., Field trips in the southern Rocky Mountains, USA: Geological Society of America Field Guide 5, p. 69–88. For permission to copy, contact editing@geosociety.org. © 2004 Geological Society of America

1999). Although material remains of Paleo-Indians have been discovered throughout the Great Plains, many of the significant sites, including Clovis, Plainview, Lubbock Lake, Lindenmeir, Hell Gap, Scottsbluff, Olsen-Chubbuck, Lime Creek, Dutton, Lange-Ferguson, and Agate Basin, are on the High Plains. The Southern High Plains of Texas and New Mexico and the Western High Plains of Colorado and Wyoming have especially high concentrations of recorded early sites (Holliday, 1997; Albanese, 2000). However, this pattern does not hold up on the High Plains of western Kansas. Despite the numerous finds of Paleo-Indian projectile points on uplands and in streambed contexts across Kansas, few camps and kill sites predating 9 ka[1] have been documented in the state (Hofman, 1996). The dearth of recorded stratified early sites is especially apparent on the High Plains of Kansas, a region that should have been attractive to the early human inhabitants of North America.

The paucity of recorded early sites on the Kansas High Plains is partially a result of insufficient archaeological research in the region. Numerous archaeological surveys, many associated with reservoir and highway construction projects, have been conducted in eastern Kansas, whereas western Kansas has witnessed few systematic surveys. Furthermore, there is strong evidence suggesting that the low number of recorded early sites in western Kansas is a product of the filtering effects of geomorphic processes on the regional archaeological record (Mandel, 1992, 1995, 2000b, 2001a, 2005; Mandel and Hofman, 2003; Mandel et al., 2004). Specifically, the two geomorphic settings and associated micro-environments that would have been most attractive to the early residents of the High Plains—stream valleys and playas—were zones of sedimentation and episodic soil development during the terminal Pleistocene and early Holocene (ca. 13–9 ka) (Mandel, 1995, 2005). Consequently, Paleo-Indian and pre-Clovis landscapes that may harbor in situ cultural deposits are deeply buried and are rarely detected using traditional archaeological survey techniques.

The primary objective of this one day field trip is to demonstrate how late Quaternary landscape evolution has affected the distribution and detection of Paleo-Indian and earlier sites on the Kansas High Plains. Recent research has focused on temporal and spatial patterns of late Quaternary erosion, sedimentation, and landscape stability in stream valleys and playas throughout western Kansas (Mandel, 1995, 2005). Information gleaned from this research provides the basis for determining whether sedimentary deposits of certain ages are systematically preserved in valley and upland landscapes. From an archaeological perspective, it is reasonable to assume that sites predating 9 ka will be found only where there are sedimentary deposits old enough to contain them. A corollary is that where sufficiently thick deposits postdating 9 ka are present, evidence of these sites will not be found on the modern land surface.

In order to demonstrate our landscape/soil-stratigraphic model for predicting where buried Paleo-Indian and pre-Clovis sites may occur on the Kansas High Plains, five localities in the upper Beaver Creek drainage basin of northwestern Kansas will be visited (Fig. 1). The first stop is the Kanorado locality in Middle Beaver Creek valley near the town of Kanorado, Kansas. Three stratified Paleo-Indian sites at this locality are associated with alluvial deposits stored in a draw. Site 14SN105 is an especially important and intriguing find because it contains

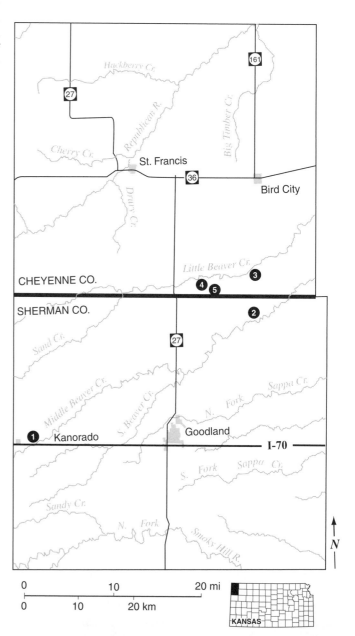

Figure 1. Map of Sherman and Cheyenne counties, northwestern Kansas. Field trip stops: 1—Kanorado locality; 2—the Laird Paleo-Indian site; 3—Steve Busse Sand and Gravel Pit; 4—Powell Clovis site and locality; and 5—Busse playa locality.

---

[1]All ages are given in uncalibrated ^{14}C yr before present (B.P.), unless otherwise noted. Calibrated ages are denoted with "cal."

a possible lithic artifact with a concentration of mammoth and *Camelops* bones more than 2.5 m below a likely Clovis horizon. Stratigraphic sections and archaeological excavation units will be examined during the field trip. Also, faunal remains and lithic artifacts recovered during excavations will be on display at the sites. A radiocarbon chronology for site 14SN105 should be available by the time of the field trip.

The second stop is at the Laird site (14SN2), a Late Paleo-Indian bison kill in Beaver Creek valley northeast of Goodland, Kansas (Fig. 1). The site is represented by a bison bone bed contained in gully fills on the side-slope of the valley wall. One of the artifacts found in association with the bone bed is believed to be a Dalton type point. This point and other artifacts will be shown during the site visit, and field-trip participants will have an opportunity to examine the soil-stratigraphic context of the bison bone bed. Although Dalton sites are common in the forested Southeast United States and have been documented along the forest-prairie border, they are rarely recorded in the High Plains region. Ongoing investigations at Laird are shedding new light on the nature of Dalton subsistence strategies on the open grasslands of the High Plains.

The third, fourth, and fifth stops are in Little Beaver Creek basin south of Bird City, Kansas (Fig. 1). Stop 3 is at the Steve Busse Sand and Gravel Pit, where a thick package of late Wisconsinan and early Holocene valley fill is stored in a draw. Deeply buried soils representing former stable landscapes that may have been occupied by Paleo-Indians are exposed in a 6-m-high section at this locality. The field trip participants will have an opportunity to examine this section, as well as a late Holocene alluvial section at a lower position in the valley landscape. At the fourth stop (Powell site), ~10 km upstream from Stop 3, alluvial deposits and associated buried paleosols dating to the Pleistocene-Holocene transition will be examined along Little Beaver Creek. A Clovis point and *Camelops* and mammoth bones were discovered on the floor of the draw at Stop 4. This find underscores the geologic potential for buried Paleo-Indian cultural deposits in draws on the High Plains. The fifth stop is a locality with several playas on the uplands overlooking Little Beaver Creek valley. Playas were undoubtedly "magnets" for game and the earliest human occupants on the High Plains. Therefore, an understanding of the soil-stratigraphy and geochronology of playa deposits is essential to locating the region's earliest sites. At Stop 5, a ^{14}C-dated soil-stratigraphic sequence will be examined in a backhoe trench on the floor of a playa.

By the end of the field trip, participants should have a good understanding of the late Quaternary soil-stratigraphic record in the upper Beaver Creek drainage basin. This knowledge provides the basis for interpreting temporal and spatial patterns in the regional archaeological record. It is also invaluable to the search for stratified Paleo-Indian and pre-Clovis cultural deposits on the High Plains.

A road log is not included with this field trip. Unfortunately, archaeological sites, especially early ones, are threatened by artifact collectors who turn to the literature for information. The recent destruction of large portions of the Burntwood Creek site in northwestern Kansas is a case in point. Hence, we do not want to reveal the precise locations of the sites that will be visited during this trip.

PHYSIOGRAPHY AND REGIONAL GEOLOGY

The study area for this field trip is in the semiarid High Plains physiographic subprovince of Fenneman's (1931) Great Plains physiographic province. The High Plains region forms most of the western one-third of Kansas. Although the edge of the High Plains is heavily dissected, this region is essentially a plateau characterized by broad reaches of flat uplands with poorly developed surface drainage.

The geologic history of the High Plains can be traced back to the evolution of the Rocky Mountains. As this mountain system was slowly uplifted during the Tertiary period, large volumes of rock were eroded from its slopes and transported eastward by streams. These streams eventually carried the sediments into eastern Colorado and western Kansas. Wilson (1984, p. 33) wrote,

So great was the mass of eroded material that it literally overflowed the stream valleys and spread out over the uplands. By the end of the Tertiary period, the upper surface of this immense sheet of sand and gravel formed a gently eastward-sloping plain extending from the eastern front of the Rockies to the west slopes of the Flint Hills in central Kansas.

The High Plains of western Kansas represent the uneroded remnants of this extensive plain, and deposits of pre-Quaternary sand and gravel underlying the surface comprise the Ogallala Formation (Frye et al., 1956).

In Kansas, the Ogallala Formation mostly consists of Miocene and Pliocene deposits of fluvial sand and gravel (Merriam, 1963, p. 22). There are multiple carbonate-rich horizons in the Ogallala Formation; the most distinct of these is the resistant "caprock caliche" forming the upper surface of the Ogallala (Frye et al., 1956). The caprock typically is several meters thick and preserves the plateau topography of the High Plains. Because of the thick package of Quaternary deposits (loess and alluvium) on the High Plains, surface exposures of the Ogallala Formation are confined to deeply dissected or eroded areas, especially along the High Plains escarpment. The Ogallala also is exposed in the valley walls of streams, including Beaver Creek.

The Ogallala Formation is a major aquifer; hence, springs are common along the High Plains escarpment and where the Ogallala is exposed in the walls of stream valleys (Sawin et al., 2002). Because many of these springs were reliable sources of water for game and people during prehistoric times, archaeological sites are often located at or near them.

The High Plains surface is mantled by a sheet of late Quaternary loess that is generally 2–3 m thick, but is more than 10 m thick in some areas near major river valleys. Peoria Loess, the dominant surface deposit on uplands in the field trip area, directly overlies Pleistocene alluvium or the Ogallala Formation. Radiocarbon and optically stimulated luminescence (OSL) ages for the Peoria Loess in the Great Plains range from ca. 24,000 cal. yr B.P. at its base to ca. 11,200–11,300 cal. yr B.P. near the top (Bettis et al. 2003; Roberts et al., 2003).

Shallow, closed depressions, with diameters ranging from a few meters to several kilometers, are common on the Kansas High Plains (Frye, 1950; Mandel, 2000c). Most of these depressions, or playas, are <3 m deep, but some of the larger ones are as much as 15–20 m deep. Radiocarbon ages determined on soils developed in playa fills suggest that many of these depressions formed during the early Holocene (Mandel, 2000c). Their development has been attributed to several causes, including wind deflation, solution subsidence, and wallowing of buffalo. Frye (1950) suggested that most, if not all, of these processes played a role in the development of depressions on the High Plains. There is also evidence indicating that some playas formed when intermittent drainage systems were blocked by sand dunes (Mandel 2000c). Like springs, the playas attracted game and people during prehistoric times. The Winger site (14ST401), a Late Paleo-Indian bison bone bed contained in playa fill on the High Plains of southwestern Kansas, is a case in point (see Mandel and Hofman, 2003).

STOP 1: THE KANORADO LOCALITY

The Kanorado locality is in Middle Beaver Creek valley near the Kansas-Colorado state line (Fig. 1). This locality is high in the drainage network of Beaver Creek, and like other draws in the region, upper Middle Beaver Creek is an intermittent stream; it only carries water immediately after heavy rainfalls. Late Quaternary alluvium is stored beneath two geomorphic surfaces in the draw: a low, narrow floodplain (T-0) and a broad, flat terrace (T-1). The T-1 and T-0 surfaces are separated by a 2–3-m-high scarp, and the T-1 surface gradually merges with colluvial slopes at the foot of the valley wall.

At Kanorado, three stratified Paleo-Indian sites—14SN101, 14SN105, and 14SN106—are within several hundred meters of each other. Artificial channels excavated by the Kansas Department of Transportation nearly 30 yr ago exposed lithic artifacts and the remains of extinct late Pleistocene fauna in alluvium beneath the T-1 terrace of Middle Beaver Creek. The cluster of sites at Kanorado represents one of the most important archaeological finds in the High Plains and perhaps North America.

History of Investigations

The Kanorado locality was first investigated by the Denver Museum of Natural History (now the Denver Museum of Nature & Science) in 1976 under the direction of K. Don Lindsey, former Curator of Paleontology. Lindsey was contacted by the landowner's son, who reported the discovery of several large bones in the area of the channelized stream. Lindsey visited the locality and recovered a femur, some fractured ribs, and a partial pelvis of a mammoth from sandy alluvium near the base of the fresh channel cut. According to Lindsey (1981), one large cobble found with mammoth bone appeared to be out of place in the fine sand. He also noted that spiral fractures and wear patterns on some mammoth elements did not appear to be caused by natural processes. During the course of his investigation, a mammoth tooth was found at a higher level in the section, indicating the presence of at least two mammoths at the locality.

In 1981, Lindsey brought Robin Boast, an archaeology graduate student and employee of the Denver Museum's anthropology department, to the Kanorado locality. Salvage excavations in the upper mammoth level produced spiral fractured limb-bone fragments but no lithic artifacts. However, one piece of cortical limb bone has clear evidence of three facets, probably caused by cutting with stone tools. Camelops vertebrae were also found in the upper level during this excavation. Lindsey (1981, p. 22) noted that, "Salvage work at a late Pleistocene locality in western Kansas is of special interest due to the uncommon distribution of the bone, unnatural breakage and the presence of polished fragments."

In 2001, Steven Holen, Curator of Archaeology at the Denver Museum of Nature & Science, noticed the unusual fracture and wear patterns on some mammoth elements in the Kanorado paleontology collection. Holen and Jack Hofman visited the locality in February 2002 and found mammoth bones, and some burned bone fragments, on a pile of talus where Lindsey had excavated mammoth remains. Also, Camelops elements and one unidentifiable bone fragment were observed in situ in a cutbank exposing valley fill beneath the T-1 terrace. The Camelops elements were ~40 cm below the top of the A horizon of a buried paleosol, or ~1.4 m below the T-1 surface (Fig. 2). The Camelops bones were estimated to be of Clovis age (ca. 11,000–11,500 ^{14}C yr B.P.). However, no cultural material was found during the 2001 survey.

Holen returned to the Kanorado locality in June 2002 to conduct an additional survey. One archaeological site, 14SN101, was recorded immediately east of the original mammoth locality in another area where Middle Beaver Creek had been channelized. A thin endscraper made of Alibates flint was found at 14SN101. It was discovered on talus just below the A horizon of the buried paleosol that previously yielded Camelops remains. The endscraper appeared to be an early Paleo-Indian variety. A test unit in the A horizon of the paleosol yielded additional lithic material, including three quartzite flakes and one silicified wood retouched flake. Two of the lithic artifacts were found in the lower 5 cm of the buried A horizon.

Two test units were also excavated at the original Kanorado mammoth locality. One test unit yielded the articulated remains of a Camelops distal tibia, calcaneum, and astragalus excavated from the base of the buried A horizon. The second unit produced an unidentifiable element 40 cm below the base of the buried A horizon (70 cm below the top of the paleosol) at the location of the 1976 excavation. No additional faunal elements were found in this test unit.

Figure 2. Photograph of the profile in Area B of site 14SN105. Lithic artifacts and *Camelops* bone were recovered from the lower 5 cm of the A horizon (likely Clovis horizon) of Soil 2.

In the spring of 2003, Holen revisited Kanorado and found lithic flakes eroding from the bottom of the A horizon of the buried paleosol. This find was in the cutbank ~50 m south of Lindsey's original mammoth locality and marked the first direct evidence of human occupation at this locality. Hence, the cutbank was assigned an archaeological site number (14SN105).

In June of 2003 and 2004, sites 14SN105 and 14SN101 were tested by a joint crew from the University of Kansas and the Denver Museum of Nature & Science. Test excavations were conducted to determine the extent and age of the lithic components in the buried paleosol at both sites. Test excavations were also conducted to determine if any mammoth bones remained in place at the location where the Denver Museum excavated in 1976 and 1981. As noted earlier, some of the mammoth bones in the Denver Museum's collection exhibited wear and fracture patterns that appeared to be caused by humans. Because these mammoth remains were excavated in deposits nearly 2.5 m below a likely Clovis horizon, there appeared to be potential for a pre-11.5 ka component at 14SN105.

Geomorphological investigations accompanied archaeological testing at sites 14SN105 and 14SN101. The primary objectives of these investigations were to (1) place the archaeological and paleontological deposits in a geomorphic and soil-stratigraphic context, (2) determine the geochronology of the valley fill underlying the T-1 terrace of Middle Beaver Creek, and (3) determine the site-formation processes that shaped the archaeological record at Kanorado. Results of accelerator mass spectrometry (AMS) ^{14}C dating of purified collagen from bones collected at site 14SN105 are pending and should be available at the time of the field trip.

Site 14SN105

Site 14SN105 spans a 200-m-long cutbank adjacent to the mechanically straightened channel of Middle Beaver Creek. We divided the site into three areas: A, B, and C. Area A is the location where mammoth and other faunal remains were excavated by the Denver Museum in 1976 and 1981. Area B is ~50 m south of Area A and includes the location where lithic artifacts were previously discovered in the lower 5 cm of the A horizon of the buried paleosol. Area C is ~150 m north of Area A and includes the location where *Camelops* bone was recovered from the lower 5 cm of the buried A horizon.

In June 2003, Don Lindsey's original excavation units were relocated in Area A. The entire section was cleaned with hand shovels, and five 1 m² units were excavated in undisturbed alluvium in the lower mammoth bone level. The section was described and soil and sediment samples were collected for laboratory analyses.

The surface soil (Soil 1) in Area A has an A-Bw-Bk-BC profile (Table 1 and Fig. 3). With the exception of the A horizon, the texture of Soil 1 is silt loam (Table 2). Carbonate morphology does not exceed stage I, and calcium carbonate equivalent (CCE) ranges from 0.9% in the A horizon to 6.1% in the BCk horizon.

Two buried soils (soils 2 and 3) form a pedocomplex immediately below Soil 1 (Fig. 3). Soil 2 is 88–128 cm below the T-1 surface and has an Ak-Bk profile (Table 1). Soil 3 is at a depth of 128–313 cm and has a thick, well expressed Ak-Bk-BCk-CBgk profile. The Bkb1 horizon of Soil 2 is welded onto the Akb2 horizon of Soil 3. This accounts for the subangular blocky structure and carbonate morphology (films and threads) in the Akb2 horizon. Despite welding of Soil 2 onto Soil 3, the Akb2 horizon is marked by a significant increase in organic carbon (OC) content compared to the overlying Bkb1 horizon (Table 2). Particle-size distribution is fairly consistent in Soil 3, but there is a prominent peak in CCE in the Bk2b2 and Bk3b2 horizons (Table 2). Redoximorphic features (e.g., gray matrix color, reddish mottling, and oxide concretions) appear in the CBgkb2 horizon and continue down into the C1b2 horizon. There is a dramatic change in grain-size distribution at the boundary separating the C1b2 and C2b2 horizons, with total silt content decreasing from 52.1% to 7.1% and total sand content increasing from 34.1% to 89.0%. This textural boundary marks a facies change; fine-grained vertical accretion deposits compose the upper 337 cm of the valley fill,

TABLE 1. DESCRIPTION OF THE PROFILE AT SITE 14SN105, AREA A

Depth (cm)	Soil horizon	Description
0–22	A	Dark brown (10YR 3/3) loam, very dark grayish brown (10YR 3/2) moist; weak fine granular structure; slightly hard, friable; many worm casts and open worm burrows; many fine and very fine roots; gradual smooth boundary.
22–41	Bw	Dark brown (10YR 4/3) clay loam, dark brown (10YR 3/3) moist; weak fine subangular blocky structure; hard, firm; common worm casts and open worm burrows; common fine and very fine roots; gradual smooth boundary.
41–68	Bk	Grayish brown (10YR 5/2) silt loam, dark grayish brown (10YR 4/2) moist; weak medium and fine prismatic structure parting to moderate fine subangular blocky; hard, firm; common fine threads and films of calcium carbonate; common fine and very fine pores; common worm casts and open worm burrows; common fine roots; gradual smooth boundary.
68–88	BCk	Brown (10YR 5/3) silt loam, dark brown (10YR 4/3) moist; very weak fine subangular blocky structure; hard, friable; very faint bedding; common films and fine threads of calcium carbonate; common fine and very fine pores; common worm casts and open worm burrows; common fine roots; clear, smooth boundary.
88–106	Akb1	Dark grayish brown (10YR 4/2) silt loam, very dark grayish brown (10YR 3/2) moist; weak fine subangular blocky structure parting to moderate medium and fine granular; slightly hard, friable; many films and fine threads of calcium carbonate; common worm casts and open worm burrows; common very fine roots; gradual smooth boundary.
106–128	Bkb1	Dark brown (10YR 4/3) silt loam, dark brown (10YR 3/3) moist; weak medium and fine prismatic structure parting to weak fine subangular blocky; hard, friable; common films and fine threads of calcium carbonate; common worm casts and open worm burrows; common very fine roots; common fine and very fine pores; abrupt smooth boundary.
128–160	Akb2	Dark gray (10YR 4/1) to dark grayish brown (10YR 4/2) silt loam, very dark gray (10YR 3/1) to very dark grayish brown (10YR 3/2) moist; weak fine granular structure; slightly hard, very friable; many films and fine threads of calcium carbonate; common worm casts and open worm burrows; common krotovina 6–9 cm in diameter filled with grayish brown (10YR 5/2, dry) silt loam; common fine and very fine and few medium roots; many fine and very fine pores; gradual smooth boundary.
160–170	ABkb2	Dark grayish brown (10YR 4/2) silt loam, very dark grayish brown (10YR 3/2) moist; weak fine subangular blocky structure parting to moderate fine and medium granular; slightly hard, friable; common films and fine threads of calcium carbonate; common worm casts and open worm burrows; common fine and very fine and few medium roots; many fine and very fine pores; gradual smooth boundary.
170–184	Bk1b2	Grayish brown (10YR 5/2, dry) silt loam, dark grayish brown (10YR 4/2) moist; weak medium and fine prismatic structure parting to weak fine subangular blocky; hard, friable; common very fine threads of calcium carbonate; few fine hard calcium carbonate concretions; few siliceous granules; common worm casts and open worm burrows; few krotovina 8–10 cm in diameter filled with dark brown (10YR 4/3, dry) or yellowish brown (10YR 5/4) silt loam; few very fine roots; common fine and very fine pores; clear, smooth boundary.
184–214	Bk2b2	Very pale brown (10YR 7/3) silt loam, pale brown (10YR 6/3) moist; weak medium and fine prismatic structure parting to weak fine subangular blocky; hard, friable; many films and fine threads of calcium carbonate; common fine and very fine hard calcium carbonate concretions; few worm casts and open worm burrows; few krotovina 3–4 cm in diameter filled with grayish brown (10YR 5/2, dry) silt loam; few very fine roots; many fine and very fine pores; gradual smooth boundary.
214–244	Bk3b2	Pale brown (10YR 6/3) silt loam, brown (10YR 5/3) moist; few fine faint yellowish brown (10YR 5/6) mottles; weak medium and fine prismatic structure parting to weak fine subangular blocky; hard, friable; common films and fine threads of calcium carbonate; few worm casts and open worm burrows; common krotovina 15–25 cm in diameter filled with dark brown (10YR 4/3, dry) silt loam; few very fine roots; many fine and very fine pores; gradual smooth boundary.
244–287	BCkb2	Light yellowish brown (2.5Y 6/3) silt loam, light olive brown (2.5Y 5/3) moist; common fine prominent strong brown (7.5YR 5/8) mottles, with many concentrated in macro-pores; very weak fine subangular blocky structure; slightly hard, friable; common fine threads and few fine soft concretions of calcium carbonate; few very fine roots; many fine and very fine pores; few fine soft iron and manganese oxide concretions; clear, smooth boundary.
287–313	CBgkb2	Grayish brown (2.5Y 5/2) silt loam, dark grayish brown (2.5Y 4/2) moist; common, fine, prominent strong brown (7.5YR 5/8) mottles, with many concentrated in macro-pores; very weak medium and coarse angular blocky structure; hard, firm; many films and common fine threads of calcium carbonate; few very fine roots; many fine and very fine pores; common fine soft iron and manganese oxide concretions; abrupt wavy boundary.
313–337	C1b2	Laminated light yellowish brown (2.5Y 6/3) silt loam, light olive brown (2.5Y 5/3) moist; common fine and medium prominent red (2.5YR 4/8), yellowish red (5YR 5/8), and strong brown (7.5YR 5/8) mottles; massive; hard, firm; laminae are wavy; few krotovina 12–31 cm in diameter filled with very pale brown (10YR 7/3, dry) sand and fine pebbles derived from the C2b2 horizon; many fine and very fine pores; common fine soft iron and manganese oxide concretions; abrupt wavy boundary.
337–362	C2b2	Stratified very pale brown (10YR 7/3) coarse sand, pale brown (10YR 6/3) moist; few fine and medium prominent strong brown (7.5YR 5/8) and reddish yellow (7.5YR 6/8) mottles; single grain, loose; common lenses of fine pebbles; abrupt wavy boundary.
362–387	C3b2	Light olive brown (2.5Y 5/3) loam, olive brown (2.5Y 4/3) moist; many very fine, prominent, brownish yellow (10YR 6/8) and strong brown (7.5YR 5/8) and few very fine prominent red (2.5YR 4/8) mottles; massive; soft, friable; common krotovina 10–12 cm in diameter filled with dark brown (10YR 3/3, dry) silt loam; few lenses of sand and fine pebbles; abrupt wavy boundary.
387–410+	C4b2	Coarse gravel; single grain; loose.

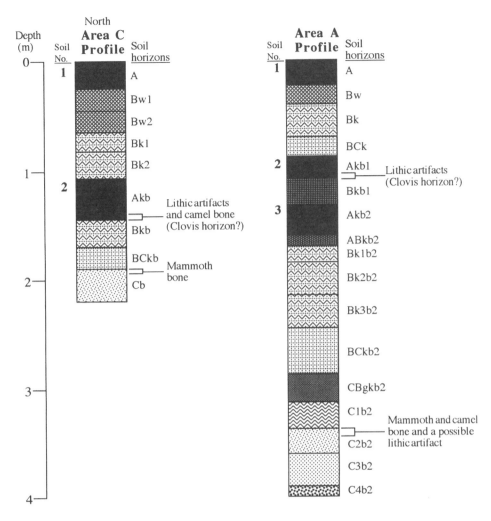

Figure 3. Stratigraphic columns of T-1 fill in areas A and C of site 14SN105. Soil horizons are designated according to criteria outlined in the U.S.D.A. *Soil Survey Manual* (Soil Survey Staff, 1993).

whereas coarse-grained lateral accretion deposits compose valley fill below a depth of 337 cm. A bed of stratified coarse sand with lenses of fine pebbles (C2b2 horizon) at a depth of 337–362 cm overlies a 25-cm-thick bed of loam (C3b2 horizon) composed of 48.4% sand (Tables 1 and 2). The C3b2 horizon is above coarse gravels (C4b2 horizon) that continue to the bottom of the section.

During the 2003 field season, a concentration of in situ mammoth bones, including a proximal fibula, a vertebra, and rib fragments, was exposed in excavation units in the lower part of the section (Fig. 4). The bones were in the loamy C3b2 horizon directly above the gravels. Spiral fractures were observed on the fibula and rib fragments and on one piece of cortical limb bone. The vertebra, however, showed no evidence of breakage. A phalanx and fragment of pelvis from a *Camelops* were also excavated from the C3b2 horizon. These elements were found among the mammoth bones. However, no lithic artifacts were found in this lower level.

In addition to focusing on intact deposits, seven 1 m² units were excavated in talus at the foot of the cutbank above the lower mammoth/*Camelops* bone level in Area A. The disturbed sediments yielded a highly fragmented mammoth tibia and two lithic flakes.

In June 2004, three 1 m² units were excavated in the lower mammoth/*Camelops* bone level (C3b2 horizon) in Area A. Two mammoth rib elements and two unidentifiable bone fragments were found. However, a more notable discovery was made: a quartzite flake was found in the mammoth bone level. The flake exhibits a prominent bulb of percussion and dorsal flake scars, but exhibits some wear probably caused by movement in alluvium. Quartzite occurs naturally in local gravels, but the morphology of the flake suggests that it was produced by human action. Analysis of the flake is under way.

In addition to the finds in the lower mammoth/*Camelops* bone level, one flake was discovered in the middle of the 2Akb1 horizon in Area A. This represents the first in situ lithic artifact recorded in the buried paleosol in Area A. This cultural horizon may be the source of lithic flakes excavated with a mammoth tibia in talus on the cutbank.

TABLE 2. LABORATORY DATA FOR SOILS AT SITE 14SN105, AREA A

| Depth (cm) | Soil horizon | Sand (mm) | | | | | | Silt (µm) | | | | Clay (µm) | | OC[†] (%) | CCE[§] (%) |
		VCS 2.00–1.00	CS 2.00–0.50	MS 0.50–0.25	FS 0.25–0.10	VFS 0.10–0.05	TS 2.00–0.05	CSI 50.0–20.0	MSI 20.0–5.0	FSI 5.0–2.0	TSI 50.0–2.0	TC <2.0	Textural class*		
0–22	A	.9	3.0	4.7	3.5	14.6	26.7	31.4	9.8	5.6	46.8	26.5	l	0.80	0.9
22–41	Bw	.2	0.7	1.1	1.1	19.0	22.1	33.7	10.2	4.9	48.8	29.1	cl	0.48	1.5
41–68	Bk	0.0	0.1	0.2	0.5	27.1	27.9	35.8	11.0	6.6	53.4	18.7	sil	0.32	6.2
68–88	BCk	0.2	0.4	0.6	0.9	15.3	17.4	44.7	15.4	8.3	68.4	14.2	sil	0.25	6.1
88–106	Akb1	0.1	0.8	0.9	1.8	15.0	18.6	40.4	23.1	7.7	71.2	10.2	sil	0.47	2.1
106–128	Bkb1	0.1	1.0	1.0	1.9	15.7	19.7	36.3	15.6	6.2	58.1	22.2	sil	0.42	1.7
128–160	Akb2	0.2	0.9	1.6	2.7	14.6	20.0	33.1	16.8	8.5	58.4	21.6	sil	0.73	3.3
160–170	ABkb2	0.5	1.0	1.7	3.1	17.3	23.6	34.3	12.8	6.4	53.5	22.9	sil	0.55	2.7
170–184	Bk1b2	0.3	1.0	1.4	3.2	19.0	24.9	36.9	12.0	6.8	55.7	19.4	sil	0.55	1.1
184–214	Bk2b2	0.0	0.2	0.4	1.8	18.8	21.2	41.3	14.6	6.2	62.1	16.7	sil	0.44	15.0
214–244	Bk3b2	0.0	0.3	0.5	3.8	24.3	28.9	37.1	10.6	5.5	53.2	17.9	sil	0.21	11.9
244–287	BCkb2	0.1	0.3	0.8	4.3	25.5	31.0	37.7	11.6	4.8	54.1	14.9	sil	0.12	8.0
287–313	CBgkb2	0.1	0.3	0.9	2.2	17.8	21.3	41.1	15.1	4.8	61.0	17.7	sil	0.08	6.3
313–337	C1b2	0.1	0.1	0.1	3.0	30.8	34.1	36.1	11.3	4.7	52.1	13.8	sil	0.00	5.8
337–362	C2b2	24.6	42.1	16.8	3.9	1.6	89.0	4.0	1.3	1.8	7.1	3.9	cs	0.00	0.5
362–387	C3b2	0.1	0.2	2.2	11.4	34.5	48.4	30.4	6.0	3.4	39.8	11.8	l	0.00	2.4

Note: Bulk samples were collected from each soil horizon. Air-dry bulk samples were crushed with a wooden rolling pin and passed through a No. 10 sieve with 2 mm square openings. Total carbon (TC) was determined using a high-frequency induction furnace (Leco Model CNS-2000, St. Joseph, MI) following the procedure of Tabatabai and Bremner (1970). Particle size distribution was determined using a modification of the pipet method of Kilmer and Alexander (1949) and method 3A1 from the Soil Survey Laboratory Method Manual (Soil Survey Laboratory Staff, 1996). Organic matter was removed from samples containing greater than 1.4% total C with 30% hydrogen peroxide. Calcium carbonate (CaCO₃) equivalent (CCE) was estimated by the Piper method (Hesse, 1971). Soil organic carbon (SOC) was calculated by adjusting the total carbon content using the CCE to account for C contained in calcium carbonate. Abbreviations for the sand, silt, and clay fractions: VCS—very coarse sand; CS—coarse sand; MS—medium sand; FS—fine sand; VFS—very fine sand; TS—total sand; CSI—coarse silt; MSI—medium silt; FSI—fine silt; TSI—total silt; TC—total clay.
*Textural classess: l—loam; cl—clay loam; sil—silt loam; cs—coarse sand.
[†]OC—Organic carbon,
[§]CCE—Calcium carbonate equivalent.

Excavations were conducted in Area C during the 2003 and 2004 field seasons. Soils 2 and 3 merge to the north of Area A, forming one buried paleosol (Soil 2) in Area C (Fig. 3). Soil 2 is at a depth of 106–191 cm below the T-1 surface and has an Ak-Bk-BCk profile. Silty overbank alluvium gives way to sandy lateral accretion deposits at a depth of 191 cm.

In 2003, three Alibates flint flakes and one Flattop Chalcedony blade fragment were found on the surface of a two-track ranch road that cuts into the Akb horizon of Soil 2. Also, one Alibates retouched flake was found eroded out of the Akb horizon in a cutbank adjacent to the ranch road. A 1 m² unit was excavated in the Akb horizon where in situ bone fragments were observed. Spirally fractured bone fragments and two lithic flakes were documented in the lower 5 cm of the Akb horizon. One of the flakes was Alibates; hence, this archaeological component probably yielded the Alibates flakes found on the ranch road. Also, one in situ spirally fractured lateral fragment of a large ungulate limb bone, probably *Camelops* or bison, was excavated from the lower 5 cm of the Akb horizon on the ranch road.

In 2004, 2 m² were excavated in the upper 50 cm of Soil 2 in Area C. Ungulate bone fragments were found in the lower 5 cm of the Akb horizon, and a few, very small lithic flakes were found 5–10 cm above the bone fragments. The flakes either represent a separate archaeological component or they were moved upward from the bone level by bioturbation.

Area B of site 14SN105 has received relatively little attention. However, in June 2003, three 1 m² units were excavated in Soil 2. Three flakes were found in the Akb1 horizon, but the area was heavily disturbed by burrowing animals, and the exact context of the flakes within the soil could not be determined.

Site 14SN101

Site 14SN101 is located ~300 m east of site 14SN105 and borders a segment of Middle Beaver Creek that was straightened in the late 1960s. Valley fill beneath the T-1 surface is exposed in 100-m-long cutbanks on both sides of the channelized reach. The stratigraphy of the T-1 fill at 14SN101 is remarkably similar to the stratigraphy in Area C of site 14SN105. There is a single buried soil (Soil 2) with an Ak-Bk-BCk profile developed in silty vertical accretion deposits. The package of fine-grained alluvium is 3.5 m thick and overlies coarse-grained lateral accretion deposits.

In June 2003, archaeological investigations at 14SN101 consisted of test excavations and survey work. Two 1 m² test units were excavated next to the 2002 test pit on the south bank. One sandstone abrader and several flakes were recorded in the middle of the Akb horizon of Soil 2. Also, a piece of thick, spirally-fractured cortical bone was found in situ in the lower 5 cm of the Akb horizon ~4 m west of the excavation. An Alibates flake knife was found eroded out of place just below the Akb horizon 60 m west of the excavations.

In October 2004, an additional survey was conducted at 14SN101 on the north bank of the channelized Middle Beaver Creek (immediately across from the 2003 test excavations in the south bank). This survey produced faunal and lithic evidence indicating that the component in the Akb horizon extends to the north cutbank. It also revealed that an area of the site ~70 m wide was destroyed by the channelization of the creek in the early 1970s. Survey of the north cutbank produced one Smoky Hill Jasper biface found eroded out just below the base of the Akb horizon of Soil 2. In addition, large ungulate limb-

Figure 4. Photograph of the lower mammoth/*Camolops* horizon in Area A of site 14SN105. The photo scale (arrow) is 20 cm long.

bone fragments and one *Camelops* patella were found in the lower 5 cm of the Akb horizon 1 m east and ~70 m west of the biface, respectively.

In June 2004, 3 m² were excavated in the Akb horizon of Soil 2 on the north bank at site 14SN101. A single 1 m² test unit was placed at the location where a *Camelops* patella was excavated from the base of the Akb horizon in 2003. Small unidentifiable bone fragments were found just below the Akb horizon, but no cultural material was found. Two contiguous 1 m² units were excavated at the location where a broken biface was discovered eroded out of place just below the Akb horizon. However, no cultural material and only a small amount of fragmented bone were found in these units.

Site 14SN106

Site 14SN106 was recently recorded in a channelized segment of the Middle Beaver Creek ~300 m upstream from site 14SN105. A detailed geomorphological investigation has not been conducted at this site, but some general observations were made during the June 2004 field season. There is a buried paleosol (Soil 2) with Ak-Bk horizonation ~1 m below the T-1 surface. Soil 2 is developed in a 1.5-m-thick unit of silty vertical accretion deposits that overlie sandy and gravelly alluvium. Bison bone and one Hartville chert endscraper were discovered on talus immediately below the Akb horizon of Soil 2. A 0.5 m² test unit in the upper 50 cm of Soil 2 yielded three lithic flakes, all from the middle of the Akb horizon.

Summary of the Kanorado Locality

The Kanorado locality consists of three archaeological sites with cultural components in a buried paleosol (Soil 2) developed in the upper 1.5 m of the T-1 valley fill. At sites 14SN101 and 14SN105, lithic artifacts are in the lower 5 cm of the Akb horizon of Soil 2. *Camelops* elements were also found in the lower 5 cm of the Akb horizon at both sites, but the bones were not directly associated with lithic artifacts. *Camelops* extinction on the Great Plains is thought to date ca. 11,200 ¹⁴C yr B.P. based on a radiocarbon-dated *Camelops* bone from the Casper Site (Frison, 2000). This represents the most recent occurrence of *Camelops* on the Great Plains. The Casper site may contain evidence of an association of dynamically impacted and spirally fractured *Camelops* bone with the Clovis projectile point found at the site. Suggestions of *Camelops* surviving into the Holocene in North America have not been supported by radiocarbon evidence. Hence, it is likely that the cultural horizons in the base of the buried A horizon at both 14SN101 and 14SN105 are Clovis and date to the late Pleistocene, not the Holocene.

Archaeological evidence suggesting that Clovis people hunted *Camelops* anywhere in North America is scant (Frison et al., 1978; Haynes and Stanford, 1984), although the Casper site in east-central Wyoming may represent a human-*Camelops* association (Frison, 2000). The Colby mammoth site in northwestern Wyoming also contained a possible tool made from *Camelops* bone in the second mammoth bone pile (Frison and Todd, 1986). The presence of *Camelops* bone in deposits of the same age as those containing probable Clovis lithic tools, and the occurrence of large, spirally fractured ungulate limb bone fragments with lithics, suggest that sites 14SN101 and 14SN105 may contain evidence of Clovis *Camelops* procurement.

Preliminary lithic procurement data support the interpretation that the components at 14SN101 and 14SN105 are Clovis. Alibates artifacts from outcrops in the Texas Panhandle occur at both sites. Holen (2001) demonstrated that Clovis people moved significant amounts of Alibates into northeastern Colorado and southwestern Nebraska, probably during long distance seasonal rounds. Morphology and technological attributes of the Alibates endscraper and flake knife found at site 14SN101 are both consistent with Clovis tools, although other Paleo-Indian groups used similar types. Based on the stratigraphic evidence, the presence of *Camelops*, and the lithic procurement patterns, it is likely that the cultural deposits in the lower 5 cm of the Akb horizon of Soil 2 at 14SN105 and 14SN101 represent Clovis hunting camps. The low density of artifact and faunal elements are consistent with temporary occupation of the sites. The radiocarbon age pending on *Camelops* bone from the lower 5 cm of the Akb horizon is crucial for determining the cultural affiliation of the artifacts at that level.

There may be one or two cultural components below the suspected Clovis horizon at site 14SN105. Based on photographs and field notes from the 1981 Denver Museum excavation, a mammoth appears to be eroding 70–80 cm below the top of Soil 2. This component has produced spirally fractured mammoth limb bone. One bone fragment has several cut facets that appear to have been produced by stone tools. Lithic flakes were found in a slump near a displaced mammoth tibia, although these could have eroded from the component in the buried A horizon of Soil 2.

The lower mammoth/*Camelops* component in Area A of site 14SN105 is especially intriguing. Mammoth remains from this horizon include spirally fractured ribs and one limb bone fragment with three intersecting spiral fractures. Taphonomic evidence suggests that small limb bone fragments with numerous intersecting spiral fractures were probably caused by human actions rather than by natural causes. One rib fragment from the 1976 excavation exhibits heavy wear on one end. This wear pattern may have been produced by humans using the rib as a tool. The 1976 excavation also recorded a large cobble associated with mammoth bone in fine sand. This large cobble is out of place in relation to the surrounding sediment; hence, it may be a hammer stone used by people involved in breaking mammoth bone. Finally, the 2004 excavations recovered an in situ quartzite flake from the lower mammoth/*Camelops* component. The flake was not in a visibly disturbed context, but the possibility of vertical displacement from overlying sediments cannot be ruled out entirely. Also, we cannot rule out the possibility that the flake is a product of natural breakage, though it has all the characteristics of an artifact, including a prominent bulb of percussion and dorsal flake scars indicating prior flake removals. The flake and fractured mammoth limb bone elements suggest the lower mammoth/*Camelops* component is a cultural deposit. The numerical age of this component is unknown (radiocarbon dates are pending), but we suspect it is considerably older than Clovis. In sum, the Kanorado locality has great potential for shedding new light on the timing of human entry into the Western Hemisphere. It may also provide information concerning Clovis and possibly pre-Clovis subsistence and mobility strategies on the High Plains.

STOP 2: THE LAIRD SITE

The Laird site (14SN2) is located on a side-slope in Beaver Creek valley in northeastern Sherman County, Kansas (Fig. 1). The site was discovered in July 1990 by Rod Laird and Dan Busse, and was recognized at that time as a small, dense concentration of bison bone and teeth fragments covering an area of ~18 m². A distinctive projectile point was found in this bone concentration and is of particular interest because it is believed to be a Dalton type point. The primary geographic region for the Dalton cultural/technological complex is the Southeast United States, though Dalton sites have been recorded in the Eastern Plains. Occasional finds of Dalton artifacts are documented throughout the Central Plains and westward into the Rocky Mountain region. To date, however, excavated Dalton assemblages have not been documented in the High Plains, and the potential relationships between these materials and those found farther east remain unresolved issues of considerable interest. The age of the Dalton cultural complex in Missouri, Arkansas, and eastern Oklahoma is between ca. 10.2 ka and 9.2 ka. Investigations at the Laird site are in part intended to shed further light on the nature of Dalton activities on the High Plains.

Investigations at the Laird site have included repeated surface collections and excavations in 1995, 2001, and 2004. This work has been completed with the assistance of volunteers from throughout the region and students from the University of Kansas and elsewhere. In 1995, four 1 m² units were excavated and numerous post-hole tests were made (Fig. 5). These served to define the general nature and extent of the bison bone bed (Hofman and Blackmar, 1997). Bison skeletal elements, with many articulated remains, are concentrated in gully fill. Bones and chipped-stone pieces were found from the surface to a depth of ~50 cm, but there was a dramatic increase in the number of complete and articulated bones below a depth of 15 cm. Considerable weathering and fragmentation of the bones had occurred near the surface. Modern cultivation (until the 1960s) and the action of plant roots, insects, and rodents have severely impacted the upper portion of the original bone bed. The original depth of the bone bed and width of the gully at the time of the kill remain unknown.

In June 1999, Mandel used a Giddings hydraulic soil probe to collect cores at Laird. Core 3, which was taken in gully fill that yielded bison bone, mostly consisted of silty sediment derived from Peoria Loess on the adjacent uplands. The surface soil has a weakly expressed A-Bw profile spanning the upper 73 cm of

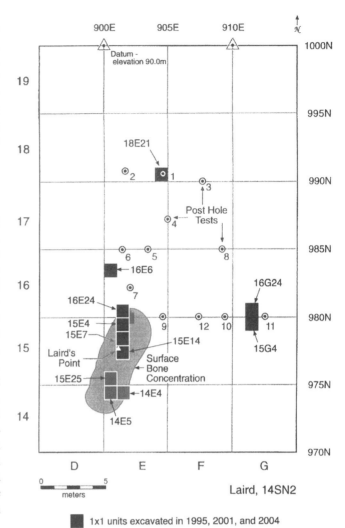

Figure 5. Map of the excavation grid at the Laird site (14SN2).

the gully fill (Table 3). A buried soil with A-Bw horizonation is developed in silty gully fill 73–110 cm below the surface, and gravel was intercepted at a depth of 110 cm. The Ab horizon of the buried soil contained a few bone fragments. Hence, in the area of Core 3, the bison bone bed is associated with a buried early Holocene soil. However, erosion, most of which is a product of Historic land use, has exhumed or nearly exhumed the bone bed in many areas.

Though smaller in size, the nature of the Laird site bison kill was apparently very similar to that at the Olsen-Chubbuck site in eastern Colorado (Wheat, 1972). At least two to three layers of bison carcasses were present in the gully at Laird, and some animals at the bottom appear to have been inaccessible, or were not completely butchered and processed by the hunters. Numerous small flakes were found during the 1995 excavation, and these primarily represent tool retouch and sharpening flakes. The northernmost excavation unit in 1995 (18E21 in Fig. 5) was

TABLE 3. DESCRIPTION OF CORE 3 AT THE LAIRD SITE

Depth (cm)	Soil horizon	Description
0–12	Ap	Brown (10YR 5/3) silt loam, dark brown (10YR 4/3) moist; weak fine subangular blocky structure parting to weak fine granular; slightly hard, friable; many fine and very fine roots; many worm casts and open worm burrows; gradual boundary.
12–30	Bw1	Brown (10YR 5/3) to yellowish brown (10YR 5/4) silt loam, dark brown (10YR 4/3) moist; weak fine subangular blocky structure; slightly hard, friable; many fine and very fine roots; many worm casts and open worm burrows; common fine and very fine pores; gradual boundary.
30–73	Bw2	Yellowish brown (10YR 5/4) silt loam, brown (10YR 5/3) moist; weak fine prismatic structure parting to weak subangular blocky; slightly hard, friable; common fine roots; common worm casts and open worm burrows; common fine and very fine pores; gradual boundary.
73–90	Ab	Dark brown (10YR 4/3) silt loam, dark brown (10YR 3/3) moist; weak fine subangular blocky structure parting to weak fine and medium granular; slightly hard, friable; common dark grayish brown (10YR 4/2) organic coatings on ped faces; common fine roots; common worm casts and open worm burrows; common fine and very fine pores; gradual boundary.
90–110	Bwb	Brown (10YR 5/3) silt loam, dark brown (10YR 4/3) moist; weak fine prismatic structure parting to weak fine subangular blocky; hard, friable; few fine roots; few worm casts and open worm burrows; common fine and very fine pores; abrupt boundary.
110+	2C	Coarse gravel; single grain, loose.

~10 m north of the surface bone exposure. The northern end of the paleo-gully was found, but it did not yield any bison bones. In 1998, a visit to the site resulted in the discovery of the tip of a second projectile point. It was found eroding from the bone bed in unit 16E24 (Fig. 5). This specimen is slightly beveled on the left edge of each face and is made of lithic material believed to have its source area in extreme eastern Nebraska or Kansas.

The 2001 investigations included additional surface mapping and completing the excavation of two units begun in 1995 (Blackmar, 2002). Excavations were expanded adjacent to these two units to expose and remove areas with high bone density. One unit (16E6) was also excavated where flakes had been found on the surface just north of the bone bed. Two units with considerable bone (15E4 and 14E5) were essentially completed in 2001, with bones removed from the bottom of the gully in these locations. Excavation of a 2 m² test located 10 m east of the bone bed was also completed in 2001. This location had yielded flakes and small bone fragments from a post-hole test made in 1995. In 2001, this test revealed numerous small pieces of bone at depths between 35 and 60 cm below the surface. One flake of Flattop Chalcedony, with a source area northwest of Sterling, Colorado, was also recovered.

A surface find of a Flattop endscraper 25 m southeast of the bone bed, an area that had also yielded a few flakes and pieces of fire-cracked cobbles, suggested the possibility of a processing area associated with the Laird bone bed. Small retouch flakes were found among the bones in 2001, but no additional artifacts were recovered from the excavation.

During the 2004 season, four 1 m² units were excavated in the bone bed area, with work in three of the units (16E24, 15E14, and 14E4) completed. Unit 16E24 was on the northwest edge of the gully, unit 15E14 was in the center area of the bone bed and

yielded a side scraper made from Niobrara Jasper (the source area probably is the Saline River valley ~100 km southeast of the Laird site), and unit 14E4 was on the southeast edge of the gully. Excavation of unit 15E25, also in the main bone bed area, yielded numerous bones but was not completed. A 1 m² unit (11F13) and a 2 m² unit (10H15 and 10H16) were excavated 20–25 m southeast of the bone bed in the potential processing area where a Flattop endscraper was found in 2001. Flakes, pieces of fire-fractured cobbles, and the tip of a Flattop knife were found in this area. Further excavation is needed to evaluate the potential relationship of this area to the bone bed.

The number of bison represented by the sample from Laird is currently under investigation, but we estimate that at least six animals are represented by the available bones and that probably twice that many are represented at the site. Butchering is indicated by a few cut marks and numerous green-bone fractures. No evidence of burning has been identified. The season of the kill is also under investigation, with a number of new molars having been recovered since the initial work. Hofman and Blackmar (1997) suggested a time of death of late winter to early spring. This interpretation is now being reevaluated.

Lithic materials from the Laird site represent an interesting variety of source areas. The initial point found by Rod Laird is manufactured from Black Forest or Elizabeth agatized wood. The primary source for this material is the Hahn divide area of east-central Colorado, ~230 km southwest of the Laird site. This material also occurs across much of eastern Colorado as cobbles in Pleistocene alluvial deposits and the Ogallala Formation. The second point tip is made from a distinctive gray speckled chert that may be Pennsylvanian material from eastern Kansas or Nebraska. No sources of this material are known to the west. The endscraper and knife tip from the area south of the bone bed are manufactured from Flattop Chalcedony, with its closest source area northwest of Sterling, Colorado, ~180 km northwest of the Laird site. Other flakes of Flattop are present from the bone bed area. The side scraper found in the bone bed in 2004 is manufactured from Niobrara Jasper, with source areas ~100 km east and northeast of Laird. Most of the small retouch flakes from the bone bed area are also made from Niobrara Jasper. Other materials represented by the lithic pieces from the site include basalt, which occurs locally as cobbles in alluvial gravels, and quartzites, which have multiple potential sources, especially to the northwest and southwest of Laird. It is likely that ongoing analysis of the small flakes recovered from the excavations at Laird will reveal additional insights about this interesting lithic assemblage.

The numerical age of the Laird site has not been firmly established. Purified collagen from a bison long-bone fragment collected in unit 16E24 yielded an AMS ^{14}C age of 8495 ± 40 yr B.P. Given the site's Dalton affiliation, this age is ~700 yr younger than the expected age of the site. However, it is not uncommon for bone dates from the region to be significantly younger than associated charcoal dates (Hofman, 1995). Additional radiocarbon dating is needed to clarify the age of the cultural deposits at Laird. Nevertheless, Laird is among the earliest kill sites recorded on the High Plains of western Kansas.

STOP 3: STEVE BUSSE SAND AND GRAVEL PIT

The Steve Busse Sand and Gravel Pit is located in Little Beaver Creek valley, ~15 km south of Bird City, Kansas (Fig. 1). Little Beaver Creek is a draw at this locality; it only carries water immediately after a heavy rainfall. The valley floor is ~300 m wide and consists of a narrow, modern floodplain (T-0) and a broad, low terrace (T-1) (Fig. 6). The T-1 surface is 3 m above T-0, and these landforms are separated by a moderately steep scarp. Valley fill underlying the T-1 terrace is exposed in a large section on the eastern edge of the sand and gravel pit (Fig. 7). There is also a good exposure of T-0 fill at this stop.

Valley fill beneath the T-0 surface consists of two upward-fining sequences of alluvium (Fig. 6). In the lowest sequence, sand at the bottom of the T-0 fill grades upward to silt loam. A

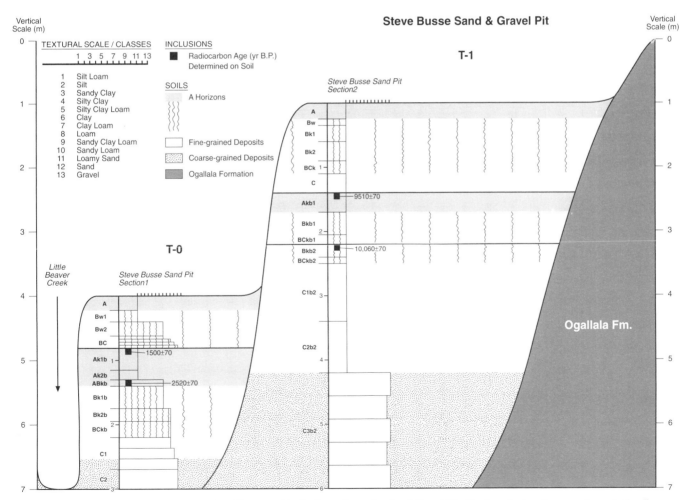

Figure 6. Cross section of late Quaternary valley fill exposed in the Steve Busse Sand and Gravel Pit. The described sections are shown in the diagram.

Figure 7. Photograph of T-1 fill exposed in the west-facing wall of the Steve Busse Sand and Gravel Pit. A description of the section is in presented in Table 4. The radiocarbon ages were determined on decalcified soil carbon from the upper 10 cm of soils 2 and 3. The section is 6 m high.

buried soil with Ak-ABkb-Bk-BCk horizonation is developed at the top of this sequence (Fig. 6). Decalcified organic carbon from the upper 10 cm of the Akb and ABk horizons yielded radiocarbon ages of 1500 ± 70 and 2520 ± 70 yr B.P., respectively. These ages indicate that the floodplain was relatively stable between ca. 2.5 and 1.5 ka, allowing soil development to occur. The late Holocene soil was buried by flood deposits soon after ca. 1500 yr B.P. The upper 80 cm of the T-0 fill consists of stratified fine sandy loam and loam grading upward to silt loam (Fig. 6). The modern surface soil at the top of this upward-fining sequence has a weakly expressed A-Bw-BC profile. In sum, the T-0 fill appears to be late Holocene in age and has high potential for containing buried Late Archaic and younger cultural deposits. The late Holocene fill is laterally inset against much older T-1 fill (Fig. 6).

A portion of the section of T-1 fill was cleaned with a hand shovel (Fig. 7) and described (Table 4). There are two buried paleosols in the 6-m-high section (Fig. 6). The lowest buried paleosol is 220–250 cm below the T-1 surface and has a Bk-BCk profile; the A horizon was stripped off by erosion before the soil was buried. This paleosol is developed at the top of a thick package of silty alluvium overlying sand and gravel.

The upper buried paleosol is at a depth of 140–220 cm and has an Ak-Bk-BCk profile. Although both paleosols have Bk horizons, the carbonate morphology is weak (stage I-I+ as defined by Birkeland [1999, his Table A1.5]). Decalcified soil carbon from the upper 10 cm of the Akb1 and Bkb2 horizons yielded radiocarbon ages of 9510 ± 70 yr B.P and $10,060 \pm 70$ yr B.P., respectively (Figs. 6 and 7). Based on these ages, there is a large volume of terminal Pleistocene and very early Holocene alluvium stored in Little Beaver Creek valley. Also, the soil-stratigraphic record indicates that alluviation during the Pleisto-

cene-Holocene transition was punctuated by landscape stability and soil formation.

We suspect that the T-1 fill at the Kanorado locality in Middle Beaver Creek valley is temporally equivalent to the T-1 fill in Little Beaver Creek. Although no faunal remains or artifacts were found in the T-1 fill at the Steve Busse Sand and Gravel Pit, the buried paleosols, like the ones at Kanorado, represent landscapes that may contain in situ Paleo-Indian cultural deposits. However, because of the depths of the buried paleosols, detection of the material record of Paleo-Indians is problematic.

STOP 4: POWELL SITE

The Powell site (14CN308) is located in Little Beaver Creek valley, ~12 km upstream from the Steve Busse Sand and Gravel Pit (Fig. 1). The site consists of a Clovis fluted point (Fig. 8) discovered in a stream bank by the landowner, Mr. Delbert Powell. The point is manufactured from Niobrara Jasper, with source areas including the Saline River valley in north-central Kansas and Medicine Creek valley in south-central Nebraska. The specimen measures $3.90 \times 2.10 \times 0.43$ cm in length, width, and thickness, respectively. The tip of this point has been heavily reworked, and the artifact was originally much larger (Hofman and Hesse, 2002).

Hofman and Mandel visited the site with Mr. Powell on July 30, 1999, and inspected the stream bank where the Clovis point was recovered. The site is associated with the modern floodplain immediately adjacent to a channelized segment of Little Beaver Creek. According to Powell, the Clovis point was removed from fine-grained sediment ~2 m below the surface of the floodplain. During Hofman and Mandel's visit to the site, a 3-m-deep trench

TABLE 4. DESCRIPTION OF THE T-1 FILL EXPOSED IN SECTION 2 AT THE STEVE BUSSE SAND AND GRAVEL PIT

Depth (cm)	Soil horizon	Description
0–25	A	Grayish brown (10YR 5/2) silt loam, dark grayish brown (10YR 4/2) moist; weak fine subangular blocky structure parting to weak fine granular; hard, friable; many fine and very fine roots; many worm casts and open worm burrows; gradual smooth boundary.
25–35	Bw	Brown (10YR 5/3) silt loam, dark brown (10YR 4/3) moist; weak fine and medium prismatic structure parting to weak fine subangular blocky; hard, friable; many fine and very fine roots; many worm casts and open worm burrows; gradual smooth boundary.
35–60	Bk1	Brown (10YR 5/3) silt loam, dark brown (10YR 4/3) moist; moderate medium prismatic structure parting to moderate medium and fine subangular blocky; hard, friable; common films and threads of calcium carbonate; common fine and very fine roots; common worm casts and open worm burrows; gradual smooth boundary.
60–90	Bk2	Brown (10YR 5/3) silt loam, dark brown (10YR 4/3) moist; moderate medium prismatic structure parting to moderate fine subangular blocky; hard, friable; many films and threads of calcium carbonate; common fine and very fine pores; few fine and very fine roots; common worm casts and open worm burrows; gradual smooth boundary.
90–110	BCk	Brown (10YR 5/3) silt loam, dark brown (10YR 4/3) moist; weak fine subangular blocky structure; hard, friable; common fine films and very fine threads of calcium carbonate; common fine and very fine pores; few fine and very fine roots; few worm casts and open worm burrows; gradual smooth boundary.
110–140	C	Pale brown (10YR 6/3) silt loam, brown (10YR 5/3) moist; massive; soft, very friable; many fine and very fine pores; few very fine roots; abrupt smooth boundary.
140–170	Akb1	Dark brown (10YR 4/3) silt loam, dark brown (10YR 3/3) moist; weak medium prismatic structure parting to weak very fine subangular blocky; hard, friable; few films and common very fine threads of calcium carbonate; common fine and very fine pores; few fine roots; few worm casts and open worm burrows; gradual smooth boundary.
170–205	Bkb2	Brown (10YR 5/3) silt loam, dark brown (10YR 4/3) moist; weak medium prismatic structure parting to moderate fine subangular blocky; hard, friable; few films and common very fine threads of calcium carbonate; common fine and very fine pores; few fine roots; few worm casts and open worm burrows; gradual smooth boundary.
205–220	BCkb1	Pale brown (10YR 6/3) silt loam, brown (10YR 5/3) moist; very weak medium prismatic structure parting to very weak fine subangular blocky; hard, friable; few films and common very fine threads of calcium carbonate; common fine and very fine pores; few fine roots; clear, smooth boundary.
220–240	Bkb2	Brown (10YR 5/3) to dark brown (10YR 4/3) silt loam, dark brown (10YR 4/3) moist; moderate medium prismatic structure parting to moderate medium and fine subangular blocky; hard, friable; common films and threads of calcium carbonate; few fine roots; common fine and very fine pores; gradual smooth boundary.
240–250	BCkb2	Pale brown (10YR 6/3) silt loam grading downward to very fine sandy loam, brown (10YR 5/3) moist; very weak fine prismatic structure parting to very weak fine subangular blocky; hard, friable; few very fine threads of calcium carbonate; many fine and very fine pores; gradual smooth boundary.
250–340	C1b2	Laminated pale brown (10YR 6/3) fine sandy loam interbedded with loamy fine sand, brown (10YR 5/3) moist; massive; soft, very friable; few lenses of fine pebbles; gradual smooth boundary.
340–420	C2b2	Laminated pale brown (10YR 6/3) fine sandy loam interbedded with loamy fine sand, brown (10YR 5/3) moist; few fine prominent yellowish red (5YR 5/8) mottles; massive; soft, very friable; few lenses of fine pebbles; common gastropods; gradual smooth boundary.
420–500+	C3b3	Coarse sandy gravel; single grain, loose.

was excavated behind the stream bank that yielded the point. It revealed that the upper 2.2 m of the floodplain fill consists of laminated silt loam and loam. The fine-grained vertical accretion deposits overlie coarse gravel. A 35-cm-thick surface soil with A-AC horizonation is developed at the top of the fine-grained fill. Such weak soil development suggests that the floodplain surface at this locality has been stable for <100 yr. A more significant find was the discovery of a strand of barbed wire at a depth of 1.02 m. Based on this evidence, the unit of fine-grained alluvium that contained the Clovis point is Historic in age; hence, the point was eroded from its original context and redeposited.

While searching for landform sediment assemblages that may have yielded the Clovis point, Hofman and Mandel discovered a spirally fractured distal metapodial of a *Camelops*. This bone was on the floor of the channel ~50 m upstream from the Clovis find and, like the point, was not in situ. However, valley fill that may be the source of the bone and/or point is only ~50 m upstream from the *Camelops* find. The fill underlies the T-1 terrace and is exposed in a long, high cutbank on the north side of the channelized creek (Fig. 9). A backhoe was used to remove talus from the bottom of the cutbank and to clean off a profile for study (Figs. 9 and 10).

Figure 8. Photograph of the Clovis point from the Powell site (14CN308). The photo scale is 5 cm long.

Figure 10. Photograph of the described section of T-1 fill at the Powell locality. The radiocarbon ages were determined on decalcified soil carbon from the upper 10 cm of soils 2, 3, and 4. The pole is 1.5 m long.

Figure 9. Photograph of the cutbank at the Powell locality. The arrow indicates the location of the described section (Fig. 10)

There are three buried paleosols (soils 2, 3, and 4) in the 4.7-m-thick section of T-1 fill at the Powell locality (Fig. 11). They form a dark, organic-rich pedocomplex 165–330 cm below the surface (Figs. 10 and 11). The surface soil at the top of the section has A-AB-Bw-Bk-BCk horizonation and is welded onto Soil 2, which is represented by a 32-cm-thick ABkb1 horizon (Fig. 11). Soils 3 and 4 have Ak-Bk and ABk-BCk profiles, respectively. Soil 4 is developed at the top of an upward-fining sequence, with sandy gravel at the bottom of the section grading upward to silt loam in the ABkb3 horizon.

Soil development is not particularly strong among any of the buried paleosols at the Powell locality; there are no argillic (Bt) horizons, and carbonate morphology does not exceed stage I. Hence, the individual paleosols do not appear to represent long periods of pedogenesis. This interpretation is supported by the radiocarbon chronology. Decalcified organic carbon from the upper 10 cm of soils 2, 3, and 4 yielded ^{14}C ages of 9160 ± 70 yr B.P., 9800 ± 70 yr B.P., and $10,800 \pm 70$ yr B.P., respectively. The ^{14}C age for Soil 4 indicates that pedogenesis was occurring on the late Wisconsinan floodplain (now the T-1 terrace) at ca. 10.8 ka and may have been under way long before that time. Soil 4 was buried by vertical accretion deposits soon after 10.8 ka, but floodplain aggradation was slow and accompanied by soil development between ca. 9.8 ka and 9.1 ka.

The suite of radiocarbon ages indicates there is high potential for Early through Late Paleo-Indian cultural deposits in the buried pedocomplex at the Powell locality. Although the origin of the Powell Clovis point is unknown, it probably was derived from the T-1 fill, and more specifically, from Soil 4 or the underlying alluvium. It is notable that recent borrowing of fill for construction of a small earth dam ~50 m downstream from the described section brought a *Proboscidian* long bone and several bone fragments to the surface. Much of the fill was derived from the buried pedocomplex, but the original stratigraphic context of the bone

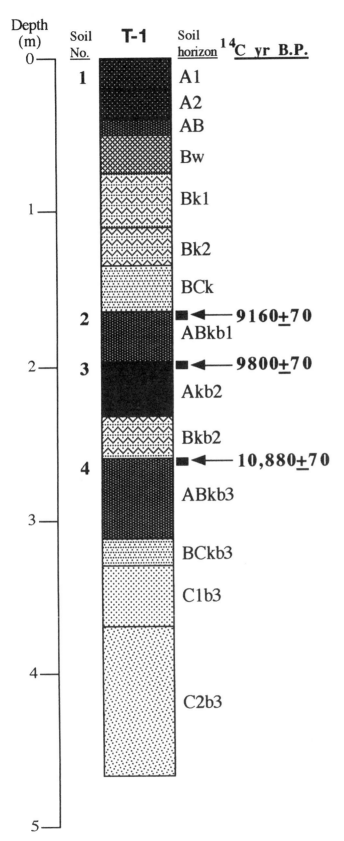

Figure 11. Stratigraphic column of T-1 fill at the Powell locality (see Fig. 10). Soil horizons are designated according to criteria outlined in the U.S.D.A. *Soil Survey Manual* (Soil Survey Staff, 1993).

is unknown. We continue to monitor exposures of T-1 fill at the Powell locality and anticipate that they will eventually yield in situ Paleo-Indian cultural deposits.

STOP 5: BUSSE PLAYA LOCALITY

The Busse playa locality is on the uplands overlooking Little Beaver Creek in southern Cheyenne County (Fig. 1). There is a cluster of four playas in an ~3 km² area. Two of the basins, informally named the Busse west playa and Busse east playa, have been studied (Mandel, 2000c). The Busse west playa is ~0.7 km in diameter and has 8 m of relief from its floor to rim, making it one of the largest basins in northwest Kansas. This playa formed behind a large dune blocking an intermittent drainage. Despite the impressive size of the basin, the playa fill is only 1 m thick and <3000 ¹⁴C yr old (Mandel, 2000c). In contrast, the Busse east playa is ~150 m in diameter and has 3 m of relief, but it contains nearly 2 m of fill consisting of early and late Holocene pond deposits. This playa is the focus of Stop 5.

Excavation of a 2.3-m-deep trench on the floor of the Busse east playa exposed clayey, organic-rich pond deposits beneath a thin veneer of Historic slope wash (Figs. 12 and 13). There are

Figure 12. Photograph of the profile exposed in the trench on the floor of Busse east playa. The radiocarbon ages were determined on decalcified soil carbon from the lower 10 cm of the 2Bt2b1 horizon of Soil 2 and the upper 10 cm of Soil 3. The trench is 2.3 m deep.

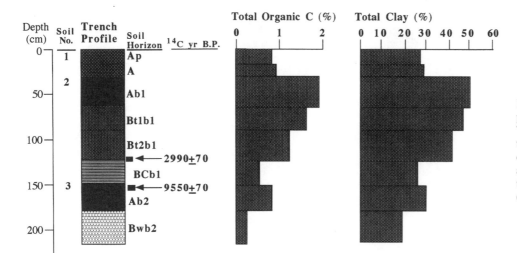

Figure 13. Stratigraphic column of playa fill in the Busse east playa. The histograms show depth functions of organic carbon and total clay. Soil horizons are designated according to criteria outlined in the U.S.D.A. *Soil Survey Manual* (Soil Survey Staff, 1993).

two buried soils developed in the fine-grained playa deposits: one at a depth of 30–150 cm (Soil 2), and the other at a depth of 150–230 cm (Soil 3) (Fig. 13). Decalcified organic carbon from the lower 10 cm of the 2Bt2b1 horizon of Soil 2 and upper 10 cm of the Soil 3 yielded ^{14}C ages of 2990 ± 70 yr B.P. and 9550 ± 70 yr B.P., respectively. The absence of sediments dating between ca. 9.5 ka and 3 ka suggests there were one or more episodes of deflation during this period. It is likely that the Busse east playa, as well as many other depressions on the High Plains, were zones of net erosion during the mid-Holocene warm/dry climatic episode (Mandel, 2000c). However, at the Busse east playa, a Paleo-Indian landscape (Soil 3) is preserved beneath late Holocene deposits. As noted earlier, playas were attractive micro-environments for game and people on the High Plains. Hence, Soil 3 may contain Paleo-Indian cultural deposits and is a good target for deep archaeological exploration.

SUMMARY

The results of recent geoarchaeological investigations on the High Plains of northwestern Kansas indicate there are buried Paleo-Indian landscapes in two geomorphic settings: draws and playas. Based on the suite of radiocarbon ages for buried paleosols developed in T-1 fills in the draws (Table 5), there is a tremendous volume of late Wisconsinan through early Holocene alluvium stored in these small, dry valleys. This pattern of late Wisconsinan through early Holocene sediment storage has been documented in draws throughout western Kansas (Mandel, 2001b, 1995; Mandel et al., 2004). The stratigraphic record indicates that although headward erosion of intermittent streams in the draws has removed most if not all middle Holocene deposits, it has not been sufficient to remove the late Wisconsinan and early Holocene valley fill. The stratigraphic record in playas is remarkably similar to the record in

the draws; middle Holocene deposits have been removed by erosion (deflation), but early Holocene deposits and associated buried soils are often preserved.

From an archaeological perspective, the presence of buried paleosols dating between ca. 11 ka and 9.1 ka in the draws, and dating back to ca. 9.5 ka in the playas, is very significant. These paleosols represent landscapes that may have been occupied by humans during and immediately after the Pleistocene-Holocene transition. It is important to note that the Clovis-age landscape is ~1–3 m below the T-1 surface in the upper Beaver Creek drainage network. This underscores the problem of finding Paleo-Indian archaeological deposits in a stratified context in the region. Artifact collectors have discovered Early through Late Paleo-Indian artifacts in the dry channels of draws, and we now know the likely source of these materials: the T-1 fill. The task is to find "windows" into the T-1 fill and thereby systematically search for in situ Paleo-Indian and pre-Clovis cultural deposits.

ACKNOWLEDGEMENTS

We want to thank Eric Nelson, Field Guide Editor, for providing helpful suggestions that improved the manuscript. We are also grateful to Jennifer Sims and John Charlton for preparing many of the illustrations and to Janice McLean for producing Fig. 6. Numerous people supported our efforts in the field, and we especially want to thank the Dan Busse family, the Keith Coon family, Rob Laird, Delbert Powell, Mark Muniz, Mel Grantham, and the many volunteers and University of Kansas students. Research support was provided by the University of Kansas Odyssey Archaeological Research Fund, the U.S. Department of Transportation, the Kansas Department of Transportation, the University of Kansas General Research Fund, the University of Kansas Museum of Anthropology, and the Carlyle Smith Fund.

TABLE 5. ¹⁴C DATES FOR UPPER BEAVER CREEK DRAINAGE BASIN

Locality	Landform	Material	Depth (cm)	δ¹³C (per mil)	¹⁴C age (yr B.P.)	Lab no.
Laird site (14SN2)	Colluvial slope	Purified collagen	20–30	−8.5	8495 ± 40	CAMS 82397
Busse Sand & Gravel Pit	T-0	Soil OC*	81–91	−15.4	1500 ± 70	ISGS-4678
Busse Sand & Gravel Pit	T-0	Soil OC*	130–140	−16.0	2520 ± 70	ISGS-4669
Busse Sand & Gravel Pit	T-1	Soil OC*	140–150	−17.5	9510 ± 70	ISGS-4671
Busse Sand & Gravel Pit	T-1	Soil OC*	220–230	−16.7	10,060 ± 70	ISGS-4670
Powell locality	T-1	Soil OC*	165–175	−16.1	9160 ± 70	ISGS-4421
Powell locality	T-1	Soil OC*	197–202	−17.4	9800 ± 70	ISGS-4428
Powell locality	T-1	Soil OC*	259–269	−19.9	10,800 ± 70	ISGS-4426
Busse west playa	Playa	Soil OC*	32–42	−20.5	1730 ± 70	ISGS-4422
Busse west playa	Playa	Soil OC*	85–95	−21.3	2890 ± 70	ISGS-4430
Busse east playa	Playa	Soil OC*	110–120	−20.1	2990 ± 70	ISGS-4449
Busse east playa	Playa	Soil OC*	150–160	−17.0	9550 ± 70	ISGS-4448

*Decalcified organic carbon.

REFERENCES CITED

Albanese, J., 2000, Resume of geoarchaeological research on the northwestern Plains, *in* Mandel, R.D., ed., Geoarchaeology in the Great Plains: Norman, University of Oklahoma Press, p. 199–249.

Bettis, E.A., III, Muhs, D.R., Roberts, H.M., and Wintle, A.G., 2003, Last glacial loess in the conterminous USA: Quaternary Science Reviews, v. 22, p. 1907–1946, doi: 10.1016/S0277-3791(03)00169-0.

Birkeland, P.W., 1999, Soils and geomorphology (3rd edition): Oxford, Oxford University Press, 430 p.

Blackmar, J.M., 2002, The Laird site: University of Kansas Museum of Anthropology Cultures Newsletter, v. 65, p. 3.

Fenneman, N.M., 1931, Physiography of Western United States: New York, McGraw-Hill, 534 p.

Frison, G.C., 2000, A ¹⁴C date on a late-Pleistocene *Camelops* at the Casper-Hell Gap site, Wyoming: Current Research in the Pleistocene, v. 17, p. 28–29.

Frison, G.C., and Todd, L.C., 1986, The Colby mammoth site: Taphonomy and archaeology of a Clovis kill in northwestern Wyoming: Albuquerque, New Mexico, University of New Mexico Press, 238 p.

Frison, G.C., Walker, D.N., Webb, S.D., and Zeimens, G.M., 1978, Paleo-Indian procurement of *Camelops* on the Northwestern Plains: Quaternary Research, v. 10, p. 385–400.

Frye, J.C., 1950, Origin of Kansas Great Plains depressions: Kansas Geological Survey Bulletin 86, pt. 1, 20 p.

Frye, J.C., Leonard, A.B., and Swineford, A., 1956, Stratigraphy of the Ogallala Formation (Neogene) of northern Kansas: Kansas Geological Survey Bulletin 118, 92 p.

Haynes, G., and Stanford, D., 1984, On the possible utilization of *Camelops* by Early Man in North America: Quaternary Research, v. 22, p. 216–230.

Hesse, P.R., 1971, Piper Method: Textbook for soil chemical analysis: New York, Chemical Publishing Co., 520 p.

Hofman, J.L., 1995, Dating the Folsom occupation of the Southern Plains: The Lipscomb and Waugh sites: Journal of Field Archaeology, v. 22, p. 421–437.

Hofman, J.L., 1996, Early hunter-gatherers of the Central Great Plains: Paleo-Indians and Mesoindian (Archaic) cultures, *in* Hofman, J.L, ed., Archaeology and paleoecology of the Central Great Plains: Fayetteville, Arkansas Archaeological Survey Research Series No. 48, p. 41–100.

Hofman, J.L., and Blackmar, J.M., 1997, The Paleoindian Laird bison bone bed in northwestern Kansas: The Kansas Anthropologist, v. 18, p. 45–58.

Hofman, J.L., and Graham, R.W., 1998, The Paleoindian cultures of the Great Plains, *in* Wood, R.E., ed., Archaeology of the Great Plains: Lawrence, University of Kansas Press, p. 87–139.

Hofman, J.L., and Hesse, I.S., 2002, Clovis in Kansas, *in* Dort, W., Jr., ed., Institute for Tertiary-Quaternary Studies (TER-QUA) Symposium Series: Lincoln, Nebraska Academy of Science, v. 3, p. 15–35.

Holen, S.R., 2001, Clovis mobility and lithic procurement on the Central Great Plains of North America [Ph.D. Dissertation]: Lawrence, University of Kansas, 273 p.

Holliday, V.T., 1997, Paleo-Indian geoarchaeology of the Southern High Plains: Austin, University of Texas Press, 297 p.

Kilmer, V.J., and Alexander, L.T., 1949, Methods of making chemical analyses of soils: Soil Science, v. 68, p. 15–24.

Lindsey, K.D., 1981, Paleontology division, *in* Denver Museum of Natural History Annual Report 1981: Denver, Colorado, Denver Museum of Natural History, p. 22.

Mandel, R.D., 1992, Soils and Holocene landscape evolution in central and southwestern Kansas: Implications for archaeological research, *in* Holliday, V.T., ed., Soils in archaeology: landscape evolution and human occupation: Washington, D.C., Smithsonian Institution Press, p. 41–100.

Mandel, R.D., 1995, Geomorphic controls of the Archaic record in the Central Plains of the United States, *in* Bettis, E.A., ed., Archaeological geology of the Archaic period in North America: Geological Society of America Special Paper 297, p. 37–66.

Mandel, R.D., editor, 2000a. Geoarchaeology in the Great Plains: Norman, University of Oklahoma Press, 306 p.

Mandel, R.D., 2000b. The history of geoarchaeological research in the Central Plains of Kansas and northern Oklahoma, *in* Mandel, R.D., ed., Geoarchaeology in the Great Plains: Norman, University of Oklahoma Press, p. 79–136.

Mandel, R.D., 2000c, Lithostratigraphy and geochronology of playas on the High Plains of western Kansas: Geological Society of America Abstracts with Programs, v. 32, no. 7, p. 223.

Mandel, R.D., 2001a, Archaeological geology of the Pleistocene-Holocene boundary in stream valleys of the Central Great Plains, U.S.A: Geological Society of America Abstracts with Programs, v. 33, no. 6, p. 287–288.

Mandel, R.D., 2001b, The effects of Holocene climatic change on river systems in the Central Great Plains of North America: Lincoln, University of Nebraska, Program and Abstracts of the 7th International Conference on Fluvial Sedimentology, p. 191.

Mandel, R.D., 2005, Late Quaternary landscape evolution in western Kansas: implications for archaeological research: Lawrence, Kansas Geological Survey Bulletin (in press).

Mandel, R.D., and Hofman, J.L., 2003, Geoarchaeological investigations at the Winger Site: A Late Paleoindian bison bone bed in southwestern Kansas, U.S.A.: Geoarchaeology: International Journal (Toronto, Ontario), v. 18, p. 129–144, doi: 10.1002/GEA.10054.

Mandel, R.D., Hofman, J.L., Bement, L., and Carter, B., 2004, Late Quaternary alluvial stratigraphy and geoarchaelogy in the Central Great Plains, *in* Mandel, R.D., ed., Guidebook for field trips, 18th Biennial Meeting of the American Quaternary Association: Kansas Geological Survey Open-File Report 2004-33, p. 4-1–4-52.

Merriam, D.F., 1963, Geologic history of Kansas: Lawrence, Kansas Geological Survey Bulletin, 162, 317 p.

Roberts, H.M., Muhs, D.R., Wintle, A.G., Duller, G.A.T., and Bettis, E.A., III, 2003, Unprecedented last-glacial mass accumulation rates determined by luminescence dating of loess from western Nebraska: Quaternary Research, v. 59, p. 411–419, doi: 10.1016/S0033-5894(03)00040-1.

Sawin, R.S., Buchanan, R.C., and Lebsack, W., 2002, Kansas springs inventory: water quality, flow rate, and temperature data: Kansas Geological Survey Open-File Report 2002-46, 8 p.

Soil Survey Division Staff, 1993, Soil survey manual: Washington, D.C., Handbook No. 18, U.S. Government Printing Office.

Soil Survey Laboratory Staff, 1996, Soil survey laboratory methods manual: Lincoln, Nebraska, National Soil Survey Center, Soil Survey Investigation Report No. 42, Version 3.0, 693 p.

Stanford, D., 1999, Paleoindian archaeology and late Pleistocene environments in the Plains and Southwestern United States, *in* Bonnichsen, R. and Turnmire, K.L., eds., Ice Age people of North America: Environments, origins, and adaptations: Corvallis, Oregon State University Press, p. 281–339.

Tabatabai, M.A., and Bremner, J.M., 1970, Use of the Leco automatic 70-second carbon analysis of soils: Soil Science Society of America Proceedings, v. 34, p. 608–610.

Wheat, J.B., 1972, The Olsen-Chubbuck site: A Paleoindian bison kill: Washington, D.C., Society for American Archaeology Memoir 26, 164 p.

Wilson, F.W., 1984, Landscapes—a geologic diary, *in* Buchanan, R.C., ed., Kansas geology—an introduction to landscapes, rocks, minerals, and fossils: Lawrence, University Press of Kansas, p. 9–39.

Printed in the USA

Geological Society of America
Field Guide 5
2004

The Colorado Front Range—Anatomy of a Laramide uplift

Karl S. Kellogg
Bruce Bryant
John C. Reed Jr.
U.S. Geological Survey, Mail Stop 980, Box 25046, Denver Federal Center, Denver, Colorado 80225, USA

ABSTRACT

Along a transect across the Front Range from Denver to the Blue River valley near Dillon, the trip explores the geologic framework and Laramide (Late Cretaceous to early Eocene) uplift history of this basement-cored mountain range. Specific items for discussion at various stops are (1) the sedimentary and structural record along the upturned eastern margin of the range, which contains several discontinuous, east-directed reverse faults; (2) the western structural margin of the range, which contains a minimum of 9 km of thrust overhang and is significantly different in structural style from the eastern margin; (3) mid- to late-Tertiary modifications to the western margin of the range from extensional faulting along the northern Rio Grande rift trend; (4) the thermal and uplift history of the range as revealed by apatite fission track analysis; (5) the Proterozoic basement of the range, including the significance of northeast-trending shear zones; and (6) the geologic setting of the Colorado mineral belt, formed during Laramide and mid-Tertiary igneous activity.

Keywords: Laramide orogeny, Front Range, Colorado mineral belt, apatite fission tracks, Williams Range thrust.

INTRODUCTION

This trip provides an overview of the geology of the Colorado Front Range, from the creation of continental crust during the Early Proterozoic to Neogene deformation and erosion. The emphasis, however, will be on the Laramide history. As we leave the downtown convention center and drive west toward the Front Range, the mountain front you see represents exhumed Laramide topography, stripped of a thick sequence of Upper Cretaceous to Miocene rocks during late Neogene erosion. As used here, the Laramide orogeny describes the tectonic events that occurred between ca. 70 Ma and 45 Ma (Late Cretaceous to middle Eocene), although most of the present topography is due to post-Laramide regional uplift and erosion. At the first stop (Fig. 1), we will review the Laramide and subsequent

history. As a head start, however, perhaps no clearer or more succinct overview of the Laramide orogeny has been given than that by Ogden Tweto (1975, p. 1):

At the beginning of the Laramide orogeny, a blanket of undisturbed Cretaceous and minor older Mesozoic sedimentary rocks 1,500 to 3,000 m thick covered the Southern Rocky Mountains province, and the last of a series of Cretaceous seas was starting to withdraw northeastward across the region. Beneath the blanket was an older and inhomogeneous terrane that in some places consisted of eroded stumps of late Paleozoic mountain ranges made up of Precambrian rocks, and in other places piles of sedimentary rocks thousands of meters thick on the sites of late Paleozoic basins. At the onset of the Laramide orogeny in Late Cretaceous time, most of the buried mountain ranges were re-elevated, and adjoining Laramide basins, in part inherited from late Paleozoic basins, began to subside and receive orogenic sediments....

Kellogg, K.S., Bryant, B., and Reed, J.C. Jr., 2004, The Colorado Front Range—Anatomy of a Laramide uplift, *in* Nelson, E.P. and Erslev, E.A., eds., Field trips in the southern Rocky Mountains, USA: Geological Society of America Field Guide 5, p. 89–108. For permission to copy, contact editing@geosociety.org. © 2004 Geological Society of America

…Once started, uplift of mountain units continued through Paleocene and into Eocene time, as indicated by nearly continuous Upper Cretaceous to Eocene sedimentary sequences in the interiors of the bordering basins. Uplifts grew laterally as they rose vertically. Consequently, the major uplifts today are really larger than those that supplied the first orogenic sediments and their border structures are younger than those sediments.

…After the Laramide orogeny, the Southern Rocky Mountain region stood somewhat higher than at the beginning of orogeny—when it was at sea level—but at a much lower level than it does today…

Uplift of the Front Range began ca. 70 Ma, accommodated on the east side along a series of high-angle, mostly east-directed reverse faults, such as the Golden fault, and on the west side along low-angle, west-directed faults, including the Williams Range and Elkhorn thrusts (Fig. 1). Synorogenic conglomerates shed into the Denver basin (Raynolds, 2002) and South Park basin (Bryant et al., 1981a) record this uplift history. One of the goals of this trip is to compare the deformational styles of the east and west structural margins and consider why they are different.

Much of the following discussion is condensed and modified from Tweto (1975, 1987), Reed et al. (1987, 1993), Erslev et al. (1999), and Naeser et al. (2002). Additional useful references are the Colorado state geologic map (Tweto, 1979b), the Denver $1° × 2°$ quadrangle (Bryant et al., 1981b), and the Leadville $1° × 2°$ quadrangle (Tweto et al., 1978).

PROTEROZOIC HISTORY OF THE FRONT RANGE

The Front Range is one of numerous uplifts in the Rocky Mountain region in which Precambrian rocks are exposed. Metamorphic and plutonic rocks, including widespread migmatites, form the core of the Front Range and are part of a Proterozoic terrane called the Colorado province (Bickford et al., 1986) that is interpreted to have formed over an interval of ~130 m.y., beginning ca. 1790 Ma. This long orogenic episode is thought to accompany early Proterozoic accretion of island arcs and backarc basins to the southern margin of an Archean continent (Reed et al., 1987). The suture between the Proterozoic Colorado province and the Archean craton (Wyoming province) to the north is a deformed zone in southern Wyoming called the Cheyenne belt.

Proterozoic rocks of the Front Range include complexly folded and interlayered quartz-feldspar gneiss, amphibolite, biotite schist, and migmatite. The biotite-rich rocks, including migmatite, locally contain layers and lenses of marble, quartzite, and metaconglomerate, indicating sedimentary protoliths. Amphibolite and quartz-feldspar gneiss are abundant generally in separate areas from the metasedimentary rocks and probably originated as volcanic complexes. Both the metasedimentary and metavolcanic packages are complexly interlayered on a regional scale. The rocks are commonly metamorphosed to high-T, low-P, upper amphibolite mixed assemblages (sillimanite–K-feldspar, $±$ garnet, $±$ cordierite migmatites). Evidence for earlier high-T, high-P

conditions locally have been reported (Selverstone et al., 1997; Munn and Tracy, 1992; Munn et al., 1993).

Details of the history of the Proterozoic rocks in Colorado are only locally well known. Small regions near Gunnison, west of the Sawatch Range, and Salida, at the south end of the Mosquito Range, where the metavolcanic rocks are somewhat lower metamorphic grade, are better understood than regions elsewhere (Bickford et al., 1989). On the route of our transect across the Front Range, no metavolcanic rocks have been dated despite considerable 1:24,000 scale mapping, and few convincing facing indicators or regional marker horizons have been found in the supracrustal rocks. One exception is in Big Thompson Canyon, ~75 km north of our route, where relatively low-grade, low-strain metasediments are preserved, and metamorphic isograds have been mapped successfully (Braddock and Cole, 1979). At another location only 6 km north of our route, near the eastern margin of the range, steeply dipping metasandstones and metaconglomerates (Sheridan et al., 1967) contain clear north-facing, "stratigraphic-up" indicators.

At most places in the Front Range, however, the rocks have been severely deformed in a ductile fashion, recrystallized, and partially melted. In Big Thompson Canyon, peak metamorphism, presumably during crustal accretion, occurred ca. 1750 Ma. Compositions of cores and inclusions in garnet and plagioclase grains suggest pressures as great as 10 kb, followed by crystal growth under decreasing pressures down to 4–6 kb (Selverstone et al., 1997). However, the mechanism by which these rocks descended to over 30 km depth and then ascended to mid-crustal levels has not been adequately explained (J.C. Cole, 2004, personal commun.), and such high-P rocks have not been reported elsewhere in the Front Range.

Coeval with, or closely following, the ca. 1700 Ma metamorphic and deformational event, extensive batholiths and smaller bodies of mostly granodiorite and monzogranite, referred to as the Routt Plutonic Suite (Tweto, 1987), intruded the layered rocks of the Front Range. About 25 km north of our transect, the Boulder Creek Granodiorite, dated at 1715 Ma (Premo and Fanning, 2000), is synchronous with late folding of the supracrustal rocks (Gable, 2000). Numerous bodies of plutonic rock have been lithologically correlated with this batholith, but many of the larger plutons and most of the smaller plutons have not been dated.

The Colorado province was extensively modified by widespread ca. 1400 Ma intrusions, regional heating, and local deformation. Large and small plutons of granite and monzogranite of this age, called the Berthoud Plutonic Suite (Tweto, 1987), are widespread in the Proterozoic rocks of the Front Range. Near our transect, the Mount Evans batholith of metaluminous granodiorite and monzogranite superficially resembles the Boulder Creek batholith but has a U-Pb zircon age of 1442 Ma (Aleinikoff et al., 1993). Part of the Mount Evans batholith has been mylonitized along the northeast-trending Idaho Springs–Ralston shear zone, but that mylonite zone does not extend through the batholith.

The Silver Plume Granite, a peraluminous biotite-muscovite granite and monzogranite, forms a batholith that extends from

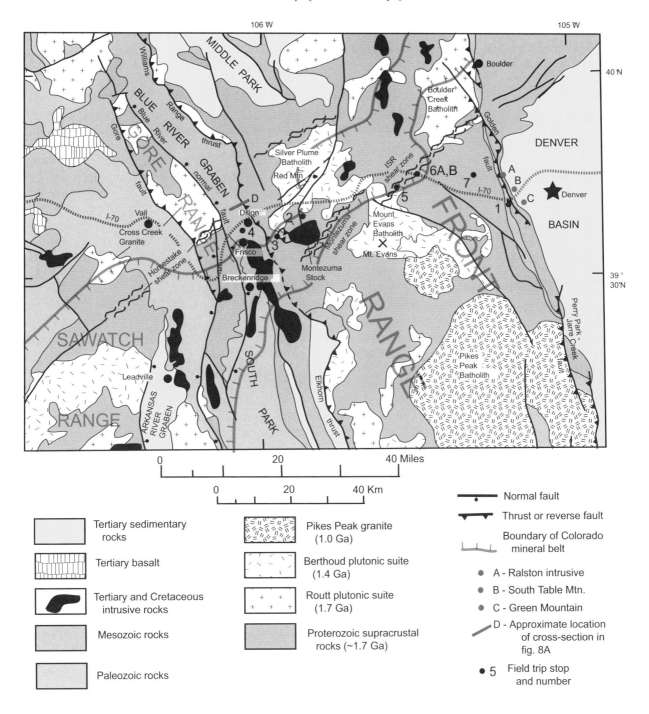

Figure 1. Sketch geologic map of the central Front Range showing field trip stops. ISR shear zone—Idaho Springs–Ralston shear zone.

the town of Silver Plume (on our transsect) to west of the Continental Divide. The Silver Plume Granite has a U-Pb zircon age of 1422 Ma (Graubard and Mattison, 1990) and was derived by limited partial melting of lower crustal material and emplaced possibly as shallow as 8 or 9 km (Anderson and Thomas, 1985). This batholith and similar ones contain many inclusions of coun-

try rock and have complex contact zones with numerous dikes and irregular bodies intruded into the country rock.

The Middle Proterozoic (ca. 1400 Ma) magmatism reset the rubidium-strontium and potassium-argon isotopic systems (Peterman et al., 1968; Shaw et al., 1999). $^{40}Ar/^{39}Ar$ dates on muscovite and biotite are all 1400–1340 Ma, reflecting cooling through closure temperature mostly after 1400 Ma. $^{40}Ar/^{39}Ar$ dates on hornblende range from 1600 Ma to 1390 Ma and represent variable retention of radiogenic argon (Shaw et al., 1999).

Diabase dikes intrude the metamorphic and granitic rocks of the Front Range. None of them has been dated in the region of our transect, but in the northeastern Front Range, similar dikes both intrude and are intruded by plutons of the Berthoud Plutonic suite with ages ca. 1420 Ma. This age, therefore, must also be the age of the dikes. East-trending lamprophyre dikes are well exposed in Clear Creek Canyon along our traverse, but their age is not known. The "Iron Dike," a distinctive north-trending gabbro dike that extends from ~10 km west of Boulder to at least as far north as the Wyoming state line, has an Rb-Sr age of 1316 Ma (Braddock and Peterman, 1989).

Emplacement of the anorogenic Pikes Peak Granite batholith at ca. 1080 Ma (Unruh et al., 1995) in the southern Front Range marks the final major Proterozoic rock-forming event.

The Proterozoic rocks of Colorado are transected by a number of northeast- to east-trending, discontinuous, en echelon shear zones consisting of steeply dipping mylonitic and non-mylonitic rocks. Our transect (Fig. 1) takes us across one of the major shear zones, the Idaho Springs–Ralston shear zone. Northeast of the Mount Evans batholith in the Front Range, the Idaho Springs–Ralston shear zone marks a discontinuity in the trends of the major folds in the metamorphic rocks, although there is no major lithologic contrast across the zone. North of the zone, folds trend north-northeast, whereas south of the zone, they generally trend northwest (Bryant et al., 1981b).

Recent study of another of the major northeast-trending shear zones, the Homestake shear zone west of the Front Range, shows that a zone of steeply dipping, highly strained but non-mylonitic rock was formed during the main metamorphism (ca. 1720 Ma), and mylonites and ultramylonites formed locally along the shear zone during or slightly after the ca. 1400 Ma plutonic event (Shaw et al., 2001). Similar evidence was obtained on the Idaho Springs–Ralston shear zone (McCoy, 2001). Latest movement on mylonites within the shear zones suggests southeast-up reverse movement (Braddock and Cole, 1979; Selverstone et al., 2000). The shear zones are interpreted to have first developed as a system of diffuse high-strain zones related to continental assembly (Shaw et al., 2001) of terranes to form the Early Proterozoic crust of the region.

Karlstrom et al. (2002) have interpreted the rocks on the southeast side of the Idaho Springs–Ralston shear zone to be part of a 1780–1730 Ma Gunnison-Salida block and those to the northwest as part of the 1750–1700 Ma Rawah block. According to this interpretation, a hypothetical belt of steeply dipping rocks of high metamorphic grade might connect the Idaho Springs–Ralston shear zone with the Homestake Creek shear zone, forming the northwest margin of the Gunnison-Salida block; however, no such belt is yet known. To test the interpretation of Karlstrom et al. (2002), we need to know the details of the age of the volcanic rocks and the metamorphic and structural history of all the rocks on either side of the Idaho Springs–Ralston shear zone.

PALEOZOIC AND MESOZOIC HISTORY OF THE FRONT RANGE REGION

During the early Paleozoic, thin continental shelf sequences of quartz-rich sands and carbonates were deposited in shallow seas over the region. In the late Paleozoic, northwest and north-northwest–trending mountain ranges and basins formed during the basement-involved Ancestral Rocky Mountain orogeny. Over the area traversed by this field trip, erosion during uplift of the Ancestral Front Range removed the earlier Paleozoic sedimentary cover, but these strata are preserved in adjacent basins where they are overlain by as much as 5 km of Pennsylvanian and Permian, mostly clastic sediments eroded from the Ancestral Rocky Mountain uplifts. Along the east flank of the Front Range at our first stop (Fig. 1), pre-Pennsylvanian rocks were eroded before deposition of ~500 m of arkosic sandstone and conglomerate of the Fountain Formation (Fig. 2), which forms the "flatirons" at various places, such as west of Boulder and west of Denver near Stop 1. The sediment forming the Fountain Formation was shed from the east flank of the Ancestral Front Range. Just south of our Front Range traverse, near Frisco (Fig. 1), ~180 m of reddish Triassic?, Permian, and Pennsylvanian sandstone and conglomerate (tentatively correlated with the Chinle and Maroon Formations) overlie basement rock (Kellogg, 2000) and were shed off the west flank of the Ancestral Front Range.

Permian, Triassic, and Jurassic fluvial, eolian, and near-shore deposits overlie the thick clastic sequences derived from the late Paleozoic uplifts. By the middle Jurassic, the Ancestral Rockies had been eroded to low relief and were covered by fluvial and lacustrine deposits of the Morrison Formation, which are 60–100 m thick on both sides of the Front Range. Near the end of the Early Cretaceous, major subsidence coeval with a rise in sea level commenced with deposition of shoreline sediments (Dakota Group) of the western interior seaway over the entire Front Range, followed by over 2 km of marine shale and minor amounts of sandstone and limestone ("Benton Group," Niobrara Formation, and Pierre Shale) (Fig. 2).

THE LARAMIDE OROGENY

The early stirrings of the Laramide orogeny, a 20 m.y. period of crustal contraction, deformation, and igneous activity during which the present Rocky Mountains were built, were marked by renewed uplift of the Front Range region. The western interior seaway began to withdraw from the region after 69 Ma, the age of the youngest ammonite zone in the Pierre Shale (Scott and Cobban, 1965; Cobban, 1993). This age is based on $^{40}Ar/^{39}Ar$ dating of tuffs elsewhere in that ammonite zone (Obradovich, 1993). The Late Cretaceous–early Tertiary rocks overlying the Pierre Shale record the uplift history, starting with the regressive shoreline deposits of the Fox Hills Sandstone, followed by the coastal plain sandstones and coal beds of the Laramie Formation (from which the Laramide orogeny is named), in turn overlain by the fluvial conglomerates, sandstones, and claystones of the

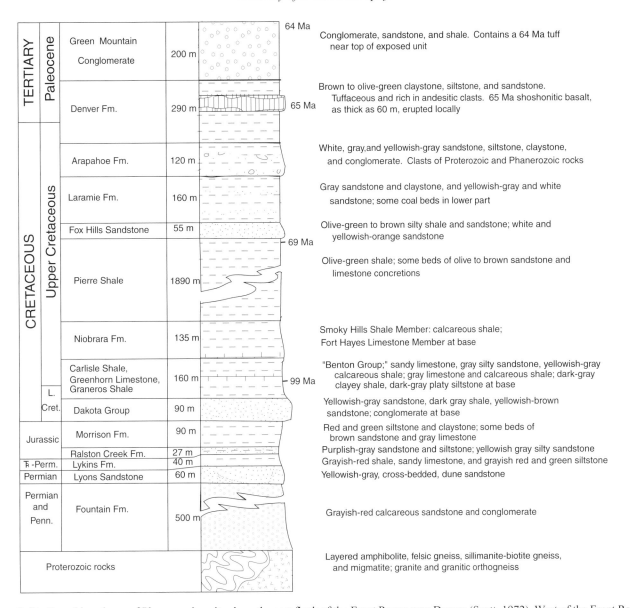

Figure 2. Stratigraphic column of Phanerozoic units along the east flank of the Front Range near Denver (Scott, 1972). West of the Front Range, units as young as the Morrison Formation locally rest directly on Proterozoic basement.

Arapaho and Denver Formations (Raynolds, 1997, 2002). Uplift in this area was geologically rapid; only a few million years separate the Late Cretaceous marine deposits of the Pierre Shale and the earliest conglomerates of the terrestrial Late Cretaceous Arapaho Formation, which contains clasts derived from Proterozoic basement rocks. During this short period, the newborn Rocky Mountains rose from the sea, and over 2 km of sedimentary rocks were eroded.

West of Denver, andesitic (field term) debris derived from volcanoes somewhere to the west of the present mountain front form a major part of the latest Cretaceous–earliest Paleocene Denver Formation. Calc-alkaline and alkalic dikes and stocks intruded into this part of the Front Range, beginning ca. 68 Ma

and continuing until ca. 27 Ma, although no Late Cretaceous or early Tertiary extrusive equivalents of any of the intrusions are preserved in this region. Volcanic rocks related to some of the younger intrusions are preserved in the northwestern part of the Front Range (O'Neill, 1981; Braddock and Cole, 1979).

Along our transect in the Golden area, the upper part of the Denver Formation contains ca. 65 Ma potassic basalt flows (shoshonites) that probably erupted from a source a few km to the north (Ralston intrusive). On South Table Mountain, ~5 km northeast of Stop 1 (Fig. 1), the K-T boundary (65.4 Ma; Obradovich, 1993) is 71 m below these basalts. Paleomagnetic directions from the Ralston intrusive (Fig. 1) are rotated, indicating that the body was emplaced before major movement on the Golden fault (Hoblitt and Larson, 1975). Near the summit of Green Mountain, ~2 km east of Stop 1 and 240 m above the basalts, the Green Mountain Conglomerate, which overlies the Denver Formation, contains a 64 Ma tuff (Obradovich, 2002). The similarity of all these ages within a relatively thick sedimentary sequence attests to rapid sedimentation, which, in turn, was due to the rapid erosion of the uplifting Front Range during the close of the Cretaceous and the opening of the Tertiary.

South of Denver, an hiatus in deposition from early Paleocene (ca. 63–64 Ma) to early Eocene (54 Ma) time separates two depositional sequences in the Denver basin (Raynolds, 2002; Obradovich, 2002), whereas in the South Park basin, on the west side of the Front Range, data suggest deposition occurred throughout the Paleocene followed by late Laramide deformation in early Eocene time (Bryant et al., 1981a). South of the latitude of Boulder, the ~80-km-wide Denver basin is highly asymmetric; the structurally deepest part, where the basement rocks are ~2,000 m below sea level (Haun, 1968), is only 8 km east of the mountain front. The elevation difference between the highest basement rocks of the Front Range (4,300 m above sea level) and the surface of the buried basement beneath the Denver basin indicates more than 6,300 m of structural relief.

On the east side of the Front Range, at the latitude of Denver, the principal Laramide structure is the Golden fault. Seismic sections and a few well data indicate that the Golden fault dips

~50°–70° to the west and has 2–3 km of eastward thrust displacement (Weimer and Ray, 1997) (Fig. 3). About 25 km north of I-70, just south of Boulder, there is a transition in the structural style. North of the transition, ENE-dipping backthrusts bring the basin side up and the range side down.

The Williams Range thrust, the western structural boundary of the Front Range, is a low-angle thrust with a minimum lateral displacement of 9 km, as shown in a window through the thrust (Stop 3) related to uplift by the 38 Ma Montezuma stock (Fig. 1) (Erslev et al., 1999). This thrust is traceable for 100 km along the west margin of the Front Range, and in South Park, a similar structure (Elkhorn thrust) is mapped along strike for 35 km. Tertiary intrusive rocks and younger deposits conceal the probable connection between the Williams Range and Elkhorn thrusts. In the western Front Range, many faults and fractures cut the Proterozoic rocks and little is known about their history, although Kellogg (2001) suggests that the fractures may be related to the large overhang along the Williams Range thrust (to be discussed at Stop 2).

Thrusts bounding both margins of the Front Range have led to many speculative models for the subsurface architecture (Fig. 4). We interpret evidence in this area to indicate that the thrust geometry is significantly different on the east and west margins of the uplift; a viable model for uplift must explain this asymmetry.

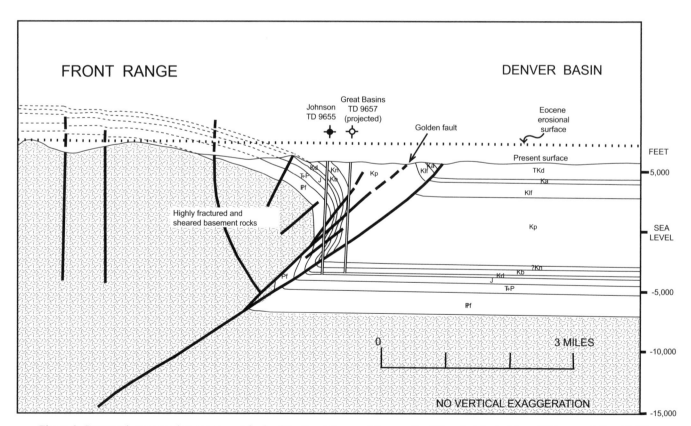

Figure 3. Structural cross section across east flank of the Front Range ~2 km south of Stop 1. Adapted from Weimer and Ray (1997).

THE COLORADO MINERAL BELT

Our transect passes through the northeastern part of the Colorado mineral belt, a northeast-trending irregular zone of Late Cretaceous and Tertiary (68–27 Ma) calc-alkalic and alkalic stocks and dikes, some of which are associated with several world-class ore deposits. The mineral belt extends northeastward across the mountainous part of Colorado from the western San Juan Mountains in the southwestern part of the state to the east flank of the Front Range north of Boulder. The mineral belt contains most of the major metallic mining districts in the state (major exceptions are the gold and silver districts at Cripple Creek and Silver Cliff). The locus of intrusion of the igneous rocks of the mineral belt seems to be related to a zone of crustal weakness marked by the northeast-trending ductile shear zones in the Precambrian rocks (Tweto and Sims, 1963). The belt of intrusives has been interpreted as the expression of a large subjacent batholith or series of batholiths, a suggestion that is consistent with the

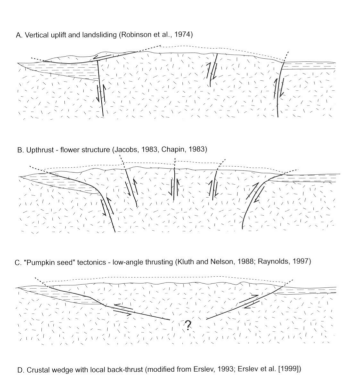

A. Vertical uplift and landsliding (Robinson et al., 1974)

B. Upthrust - flower structure (Jacobs, 1983, Chapin, 1983)

C. "Pumpkin seed" tectonics - low-angle thrusting (Kluth and Nelson, 1988; Raynolds, 1997)

D. Crustal wedge with local back-thrust (modified from Erslev, 1993; Erslev et al. [1999])

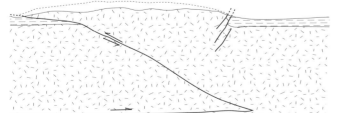

10 km

Figure 4. Models for Front Range uplift, adapted from Erslev et al. (1999).

fact that the mineral belt is nearly coincident with a major gravity low, one of the lowest Bouguer gravity anomalies in the United States (Tweto, 1975).

The Front Range portion of the mineral belt lies in the area shown on Figure 1 and is characterized by abundant stocks and dikes of Laramide and post-Laramide age emplaced into Proterozoic rocks. These intrusive bodies have a wide range of composition and a long and complicated history of emplacement. The earliest intrusions in the Front Range region are chiefly monzonites and granodiorites, followed by more alkalic intrusives, all emplaced during the interval from 68 to ca. 54 Ma (Rice et al., 1982). Ore deposits at Central City and Idaho Springs may have developed from a short-lived (~1 m.y.), complex hydrothermal system that developed 62 Ma (Nelson et al., 2003) at the end of the monzonitic and granodioritic sequence of intrusive activity. Farther to the southwest, much larger granitic plutons emplaced at ca. 40 ± 5 Ma (late Eocene) include the Montezuma stock, intrusions in the Breckenridge district, and the Mount Princeton batholith in the Sawatch Range. Hydrothermal systems that formed the base and precious metal deposits at the Georgetown, Silver Plume, Montezuma, and Breckenridge districts are associated with these younger granitic intrusives.

Some of the largest ore bodies in the mineral belt are associated with the youngest intrusives (30–25 Ma) in the Front Range region, many of which are high-silica alkali granite or rhyolite porphyries (rhyolite-A suite) (White et al., 1981). These include the world-class molybdenum deposits at Climax and Henderson.

The Central City district near Idaho Springs (Stop 6) consists of a ca. 62 Ma zoned hydrothermal system having a core of gold-bearing pyrite-quartz veins, an intermediate zone of pyrite veins carrying copper, lead, and zinc sulfide minerals, and a peripheral zone of galena-sphalerite-quartz-carbonate veins (Sims et al., 1963; Sims, 1988; Rice et al., 1982). The ca. 37 Ma Silver Plume–Georgetown district, west of the Central City district (our route traverses this district), has silver-lead-zinc–bearing quartz-carbonate veins and some gold-silver veins trending east to northeast and controlled by several sets of steeply dipping fractures (Bookstrom, 1988; Bookstrom et al., 1987).

THERMAL AND UPLIFT HISTORY AS REVEALED BY APATITE FISSION TRACKS

Apatite fission-track (AFT) studies (Bryant and Naeser, 1980; Naeser et al., 2002; Kelley and Chapin, 2004) have shown that southeast of the Colorado mineral belt the top of the Laramide apatite annealing zone can be located. (The top of the zone is a "fossil isotherm," where $T \approx 120$–$130\ °C$ during the Laramide, and which yield AFT ages of 70–50 Ma. Ages above the top of the Laramide apatite annealing zone, in the Laramide partial annealing zone and above, are older). Laramide or post-Laramide movements on faults can be determined by displacements of that zone. North of the mineral belt, most of the Front Range is in the Laramide annealing zone. This may be due to a combination of heat introduced by magma into the crust along the mineral belt and by a northward increase in the thickness of the Pierre Shale (Kelley and Chapin, 2004). East and south of Mount Evans, south of the mineral belt, the top of the Laramide annealing zone is at ~3.5 km altitude (Bryant and Naeser, 1980; Kelley and Chapin, 2004). Assuming a normal geothermal gradient, rough calculations project the top of the Precambrian basement above Mount Evans (the top of which is 4.3 km above sea level) at ~5 km and the Fox Hills Sandstone at 7.5 km altitude (Bryant and Naeser, 1980). This is ~6.7 km above the Fox Hills in the deepest part of the present Denver basin. South of Mount Evans, the top of the Laramide annealing zone dips gently east, suggesting eastward tilting. In the western Front Range north of the mineral belt, no Laramide annealing zone was found (ages are younger than Laramide). Anomalously young (as young as 5 Ma) Oligocene and Miocene annealing ages in the western Front Range and the Gore Range are interpreted to be due to heating, uplift, and cooling along the northern extension of the Rio Grande rift (Blue River graben) (Naeser et al., 2002). We will discuss additional evidence from AFT studies at Stop 4.

POST-LARAMIDE MODIFICATIONS OF THE FRONT RANGE

By the close of the Laramide orogeny, during the early Eocene (ca. 50 Ma), erosional debris derived from the Laramide Front Range uplift formed a sedimentary apron, now largely eroded, that lapped onto the flanks of the range. By the late Eocene, a widespread erosional surface formed across the Front Range (Epis and Chapin, 1975). The surface was not a peneplain, but consisted of hilly topography, presently between ~2200 and 2600 m, that forms a bench-like surface visible from Denver. High peaks similar to those along the Continental Divide protruded above this erosional surface, as they do now.

Renewed uplift of the western Front Range and probable eastward tilting during the Miocene (Naeser et al., 2002) led to erosion of the range and deposition on the plains of the Ogallala Formation, an important aquifer in the region. High-level gravel deposits found on numerous ridges in the Front Range east of the continental divide are probably coarse-grained proximal equivalents of the Ogallala Formation. These resistant gravel-capped ridges, in some cases forming sinuous east-trending patterns, represent an "inverted topography." Between the late Miocene and present, widespread erosion of the west side of the Denver basin stripped away most of the Tertiary sedimentary cover, as well as large volumes of the late Paleozoic and Mesozoic rocks, leaving a deeply incised topography. Deep canyons were cut during this time into the early and middle Proterozoic crystalline rocks along the eastern margin of the Front Range.

One of the current controversies concerns the Neogene uplift history of the Front Range. One school of thought holds that the range has been at essentially its present elevation since at least Eocene time and that the profound erosion and canyon cutting along the eastern flank is an outcome of Pliocene to Recent climate change (Gregory and Chase, 1992; Molnar and England,

1990). Evidence for this model is from paleobotanical observations on mid-Tertiary leaf morphology, which suggests paleotemperatures and, therefore, lapse rates (change of temperature with altitude) (Gregory and Chase, 1992). Another school, favored by us, advocates that a broad region, centered on the Rio Grande rift, was uplifted by as much as a thousand meters during late Miocene to Recent time (Eaton, 1986, 1987), leading to eastward tilting of the Front Range (Steven et al., 1997; McMillan et al., 2002), and that canyon cutting along the east side of the Front range is an outcome of this uplift.

How much of the Front Range has eroded since Laramide uplift? The mining industry has provided important data that indicate that the western Front Range either rose significantly during the late Tertiary or contained mid-Tertiary peaks significantly higher than they are today (Geraghty et al., 1988). A plug of porphyritic rhyolite forms the 3700 m summit of Red Mountain (Fig. 1) and is underlain at depth by the 27–30 Ma Red Mountain intrusive system, now being exploited by the Henderson molybdenum mine. Minimum homogenization pressures were obtained from vapor-rich fluid inclusions collected underground at 2500 m altitude in distal parts of extensive open-space veins. These pressures indicate 2900 m of overburden at the time the inclusions formed, which suggests that at the time of intrusion the ground surface was ~1700 m above the present top of Red Mountain. Either the Front Range was significantly higher during the Oligocene or, more likely, significant uplift has occurred since the Oligocene. In either case, ~1.7 km was removed from above the summit of Red Mountain during Neogene erosion. Detailed paleomagnetic study also shows that the intrusions and associated ore system in the Henderson mine were tilted east 25° between two faults (Geissman et al., 1992).

Directly west of the Front Range, the Blue River half graben represents a northern portion of the Rio Grande rift system (Fig. 1); extensional faults coeval with the rift system extend to the Wyoming border. The Gore Range bounds the half graben on the west. Remnants of valley fill in tilted fault blocks within the graben are as old as late Oligocene. Late Miocene AFT dates from the eastern part of the Gore Range show that significant displacement occurred along that side of the graben in late Neogene time (Naeser et al., 2002), although the young ages may be due, in part, to an increased geothermal gradient.

The higher parts of the Front Range were carved by glaciers during numerous glacial episodes during the Pleistocene and perhaps the late Pliocene. The last two major glaciations were the Pinedale (16–23 ka) and Bull Lake glaciations (95 to >130 ka) (Chadwick et al., 1997). Detailed knowledge of the glacial history is sketchy for events older than the Bull Lake glaciation.

FIELD TRIP STOPS

All stops are indicated on Figure 1, and driving instructions are provided between stops. For a detailed geologic road log between Stops 1 and 2 and between Stops 4 and 5, refer to Reed et al. (1988).

Driving instructions to Stop 1: Follow 6th Avenue west out of Denver to Interstate 70 (I-70). Continue west to exit 259 (for Morrison). After exiting, turn south onto Route 26, go under the I-70 overpass, go past the eastbound on-ramp, and take an immediate left turn into the parking lot. Climb the ridge east of the parking lot for a panoramic view of the eastern Front Range.

Stop 1. Dakota Hogback North of Morrison—The Eastern Margin of the Front Range Uplift

This ridge and a spectacular road cut just below this point exposes Jurassic Morrison Formation and Cretaceous Dakota Group rocks (Fig. 2) upturned above the Golden fault, an east-directed reverse fault that lies mostly buried by surficial deposits at the eastern base of the ridge. Walkways on both sides of the roadcut offer interpretive signs to examine the section, although we will not do so on this field trip. To the west, the Pennsylvanian and Permian Fountain Formation (arkosic sandstone and conglomerate) forms the dramatic "flatirons" along portions of the eastern foothills of the Front Range. However, directly west of this location, the Cherry Gulch fault cuts out a portion of the Fountain Formation. The Fountain Formation unconformably overlies early Proterozoic supracrustal and intrusive rocks that form the core of the Front Range. To the east, Cretaceous and Tertiary rocks comprise the near-surface rocks of the Denver basin.

The Golden fault is a west-dipping reverse fault (Berg, 1962; Weimer and Ray, 1997) that carried basement rocks over the Cretaceous rocks of the interior seaway. Figure 3 shows a cross section across the Golden fault at Turkey Creek, ~6 km south of here. Just east of Stop 1, the Golden fault places Upper Cretaceous Benton Group rocks (Fig. 2) over the upper part of the Upper Cretaceous Pierre Shale, cutting out much of the Benton Group, all of the Niobrara Formation, and ~1 km of the Pierre Shale (Scott, 1972). An overturned outcrop of Fox Hills Sandstone is clearly visible. The Golden fault is one of a series of en echelon west-dipping reverse faults that mark much of the eastern flank of the Front Range. It may be connected, via a blind (subsurface) fault, with the Perry Park–Jarre Creek thrust to the south (Fig. 1). The highly asymmetrical shape of the southern and central Denver basin can be attributed to thrust loading adjacent to these and similar faults that mark the west side of the basin.

A profound change in the style of Laramide faulting, visible from Stop 1, occurs ~15 mi (24 km) to the north, near Boulder, where the eastern margin of the Front Range steps eastward, giving the margin a more northern average orientation. A series of these steps is created by northwest-striking, southwest-directed back thrusts that expose basement in the core of fault-propagation folds (Erslev and Rogers, 1993). Reflecting this change in structural style, the northern Denver basin is both shallower and more symmetric than the southern Denver basin, and the basin axis is farther from the range.

It is important to remember that the relief along the eastern margin of the Front Range is an erosional relic of Laramide uplift and thrust or reverse faulting that placed resistant Proterozoic rocks against more easily eroded sedimentary rocks. Despite the linearity of the mountain front, extension has not played a significant role in forming this topography.

A wide spectrum of models has been proposed to explain Laramide deformation (Fig. 4). Earlier vertical-tectonic models (e.g., Prucha et al., 1965; Sterns, 1978; Matthews and Work, 1978; Robinson et al., 1974; Jacobs, 1983) have largely been superseded, with some reservations, by models that invoke horizontal crustal shortening (e.g., Kluth and Nelson, 1988; Gries, 1983; Erslev, 1986, 1993; numerous papers in Schmidt and Perry, 1988; Raynolds, 1997; Erslev and Selvig, 1997).

In order to examine the western margin of the Front Range and compare it with the eastern margin, we will next travel directly to the western side of the range. On our return to Denver, the trip will examine the Proterozoic rocks that comprise the range and observe some features of the Colorado mineral belt in more detail.

Driving instructions to Stop 2: Drive 54.8 mi. (88.2 km) west on I-70 to the Loveland Pass exit (exit 216) and proceed 4.2 mi. (6.8 km) on U.S. Route 6 to the top of Loveland Pass.

Stop 2. Loveland Pass

This view from the Continental Divide provides a spectacular overlook on a large part of the western Front Range. The rocks at the pass are mostly highly fractured migmatitic biotite gneiss, and immediately north is a contact with the 1.4 Ga Silver Plume Granite. The shattering is certainly due in part to frost action, but several NNE-trending brittle faults have also been mapped here (Bryant et al., 1981b). This widespread fracturing and fault gouge caused major engineering problems during the construction of the nearby Eisenhower Tunnel, which takes Interstate 70 under the Continental Divide.

One speculative theory (Kellogg, 1999) attributes the pervasive fracturing to Laramide movement along the low-angle Williams Range thrust (Fig. 1), which forms the western structural boundary of the Front Range a few km to the west. The thrust has a demonstrable minimum overhang of 9 km (evidence for this is at the next stop—the Keystone window in the Williams Range thrust). In fact, if a vertical hole were drilled on Loveland Pass, it is reasonable to suspect that it would encounter Cretaceous shale.

The theory suggests that there is a flexure in the thrust plane, from relatively steep at depth to gentle nearer the surface, located approximately where Proterozoic rocks in the hanging wall overlie the eastern extent of Cretaceous rocks in the footwall. The hanging wall was well above the brittle-ductile transition zone (which was probably at ~15 km at the time of thrust movement), so the brittle hanging wall rocks became pervasively fractured as they ramped over the inflection in the thrust. At many places along the west side of the range, gravitational-spreading features, such as "sackungen" (deep cracks, occasionally forming open trenches and commonly with uphill-facing scarps; the singular is "sackung"), extensive landsliding, and smooth, rounded mountain tops, characterize many of the ridges and their flanks (Varnes et al., 1989; Kellogg, 1999). In contrast to the central and eastern parts of the Front Range, which contain large, rocky outcrops with relatively widely spaced fractures, the western side contains only relatively small, strongly fractured outcrops surrounded by gravelly residuum, suggesting that the underlying rocks are relatively weak.

Driving instructions to Stop 3: Continue west on Route 6 for 8.6 mi (13.8 km) and turn left on Gondola Rd. (entrance to Keystone ski area). Take an immediate left turn (still on Gondola Rd.). Cross Montezuma Rd., jogging slightly right onto North Fork Drive (note the black hornfels of the Pierre Shale in the road cuts); park on east side of circle at the end of drive. Follow a faint trail through the woods toward the prominent cliffs ~200 m to the east. The access crosses private land, so the owner should be queried and notified of your visit first.

Stop 3. Keystone Window into the Williams Range Thrust

The Williams Range thrust forms part of the west-central structural boundary of the Front Range and has been mapped from Middle Park, just north of Kremmling, to South Park, where it is probably continuous with the Elkhorn thrust to the south (Bryant et al., 1981b) (Fig. 1). North of Middle Park, the low-angle, en echelon Never Summer thrust steps east from the Williams Range thrust and defines the eastern side of North Park.

The age of thrusting on the west side of the Front Range is not precisely known, although the onset of Laramide deformation in this area has been inferred as the age of the 70 Ma Pando Porphyry near Leadville, ~40 km to the southwest (Tweto and Lovering, 1977). This is approximately the age (69 Ma) of the marine upper Pierre Shale, indicating that the surface was near sea level and that significant uplift had not yet begun. Synorogenic lower Tertiary beds in South Park (South Park Formation) are as young as upper Paleocene and are overridden by the Elkhorn thrust. If the Elkhorn and Williams Range thrusts are synchronous, movement along the Williams Range thrust probably continued into the early Eocene.

Unlike the higher-angle faults of the eastern margin of the Front Range, the Williams Range thrust is low-angle to nearly horizontal in most places. At this locality, the Montezuma stock, a quartz monzonite porphyry with an age of 38–39 Ma (Marvin et al., 1989), domed the Williams Range thrust largely along normal faults (Fig. 5), forming a thrust window with hornfels of the Pierre Shale in the footwall and Proterozoic gneiss in the hanging wall (Ulrich, 1963). A basement overhang of at least 9 km is indicated by the distance from the thrust window to the frontal exposures of the thrust to the west, combined with the additional sedimentary section that the thrust must cut through with a maximum ramp angle. The ramp angle is unknown, so the amount of overhang is conceivably much greater than 9 km. Contact metamorphism of both the cataclastic basement rocks in the hanging wall and the Pierre Shale in the footwall, forming a resistant hornfels, allows what may be the best exposure of a Laramide thrust in the entire Front Range.

On the walk up to the thrust, stop on the small ridge near an old log cabin where erratic boulders of quartz monzonite porphyry from the Montezuma stock lie on deformed hornfels. Continue up to the cliff and examine the hornfels of the Pierre Shale. Interlayered sandstone indicates that these exposures are more than 400 m above the base of the formation (Kellogg, 2000). Traverse right around the base of the cliff to a small fault-controlled gully (offset across the gully is less than a meter). The nearly horizontal Williams Range thrust is exposed on both sides of the gully. A lower, 0.3–1.0-m-thick, strongly sheared and silicified hornfels zone marks the base of the thrust zone and is structurally overlain by a 2-m-thick zone of strongly silicified and oxidized gneiss-breccia (Fig. 6). Migmatitic biotite gneiss with a gentle foliation (strike and dip of about N20E, 25E) overlies the breccia. The overall fault orientation is about N90E, 15N and several slickenside surfaces contain lineations that trend due east, consistent with west-directed thrusting. However, the slickenside surfaces may be related to late-stage stock emplacement or Neogene extension, as they contain fault-polished chlorite-epidote alteration. Pervasive chlorite-epidote alteration is probably related to contact metamorphism and hydrothermal alteration by the Montezuma stock.

Driving instructions to Stop 4: Rejoin Route 6, drive 7.3 mi (11.7 km) west to E. Anemone Rd. (just past Dillon Dam Rd.), immediately west of the town of Dillon. Turn left and proceed 0.4 mi (0.6 km) to the cul-de-sac overlooking the Blue River Valley.

Stop 4. Western Flank of the Front Range and the Blue River Half Graben

The Blue River valley follows part of a belt of mostly Cretaceous sedimentary rocks that extends along the west flank of the Front Range. These rocks lie in a 5–9-km-wide half graben, bounded on the west by the Blue River normal fault, a complex Neogene structure that defines the abrupt east margin of the Gore Range (Fig. 1). The Blue River half graben is the northernmost major structure of the Rio Grande rift, although a network of Neogene normal faults extends as far north as the Wyoming border (Tweto, 1979b; Kellogg, 1999).

Rio Grande rifting began shortly after 29 Ma (Tweto, 1979a) and was marked by a change in volcanism from the intermediate compositions that characterize the widespread Oligocene volcanic fields to the bimodal basalt-rhyolite (mostly basalt) volcanism that characterize the Rio Grande rift (Lipman and Mehnert, 1975). Consistent with this style, the main part of the Gore Range contains a few dikes of basaltic to felsic composition. These have not been dated, but they may be related to the basaltic rocks of the Yarmony Mountain area (22–24 Ma) west of the Gore Range, part of the bimodal igneous suite related to regional extension. In the Blue River valley, a 31.5 Ma trachytic lacolithic complex (Green Mountain intrusive) intrudes Cretaceous rocks, and a 27 Ma rhyolite welded tuff and 24 Ma trachyandesite flows are preserved in tilted fault blocks within the northern Blue River half graben (Naeser et

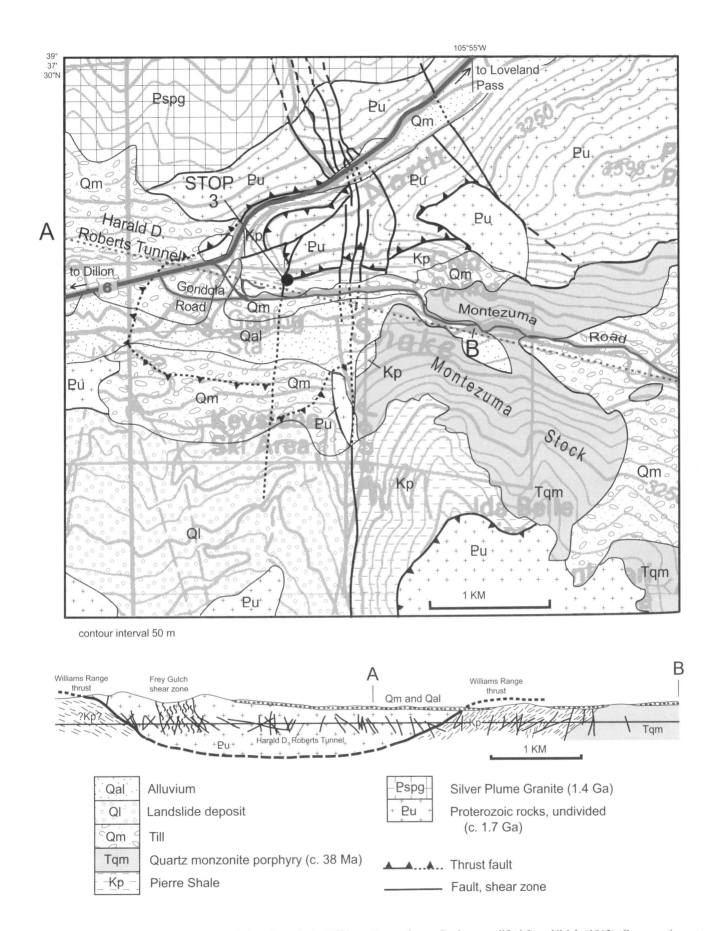

Figure 5. Geologic map of the Keystone window through the Williams Range thrust. Geology modified from Ulrich (1963). Cross section, at 0.75 scale of geologic map, is along the Harold D. Roberts Tunnel (Robinson et al., 1974), which transports water under the Continental Divide to Denver. A–B on cross section corresponds to A–B on map.

al., 2002). The tuff is near the base of the sequence, so 27 Ma (late Oligocene) represents the minimum age for initial rifting.

The Blue River normal fault along the west side of the valley has a minimum displacement of 1.2 km, based on the topographic relief between the Phanerozoic sedimentary rocks in the valley and the Proterozoic rocks that form the highest peaks in the Gore Range west of the fault. The latest movement along the fault was probably no later than Pliocene or early Pleistocene (West, 1978), although subtle scarps cutting Bull Lake till (95 to >130 ka; Chadwick et al., 1997) were mapped near Frisco, ~5 km south of here (Kellogg, 2000). An extensive apron of glacial deposits emanating from side valleys of the Blue River now covers most of the fault trace and the valley floor west of the Blue River. (The terminus of the major valley glacier that flowed north down the Blue River Valley lies under Dillon Reservoir just to the south of us.)

The half graben is cut by numerous north-striking normal faults that are almost entirely east dipping and bound west-tilted fault blocks (Fig. 7A), suggesting that the Blue River graben is a west-tilted structure above a listric fault at depth. Kellogg (1999) suggested that the west-directed Gore fault, a reverse fault along the west side of the Gore Range with significant movement in both Paleozoic and Laramide times, is listric and provided the surface along which Neogene extension beneath the Blue river half graben was accommodated (Fig. 7B).

To the northeast, on the west side of the Williams Fork Mountains (part of the western Front Range), the Williams Range thrust defines the structural margin of the Front Range. The thrust trace is buried beneath an extensive landslide complex, except along Interstate 70 ~2.5 km northeast of here (Kellogg, 2001). At that location, the thrust dips ~35° east and is marked by a 3-m-thick zone of brecciated Precambrian gneiss overlying Pierre Shale. The buried trace of the thrust climbs along the west side of the range to the north and tops the range at Ute Pass, the low point in the ridge just out of sight on the east side of the valley.

The landslide deposits that cover most of the west flank of the Williams Fork Mountains, as well as most of the trace of the Williams Range thrust, may be as thick as several hundred meters and contain blocks of Proterozoic rock tens of meters long. The deposits are deeply eroded and in most places no longer retain hummocky topography, suggesting a late Tertiary or early Pleistocene age (Kellogg, 2001, 2002).

A notable contrast in topographic form is apparent from this location: The crest of the Williams Fork Mountains on the east is rounded and not particularly steep, whereas the Gore Range on the west is rugged and steep. The rocks underlying both ranges are similar—Proterozoic gneiss and granitoid rocks—so the contrast is due to fracture density and gross rock strength underlying each range. At Stop 2 (Loveland Pass), a hypothesis for thrust-induced shattering of the hanging wall rocks was presented and is outlined in more detail in Kellogg (2001).

In late Neogene time, incision of the Blue River undercut the shattered Proterozoic rocks underlying the Williams Fork Mountains, which is inferred by Kellogg (2001) to have caused

Figure 6. View, facing northwest, showing structural zones of the Williams Range thrust fault near Keystone. Note hat for scale.

gravitational spreading of the entire mountain ridge. This spreading led to the formation of numerous sackungen ("spreading cracks") along the crest and flanks of the range (Varnes et al., 1989) and caused much of the west side of the Williams Fork Mountains to slide.

AFT ages from both sides of the Blue River valley reveal important information about the uplift and heating history of the region. The AFT ages are all younger than Laramide and range from 5 to 37 Ma in the Gore Range and 19–48 Ma in the western Front Range (Naeser et al., 2002).

A diagram showing apatite dates at 3000 m altitude from the central Front Range to the White River uplift illustrates the markedly younger dates adjacent to the Blue River valley as compared to areas several tens of km away from the valley, which yield essentially Laramide ages (Fig. 8). Similar young AFT ages along major structures related to the Rio Grande rift have been documented by Bryant and Naeser (1980), Lindsey et al. (1986), Shannon (1988), and Kelley et al. (1992). The youngest ages (5–10 Ma) are at the base of the east flank of the Gore Range west of the Blue River fault. On the tops of the high ridges near the east front of the range, AFT ages are 10–20 Ma. On the west side of the Gore Range, AFT ages are 16–25 Ma for the lower altitude samples and 26–37 Ma for the higher altitude samples. Thus, the AFT "thermochrons" (surfaces of equal age) dip west away from the Blue River valley. This indicates higher heat flow along the east side of the Gore Range, in addition to possible westward tilting of the range.

Driving instructions to Stop 5: Return to Route 6 and turn left. Proceed ~0.2 mi (0.3 km) to I-70 and take the Denver

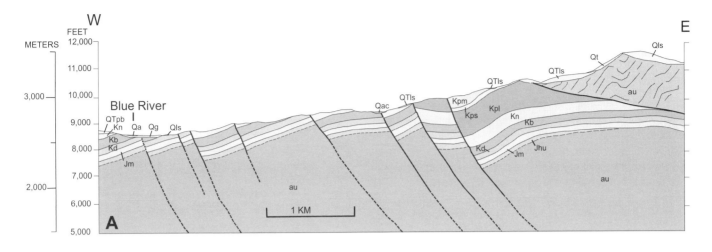

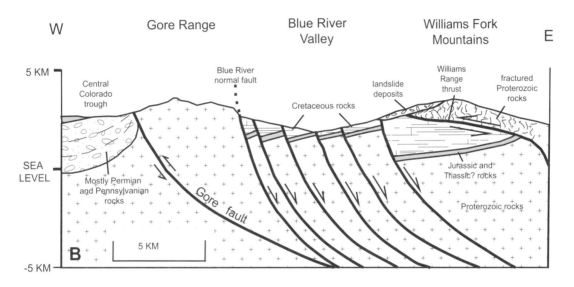

Figure 7. (A) Blue River valley cross section (from Kellogg, 1999) showing the consistent sense of normal-fault movement, which suggests a model (B) for the western margin of the Front Range in which normal faults merge into a common detachment, exploiting an older fault surface. In this case, the suggested surface is the Gore fault, which has a history at least as old as the late Paleozoic. Symbols in A: Qa—alluvium; Qg—terrace gravel; Qls—landslide deposits; QTpb—pre-Bull Lake diamicton; QTls—old landslide deposits; Kpm—Pierre Shale, sandstone and shale; Kps—Pierre Shale, sandstone; Kpl—Pierre Shale, lower shale; Kn—Niobrara Formation; Kb—Benton Shale; Kd—Dakota Group; au—Proterozoic gneiss and granite. Cross section line shown as "D" on Figure 1.

on-ramp (right turn). Drive 35.7 mi (57.4 km) east to the second Idaho Springs exit (exit 240 to Mount Evans). Turn right at the stop sign and continue 3.2 mi (5.1 km) south up the Chicago Creek valley to a wide pullout on the left (near Big Spruce Cabins). For this area, see the Idaho Springs quadrangle geologic map (Widmann et al., 2000).

Stop 5. The Idaho Springs–Ralston Shear Zone in Chicago Creek Valley

These exposures are near the southeastern side of the Idaho Springs–Ralston shear zone. The Idaho Springs–Ralston shear zone is 1–2 km wide and extends 35 km southwest from the mountain front south of Boulder to where it dies out in the Mount Evans batholith, ~7 km southwest of here (Fig. 1). This zone is typical of the numerous northeast-trending shear zones that cut Proterozoic rocks of the Front Range and adjacent ranges. These shear zones have a long and complex history. Recent studies, including monazite dating on the Homestake shear zone in the Sawatch Range (Shaw et al., 2001) and on the eastern exposures of this shear zone (McCoy, 2001), show that the zones were established during the Early Proterozoic metamorphism (ca. 1650–1700 Ma) and that they were reactivated in the Middle Proterozoic (ca. 1400 Ma) to form ductile mylonites and later brittle deformation zones.

Planar, closely spaced, steeply dipping gneiss and schist that strike consistently northeast characterize the shear zone. The younger generation of mylonitic and brittle-deformed rocks is not exposed at this stop. The rocks are layered biotite-quartz-feldspar gneiss containing some thin layers and partings of sillimanite-biotite schist and numerous stringers, lenses, and pods of pegmatite. The foliation strikes N60°E, parallels the Chicago Creek valley, and dips steeply northwest. Locally, the layers are in folds ranging from isoclinal to open, and steep northwest-plunging mineral lineations are widespread. Petrographic study of oriented thin-sections suggests that the southeast side of the shear zone moved up during formation of the down-dip lineation. West along the road is a zone where dips are gentler, and the foliation is folded. This may be a phacoid in the zone where rock escaped much of the deformation. Alternatively, it may represent the hinge zone of a large fold.

A 3-m-thick sill of Tertiary porphyry containing altered alkali feldspar phenocrysts in an aphanitic matrix is well exposed. Sills and dikes of Tertiary igneous rock are numerous in the Idaho Springs–Ralston shear zone where it crosses the Central City–Idaho Springs mining district. On the west side of the sill is a 4–5-m-thick layer of sillimanite-mica schist that has been sheared and retrogressed to form a biotite-sericite phyllonite containing porphyroclasts of biotite and muscovite, but containing relict sillimanite.

At the west end of this stop (where the road curves right), the pegmatites are thicker. At first glance, they appear to be undeformed, but close examination shows that they are foliated parallel to the shear zone, so they did not entirely escape shear-zone deformation. Farther west (around the right curve in the road) is more sillimanite-biotite-feldspar-quartz gneiss with steeply dipping lineations and pegmatite that appears to form tectonic lenses in the sheared gneiss.

Driving instructions to Stop 6A: Return to Idaho Springs. Continue over I-70 and drive two blocks through town. Turn right on Colorado Blvd. and drive 0.6 mi (1.0 km) to a grassy park. The impressive mill and dumps for the Argo Tunnel are visible across Clear Creek.

Stop 6A. The Colorado Mineral Belt and the Central City and Idaho Springs Mining Districts

The following discussion of the Central City and Idaho Springs mining districts (Fig. 9) is slightly modified from Sims (1988), which, in turn, was condensed from Sims (1983) and Sims et al. (1963). Both districts are parts of the same ore system. The Central City district produced ~70,000 metric tons of lead, 4300 metric tons of copper, 8700 metric tons of zinc, 28,350 kg of gold, 153,630 kg of silver, and small amounts of uranium. Gold accounted for ~85% and silver 10% of the value of the ore. Production in the Idaho Springs district was about two thirds of that in the Central City district, and silver was more important than gold. Production has been low since 1914.

Mining in the area played an important role in the development of the Rocky Mountain region. Placer gold was discovered

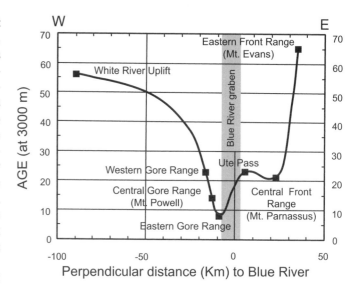

Figure 8. Regional east-west distribution of apatite fission-track ages at 3000 m altitude from the White River uplift to the Front Range showing the relation of the younger ages near the Blue River half graben (from Naeser et al., 2002).

near the present site of Idaho Springs in January 1859, and in May of the same year the first lode discovery was made in Gregory Gulch, between the present towns of Black Hawk and Central City (~6 km north of here). During the early years of the camp, mining was confined to placer gravels and surface gossans that developed on weathered veins. Later, with the development of better milling and smelting processes, mining took place below the oxidized zone. In 1900, it reached a maximum depth of 685 m in Central City. The Argo tunnel was started at Idaho Springs in 1893 to intersect the veins at depth and provide drainage and easy haulage to the mill at Idaho Springs. The 7.36-km-long tunnel was essentially completed by 1907, but for many reasons it did not stimulate significant new ore production. In 1942, the tunnel was closed because of accidental breakage into old water-filled stopes on a major vein. This resulted in a disastrous flood, and the tunnel was largely blocked by stope fill. It has never been reopened.

The host rocks in the area consist of Early Proterozoic biotite-quartz gneiss, microcline-quartz gneiss, amphibolite, migmatite, and intrusive rocks. Granodiorite and quartz diorite (ca. 1700 Ma) and two-mica granite (ca. 1400 Ma) intruded the layered rocks. A slightly younger, ~2-km-wide, northeast-trending zone of ductile deformation (Idaho Springs–Ralston shear zone) traverses the southeastern edge of the district. The country rocks in the area are cut by abundant faults, some of which exploited Early Proterozoic structures that were reactivated during the Laramide orogeny. Most of the faults, therefore, are Laramide in age and formed shortly before the mineralization.

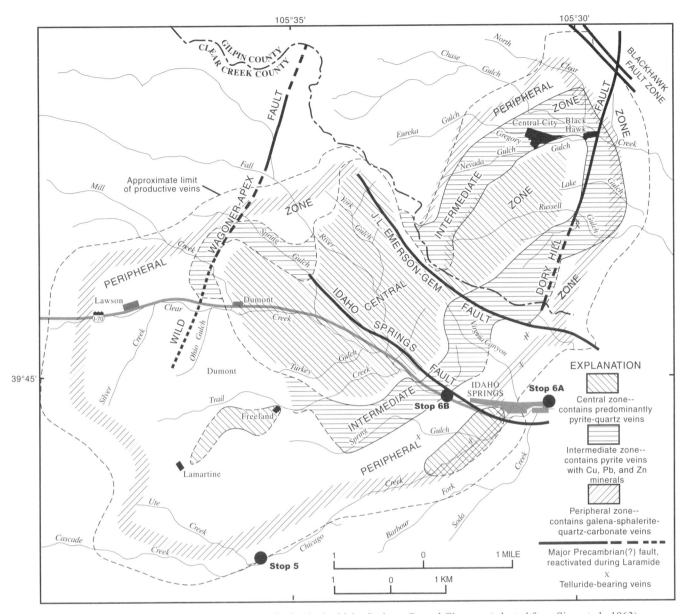

Figure 9. Map showing the zonal arrangement of veins in the Idaho Springs–Central City area (adapted from Sims et al., 1963).

The ore deposits in the Central City and Idaho Springs areas are sulfide-bearing quartz veins that contain precious metals, base metals, and sparse uranium. Formerly thought to have formed ca. 59 Ma (Rice et al., 1982), the ore deposits have been redated at ca. 62–66 Ma (Nelson et al., 2003). The deposits are related to dikes, sills, and small stocks of porphyritic igneous rocks, although they postdate intrusion. The veins are hydrothermal fillings in faults and are similar in mineralogy, texture, and structure to the deposits classified by Lindgren (1933) as mesothermal. The principal ore minerals are pyrite, sphalerite, galena, chalcopyrite, and tennantite. Less abundant are enargite, telluride minerals, and molybdenite (Rice et al.,

1985). The gangue minerals include quartz, barite, fluorite, and rhodochrosite. A portion of the area may represent the upper part of a younger, alkaline porphyry molybdenum system (Rice et al., 1985).

The ores of the area have a well-defined, concentric hypogene mineral zoning (Fig. 9). A large, central zone containing pyrite-quartz veins with gold is surrounded by an area (intermediate zone) of pyrite-type veins that carry copper, lead, and zinc sulfide minerals. The peripheral zone contains predominantly galena and sphalerite in quartz-carbonate veins.

Driving instructions to Stop 6B: Drive west on Colorado Blvd. 1.2 mi (1.9 km) back through Idaho Springs to a large parking area on the left (south) side of the road just before the entrance to I-70; do not merge onto I-70. We will visit the large road cut just to the north.

Stop 6B. Hands on the Mineral Belt—West End of Idaho Springs

The rocks along the road cut (Fig. 10) are mainly migmatitic quartz-microcline-plagioclase-biotite gneiss ("felsic gneiss"), amphibolite, and pegmatite. The layered rocks may represent a metamorphosed bimodal (rhyolite-basalt) volcanic sequence and are folded about northeast-trending axes, generally parallel to the trend of the Idaho Springs–Ralston shear zone. Several irregular dikes of Paleocene(?) bostonite porphyry (leucocratic alkalic felsite porphyry, which contains K-feldspar or sodic plagioclase phenocrysts and very few mafic minerals), and numerous mineralized veins cut the Proterozoic rocks. Oxidation of the widespread pyrite has coated much of the outcrop with an iron-oxide stain, obscuring some of the features. This locality is in the intermediate zone of the district that contains both quartz-pyrite veins and galena-sphalerite-chalcopyrite–bearing veins. Precious metals are apparently associated with both vein types.

Figure 10 outlines a traverse across the eastern 230 ft (70 m) of the outcrop (Budge et al., 1987; refer to this work for a longer, 880 ft [267 m] traverse). The Proterozoic rocks consist of lenses of amphibolite interlayered with felsic gneiss and are cut by numerous pegmatites. The layered rocks are migmatitic to varying degrees. One irregular, gently dipping bostonite dike cuts the sequence. Several west-dipping quartz-pyrite veins are visible, and one location displays a vein of galena-sphalerite-chalcopyrite as wide as 2 cm. Tennantite [$(Cu,Fe)_{12}As_4S_{13}$, a blackish, lead-gray, isometric mineral] is reported in many veins and may be present in small amounts. Quartz is the dominant gangue mineral at this locality.

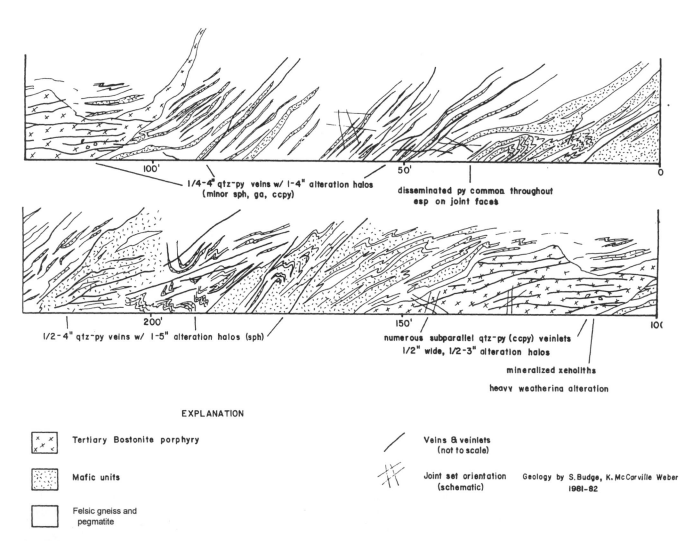

Figure 10. Detailed section across the eastern 230 ft (70 m) of an I-70 road cut at the west end of Idaho Springs (Budge et al., 1987). The "0" point is at the east end of the outcrop, near the west end of the guard rail.

Driving instructions to Stop 7: Retrace the route through Idaho Springs back to I-70 and drive 4.5 mi (7.2 km) east to the Golden exit (exit 244). Drive 3.1 mi (5.0 km) to the stop light, turn right, and continue 5.0 mi (8.0 km) to a turnout on the right, ~0.2 mi (0.3 km) past Tunnel 3 and just before Tunnel 2.

Stop 7. Ductile (and Brittle) Structures in Clear Creek

This part of Clear Creek Canyon lies in a complex, two-mile-wide transition between predominantly biotitic rocks mapped by Sheridan et al. (1967) in the Ralston Buttes quadrangle to the north and predominantly felsic gneiss and amphibolites mapped by Sheridan et al. (1972) in the Evergreen quadrangle to the south. The units mapped in this transition zone are based largely on visual estimates of relative proportions of interlayered quartz-feldspar gneiss, hornblende gneiss, amphibolite, biotite gneiss, and calc-silicate rocks, which underscores the difficulty of defining map units in this country.

Rocks exposed at this stop are chiefly fine to medium-grained biotite gneiss and medium-grained, strongly foliated migmatitic gneiss and amphibolite, all cut by dikes of pegmatite.

The general relations are well exposed on the cliff south of Clear Creek (Fig. 11). The various lithologies can be seen in outcrops and talus blocks at the east end of the parking area.

The biotite gneiss consists of various proportions of quartz, oligoclase, biotite, minor amounts of potassium feldspar, and accessory garnet, apatite, and magnetite. Some also contain appreciable amounts of hornblende. Sillimanite is found in similar rocks nearby. Migmatitic phases are characterized by wavy, 1–3-cm-thick folia of quartz, plagioclase, and pink potassium feldspar. These folia are parallel to foliation in the enclosing gneiss and generally display biotite-rich selvages a few millimeters thick. Hedge (1972) showed that the migmatites bear no direct relation to plutons of the ca. 1700 Ma Routt Plutonic Suite, and they predate the plutons. He concluded that they formed by local partial melting during high-grade regional metamorphism. Olsen (1982) found that some of the migmatites in Clear Creek Canyon formed by partial melting in a closed system, while others required some introduction of granitic or tonalitic material from external, but perhaps nearby, sources. She estimated that the migmatites formed at ~4 kb at temperatures of 650–700 °C. Less than half a mile south of here, biotite-hornblende gneiss, similar to some of the rocks

Figure 11. Sketch (by J.C. Reed) of cliff face southwest of Clear Creek at Stop 7 showing amphibolite (stippled) and pegmatite dikes (random dashed pattern) in biotite gneiss and migmatite. Cliff is ~400 ft (120 m) high.

exposed here, contains large porphyroblasts of orthopyroxene, suggesting conditions transitional between upper amphibolite and granulite facies. About 11 km to the northeast, pelitic schist contains retrograde muscovite and large, undeformed porphyroblasts of andalusite resulting from a low pressure regional metamorphism at ~525 °C (Shaw et al., 1999). The effects of this event have not yet been found in Clear Creek Canyon.

In the vicinity of this stop, layering and layer-parallel foliation strike N70°W and dip 40°–60° southwest. A strong mineral lineation defined by biotite streaks plunges ~10°W. Axes of intrafolial isoclines with amplitudes of meters to tens of meters lie approximately parallel to this lineation; one of them is exposed in the cliff at the eastern end of the parking area. Note that the sense of the small folds in the migmatite folia on the limbs of this isocline is consistent with the folds having formed as second-order, parasitic folds during formation of the isocline. The complicated pattern of amphibolite pods and layers in the cliff south of the creek is largely due to the fact that the outcrop face is subparallel to the isoclinal axes. The pattern is quite similar to the general pattern of units shown on the geologic maps of the area (Sheridan et al., 1967; Sheridan et al., 1972), suggesting that much of the complex outcrop pattern is due to early layer-parallel isoclinal folding and that lithologic sequences in these rocks may have little stratigraphic significance. A second generation of larger, more open folds of various trends commonly depicted on 1:24,000 maps affects layering and layer-parallel foliation on the limbs of isoclines of this type.

The layered gneisses are cut by dikes of pegmatite that range from sharply discordant to semi-concordant. Some are folded and contain boudins, but others seem to postdate ductile deformation of the wall rocks. The pegmatites consist of quartz, potassium feldspar, and plagioclase; some contain biotite and conspicuous knots of magnetite. There are no obvious differences in mineralogy and texture between those that are deformed and those that are not. The lack of muscovite and tourmaline in the pegmatites here (these minerals are found in pegmatites of some Silver Plume granites) and the fact that at least some pegmatites are clearly deformed suggests that they are related to the ca. 1.7 Ga Routt Plutonic Suite rather than to the ca. 1.4 Ga Berthoud Plutonic Suite.

Discordant dikes of dark hornblende lamprophyre also cut the Precambrian rocks here. An offshoot of one of these is conspicuously exposed on the north side of the highway ~15 m west of the tunnel portal. The main dike, about a meter thick, is subparallel to the face of the cut and just above it. It is exposed at road level ~50 m west of the portal. The same dike makes conspicuous notches in the ridge above Tunnel 2 and in the ridge above Tunnel 3, ~0.6 km to the west. It is not clear whether these dikes are Precambrian or Tertiary.

ACKNOWLEDGMENTS

We thank J.C. Cole, T.L. Klein, and E.P. Nelson for reviewing the manuscript and clarifying numerous points.

REFERENCES CITED

Aleinikoff, J.N., Reed, J.C. Jr., and Dewitt, E.H., 1993, The Mount Evans batholith in the Colorado Front Range: Revision of its age and reinterpretation of its structure: Geological Society of America Bulletin, v. 105, p. 791–806, doi: 10.1130/0016-7606(1993)1052.3.CO;2.

Anderson, J.L., and Thomas, W.M., 1985, Proterozoic anorogenic two-mica granites: Silver Plume and St. Vrain batholiths of Colorado: Geology, v. 13, p. 177–180.

Berg, R.R., 1962, Subsurface interpretation of the Golden fault at Soda Lakes, Jefferson County, Colorado: American Association of Petroleum Geologists Bulletin, v. 46, no. 5, p. 704–707.

Bickford, M.E., Van Schmus, W.R., and Zietz, I., 1986, Proterozoic history of the midcontinent region of North America: Geology, v. 14, p. 492–496.

Bickford, M.E., Shuster, R.D., and Boardman, S.J., 1989, U-Pb geochronology of the Proterozoic volcano-plutonic terrane in the Gunnison and Salida areas, Colorado, *in* Grambling, J.A., and Tewksbury, B.J., Proterozoic geology of the Southern Rocky Mountains: Geological Society of America Special Paper 235, p. 33–48.

Bookstrom, A.A., 1988, The Georgetown–Silver Plume district, *in* Holden, G.S., ed., Geological Society of America Field Trip Guidebook 1988: Golden, Colorado School of Mines Professional Contributions no. 12, p. 85–91.

Bookstrom, A.A., Naeser, C.W., and Shannon, J.R., 1987, Isotopic age determinations, unaltered and hydrothermally altered igneous rocks, north-central Colorado Mineral Belt: Isochron/West, no. 49, 20 p.

Braddock, W.A., and Cole, J.C., 1979, Precambrian structural relations, metamorphic grade, and intrusive rocks along the northeast flank of the Front Range in the Thompson Canyon, Poudre Canyon, and Virginia Dale areas, *in* Ethridge, F.G., ed., Field Guide, Northern Front Range and Northwest Denver Basin, Colorado: Fort Collins, Colorado, Colorado State University, Department of Earth Resources, p. 106–120.

Braddock, W.A., and Peterman, Z.E., 1989, The age of the Iron Dike—a distinctive Middle Proterozoic intrusion in the northern Front Range of Colorado: The Mountain Geologist, v. 26, no. 4, p. 97–99.

Bryant, B., and Naeser, C.W., 1980, The significance of fission-track ages of apatite in relation to the tectonic history of the Front and Sawatch Ranges, Colorado: Geological Society of America Bulletin, v. 9, p. 447–451.

Bryant, B., Marvin, R.F., Naeser, C.W., and Mehnert, H.H., 1981a, Ages of igneous rocks in the South Park-Breckenridge region, Colorado, and their relations to the tectonic history of the Front Range uplift, *in* Shorter contributions to isotope research in the western United States, 1980: U.S. Geological Survey Professional Paper 1199, Chapter C, p. 15–35.

Bryant, B., McGrew, L.W., and Wobus, R.A., 1981b, Geologic map of the Denver 1° × 2° quadrangle, north-central Colorado: U.S. Geological Survey Miscellaneous Investigations Series Map I-1163, scale 1:250,000, 2 sheets.

Budge, S., LeAnderson, P.J., and Holden, G.S., 1987, Tertiary mineralization— Idaho Springs, Colorado, *in* Beus, S.S., ed., Rocky Mountain Section of the Geological Society of America, Centennial Field Guide volume 2: Boulder, Colorado, Geological Society of America, p. 311–314.

Chadwick, O.A., Hall, R.D., and Phillips, F.M., 1997, Chronology of Pleistocene glacial advances in the central Rocky Mountains: Geological Society of America Bulletin, v. 109, p. 1443–1452, doi: 10.1130/0016-7606(1997)1092.3.CO;2.

Cobban, W.A., 1993, Diversity and distribution of Cretaceous ammonites, western United States, *in* Caldwell, W.G.E., and Kauffman, E.G., eds., Evolution of the western interior basin: Geological Society of Canada Special Publication 39, p. 435–451.

Eaton, G.P., 1986, A tectonic redefinition of the southern Rocky Mountains: Tectonophysics, v. 132, p. 163–193, doi: 10.1016/0040-1951(86)90031-4.

Eaton, G.P., 1987, Topography and origin of the southern Rocky Mountains and the Alvarado Ridge, *in* Coward, M.P., Dewey, J.F., and Hancock, P.L., eds., Continental Extensional Tectonics: Geological Society [London] Special Publication 28, p. 355–369.

Epis, R.C., and Chapin, C.E., 1975, Geomorphic and tectonic implications of the post-Laramide late Eocene erosion surface in the Southern Rocky mountains, *in* Curtis, B.F., ed., Cenozoic history of the Southern Rocky Mountains: Geological Society of America Memoir 144, p. 45–74.

Erslev, E.A., 1986, Basement balancing of Rocky Mountain foreland uplifts: Geology, v. 14, p. 259–262.

Erslev, E.A., 1993, Thrusts, back-thrusts, and detachment of Laramide foreland arches, *in* Schmidt, C.J., Chase, R., and Erslev, E.A., eds., Laramide basement deformation in the Rocky Mountain foreland of the western United States: Geological Society of America Special Paper 280, p. 125–146.

Erslev, E.A., and Rogers, J.L., 1993, Basement-cover kinematics of Laramide fault-propagation folds, *in* Schmidt, C.J., Chase, R., and Erslev, E.A., eds., Laramide basement deformation in the Rocky Mountain foreland of the western United States: Geological Society of America Special Paper 280, p. 125–146.

Erslev, E.A., and Selvig, B., 1997, Thrusts, backthrusts, and Triangle zones in the northeastern margin of the Colorado Front Range, *in* Bolyard, D.W., and Sonnenberg, S.A., eds., Geologic history of the Colorado Front Range: Denver, Rocky Mountain Association of Geologists, p. 65–76.

Erslev, E.A., Kellogg, K.S., Bryant, B., Ehlich, T.K., Holdaway, S.M., and Naeser, C.W., 1999, Laramide to Holocene structural development of the northern Colorado Front Range, *in* Lageson, D.R., Lester, A.P., and Trudgill, B.D., eds., Colorado and adjacent areas: Boulder, Colorado, Geological Society of America Field Guide 1, p. 21–40.

Gable, D.J., 2000, Geologic map of the Proterozoic rocks of the central Front Range, Colorado: U.S. Geological Survey Geologic Investigations Series I-2605, scale 1:100,000, 1 sheet.

Geissman, J.W., Snee, L.W., Grasakamp, G.W., Carten, R.B., and Geraghty, E.P., 1992, Deformation and age of the Red Mountain intrusive system (Urad-Henderson molybdenum deposits), Colorado: Evidence from paleomagnetic and $^{40}Ar/^{39}Ar$ data: Geological Society of America Bulletin, v. 104, p. 1031–1047, doi: 10.1130/0016-7606(1992)1042.3.CO;2.

Geraghty, E.P., Carten, R.B., and Walker, B.M., 1988, Rifting of Urad-Henderson and Climax porphyry molybdenum systems, central Colorado, as related to northern Rio Grande rift tectonics: Geological Society of America Bulletin, v. 100, p. 1780–1786, doi: 10.1130/0016-7606(1988)1002.3.CO;2.

Graubard, C.M., and Mattison, J.M., 1990, Syntectonic emplacement of the ca. 1440 Ma Mount Evans pluton and the history of motion and the Idaho Springs–Ralston shear zone, central Front Range, Colorado: Geological Society of America Abstracts with Programs, v. 22, no. 6, p. 12.

Gregory, K.M., and Chase, C.G., 1992, Tectonic significance of paleobotanically estimated climate and altitude of the late Eocene erosion surface, Colorado: Geology, v. 20, p. 581–585, doi: 10.1130/0091-7613(1992)0202.3.CO;2.

Gries, R.R., 1983, North-south compression of the Rocky Mountain foreland structures, *in* Lowell, J.D., and Gries, R.R., eds., Rocky Mountain foreland basins and uplifts: Denver, Rocky Mountain Association of Geologists and Denver Geophysical Society, p. 1139–1142.

Haun, J.D., 1968, Structural geology of the Denver basin—regional setting of the Denver Earthquakes, *in* Hollister, J.C., and Weimer, R.J., eds., Geophysical and geologic studies of the relationships between the Denver earthquakes and the Rocky Mountain arsenal well: Quarterly of the Colorado School of Mines, v. 63, no. 1, p. 101–112.

Hedge, C.E., 1972, Sources of leucosomes of migmatite in the Front Range, Colorado, *in* Doe, B.R., and Smith, D.K., eds., Studies in Mineralogy and Precambrian Geology: Geological Society of America Memoir 135, p. 65–72.

Hoblitt, R., and Larson, E., 1975, Paleomagnetic and geochronologic data bearing on the structural evolution of the northeast margin of the northeastern Front Range, Colorado: Geological Society of America Bulletin, v. 86, p. 237–242.

Jacobs, A.F., 1983, Mountain front thrust, southeastern Front Range and northeastern Wet Mountains, *in* Lowell, J.D., ed., Rocky Mountain foreland basins and uplifts: Denver, Rocky Mountain Association of Geologists, p. 229–244.

Karlstrom, K.E., Bowring, S.A., Chamberlain, K.R., Dueker, K.G., Eshete, T., Erslev, E.A., Farmer, G.L., Hiezler, M., Humphreys, E.D., Johnson, R.A., Keller, G.R., Kelley, S.A., Levander, A., Magnani, M.B., Matzel, J.P., McCoy, A.M., Miller, K.C., Morozova, E.A., Pazzaglia, F.J., Prodehl, C., Rumpel, H. M., Shaw, C.A., Sheehan, A.F., Shoshitaishvili, E., Smithson, S.B., Snelson, C.M., Stevens, L.M., Tyson, A.R., and Williams, M.L., 2002, Structure and evolution of the lithosphere beneath the Rocky Mountains: Initial results from the CD-ROM experiment: GSA Today, v. 12, no. 3, p. 4–10, doi: 10.1130/1052-5173(2002)0122.0.CO;2.

Kelley, S.A., and Chapin, C.E., 2004, Denudation history and internal structure of the Front Range and Wet Mountains, Colorado, based on apatite fission-track thermochronology, *in* Cather, S.M., McIntosh, W.C., and Kelley, S.A., eds., Tectonics, geochronology, and volcanism in the Southern Rocky Mountains and Rio Grande rift: New Mexico Bureau of Mines and Mineral Resources Bulletin 160.

Kelley, S.A., Chapin, C.E., and Corrigan, J., 1992, Late Mesozoic to Cenozoic cooling histories on the flanks of the northern and central Rio Grande rift,

Colorado and New Mexico: New Mexico Bureau of Mines and Mineral Resources Bulletin, v. 145, p. 40.

Kellogg, K.S., 1999, Neogene basins of the northern Rio Grande Rift—partitioning and asymmetry inherited from Laramide and older uplifts: Tectonophysics, v. 305, p. 141–152, doi: 10.1016/S0040-1951(99)00013-X.

Kellogg, K.S., 2000, Geologic map of the Frisco quadrangle, Summit County, Colorado: U.S. Geological Survey Miscellaneous Field Studies Map MF-2340, scale 1:24,000, 22 p.

Kellogg, K.S., 2001, Tectonic controls on a large landslide complex—Williams Fork Mountains near Dillon, Colorado: Geomorphology, v. 41, p. 355–368, doi: 10.1016/S0169-555X(01)00067-8.

Kellogg, K.S., 2002, Geologic map of the Dillon quadrangle, Summit and Grand Counties, Colorado: U.S. Geological Survey Miscellaneous Field Studies Map MF-2390, scale 1:24,000, 1 sheet.

Kluth, C.F., and Nelson, S.N., 1988, Age of the Dawson arkose, southwestern Air Force Academy, Colorado, and implications for the uplift history of the Front Range: The Mountain Geologist, v. 25, no. 1, p. 29–35.

Lindgren, W., 1933, Mineral deposits (4th edition): New Work, McGraw-Hill Book Company, 930 p.

Lindsey, D.A., Andriessen, P.A.M., and Wardlaw, B.R., 1986, Heating, cooling, and uplift during Tertiary time, northern Sangre de Cristo Range, Colorado: Geological Society of America Bulletin, v. 97, p. 1133–1143.

Lipman, P.W., and Mehnert, H.W., 1975, Late Cenozoic basaltic volcanism and development of the Rio Grande depression in the southern Rocky Mountains, *in* Curtis, B.F., ed., Cenozoic history of the southern Rocky Mountains, Geological Society of America Memoir 144, p. 119–154.

Marvin, R.F., Mehnert, H.H., Naeser, C.W., and Zartman, R.E., 1989, U.S. Geological Survey radiometric ages, Compilation C, part 5: Colorado, Montana, Utah, and Wyoming: Isochron/West, no. 11, 41 p.

Matthews, V., III, and Work, D.F., 1978, Laramide folding associated with basement block faulting along the northeastern flank of the Front Range, Colorado, *in* Matthews, V., III, ed., Laramide folding associated with basement block faulting: Geological Society of America Memoir 151, p. 101–124.

McCoy, A.M., 2001, The Proterozoic ancestry of the Colorado mineral belt; ca. 1.4 Ga shear zone system in central Colorado [M.S. thesis]: Albuquerque, University of New Mexico, 134 p.

McMillan, M.E., Angevine, C.L., and Heller, P.L., 2002, Post-depositional tilt of the Miocene-Pliocene Ogallala Group on the western Great Plains: Evidence of Late Cenozoic uplift of the Rocky Mountains: Geology, v. 30, p. 63–66, doi: 10.1130/0091-7613(2002)0302.0.CO;2.

Molnar, P., and England, P.C., 1990, Late Cenozoic uplift of mountain ranges and global climate change: chicken or egg?: Nature, v. 346, p. 29–34, doi: 10.1038/346029A0.

Munn, B.J., and Tracy, R.J., 1992, Thermobarometry in a migmatitic terrane, northern Front Range, Colorado: Geological Society of America Abstracts with Programs, v. 24, no. 7, p. A264–A265.

Munn, B.J., Tracy, R.J., and Armstrong, T.R., 1993, Thermobarometric clues to Proterozoic tectonism in the northern Front Range, Colorado: Geological Society of America Abstracts with Programs, v. 25, no. 6, p. 424–425.

Naeser, C.W., Bryant, B., Kunk, M.J., Kellogg, K.S., Donelick, R.A., and Perry, W.J. Jr., 2002, Tertiary cooling and tectonic history of the White River uplift, Gore Range, and western Front Range, central Colorado: Evidence from fission-track and $^{40}Ar/^{39}Ar$ ages, *in* Kirkham, R.M., Scott, R.B., and Judkins, T.W., eds., Late Cenozoic evaporite tectonism and volcanism in west-central Colorado: Geological Society of America Special Paper 366, p. 31–53.

Nelson, E.P., Beach, S.T., and Layer, P.W., 2003, Laramide dextral movement on the Colorado mineral belt interpreted from structural analysis of veins in the Idaho Springs mining district: Geological Society of America Abstracts with Programs, v. 35, no. 5, p. 13–14.

Obradovich, J.D., 1993, A Cretaceous time scale, *in* Caldwell, W.G.E., and Kauffman, E.G., eds., Evolution of the western interior basin: Geological Association of Canada Special Publication 39, p. 379–396.

Obradovich, J.D., 2002, Geochronology of Laramide synorogenic strata in the Denver basin, Colorado, *in* Johnson, K.R., Raynolds, R.G., and Reynolds, M.L., eds., Paleontology and stratigraphy of the Denver basin: Rocky Mountain Geology, v. 37, no. 2, p. 155–171.

Olsen, S.N., 1982, Open- and closed-system migmatites in the Front Range, Colorado: American Journal of Science, v. 282, p. 1596–1622.

O'Neill, J.M., 1981, Geologic map of the Mount Richthofen Quadrangle and the western part of the Fall River Pass quadrangle, Grand and Jackson Counties, Colorado: U.S. Geological Survey Miscellaneous Investigations Map I-1291, scale 1:24,000, 1 sheet.

Peterman, Z.E., Hedge, C.E., and Braddock, W.A., 1968, Age of Precambrian events in the northeastern Front Range: Journal of Geophysical Research, v. 73, p. 2277–2296.

Premo, W.R., and Fanning, C.M., 2000, SHRIMP U-Pb zircon ages for Big Creek gneiss, Wyoming and Boulder Creek batholith, Colorado: Implications for timing of Paleoproterozoic accretion of the northern Colorado province: Rocky Mountain Geology, v. 35, no. 1, p. 31–50.

Prucha, J.J., Graham, J.A., and Nickelson, R.P., 1965, Basement-controlled deformation in Wyoming province of Rocky Mountain foreland: American Association of Petroleum Geologists Bulletin, v. 49, p. 966–992.

Raynolds, R.G., 1997, Synorogenic and post-orogenic strata in the central Front Range, Colorado, *in* Bolyard, D.W., and Sonnenberg, S.A., eds., Geologic history of the Colorado Front Range: Denver, Rocky Mountain Association of Geologists, p. 43–48.

Raynolds, R.G., 2002, Upper Cretaceous and Tertiary stratigraphy of the Denver basin, Colorado, *in* Johnson, K.R., Raynolds, R.G., and Reynolds, M.L., eds., Paleontology and stratigraphy of the Denver basin: Rocky Mountain Geology, v. 37, no. 2, p. 111–134.

Reed, J.C. Jr., Bickford, M.E., Premo, W.R., Aleinikoff, J.N., and Pallister, J.S., 1987, Evolution of the Early Proterozoic Colorado province: Constraints from U-Pb geochronology: Geology, v. 15, p. 861–865.

Reed, J.C. Jr., Bryant, B., Sims, P.K., Beaty, D.W., Bookstrom, A.A., Grose, T.L.T., Wallace, S.R., and Thompson, T.B., 1988, Geology and mineral deposits of central Colorado, *in* Holden, G.S., ed., Geological Society of America field trip guidebook 1988: Golden, Colorado School of Mines Professional Contribution no. 12, p. 68–121.

Reed, J.C. Jr., Bickford, M.E., and Tweto, O., 1993, Proterozoic accretionary terranes of Colorado and southern Wyoming, *in* Reed, J.C. Jr., Bickford, M.E., Houston, R.S., Link, P.K., Rankin, D.W., Sims, P.K., and Van Schmus, W.R., eds., Precambrian Conterminous U.S.: Boulder, Colorado, Geological Society of America, The Geology of North America, v. C-2, p. 211–228.

Rice, C.M., Lux, D.R., and Macintyre, R.M., 1982, Timing of mineralization and related intrusive activity near Central City, Colorado: Economic Geology and the Bulletin of the Society of Economic Geologists, v. 77, p. 1655–1666.

Rice, C.M., Harmon, R.S., and Shepard, T.T., 1985, Central City, Colorado—the upper part of an alkaline porphyry molybdenum system: Economic Geology and the Bulletin of the Society of Economic Geologists, v. 80, p. 1786–1796.

Robinson, C.S., Warner, L.A., and Wahlstrom, E.E., 1974, General geology of the Harold D. Roberts Tunnel, Colorado: U.S. Geological Survey Professional Paper 831-B, 48 p.

Schmidt, C.J., and Perry, W.J. Jr., 1988, editors, Interaction of the Rocky Mountain foreland and the Cordilleran thrust belt: Geological Society of America Memoir 171, 582 p.

Scott, G.F., 1972, Geologic map of the Morrison quadrangle, Jefferson county, Colorado: U.S. Geological Survey Miscellaneous Investigations Map I-790-A, scale: 1:24,000, 1 sheet.

Scott, G.R., and Cobban, W.A., 1965, Geologic and biostratigraphic map of the Pierre Shale between Jarre Creek and Loveland, Colorado: U.S. Geological Survey Miscellaneous Investigations Map I-439, scale 1:48,000, 1 sheet.

Selverstone, J., Hodgins, M., Shaw, C., Aleinikoff, J.N., and Fanning, C.M., 1997, Proterozoic tectonics of the northern Colorado Front Range, *in* Bolyard, D.W., and Sonnenberg, S.A., eds., Geologic History of the Colorado Front Range, 1997 RMS-AAPG Field Trip 7: Denver, Colorado, Rocky Mountain Association of Geologists, p. 9–18.

Selverstone, J., Hodgins, M., Aleinikoff, J.N., and Fanning, C.M., 2000, Mesoproterozoic reactivation of a Paleoproterozoic transcurrent boundary in the northern Colorado Front Range—implication for ca. 1.7 and ca. 1.4 Ga tectonism: Rocky Mountain Geology, v. 35, no. 2, p. 139–162.

Shannon, J.R., 1988, Geology of the Mt. Aetna cauldron complex, Sawatch Range, Colorado [Ph.D. Thesis]: Golden, Colorado, Colorado School of Mines, 434 p.

Shaw, C.A., Snee, L.W., Selverstone, J., and Reed, J.C. Jr., 1999, ^{40}Ar/^{39}Ar thermochronology of Mesoproterozoic metamorphism in the Colorado Front Range: Journal of Geology, v. 107, p. 49–67, doi: 10.1086/314335.

Shaw, C.A., Karlstrom, K.E., Williams, M.L., Jercinovic, M.J., and McCoy, A.M., 2001, Electron-microprobe monazite dating of ca. 1.71–1.63 and ca. 1.45–1.38 Ga deformation in the Homestake shear zone, Colorado: Origin

and early evolution of a persistent intracontinental tectonic zone: Geology, v. 29, no. 8, p. 739–742, doi: 10.1130/0091-7613(2001)0292.0.CO;2.

Sheridan, D.M., Maxwell, C.H., and Albee, A.L., 1967, Geology and uranium deposits of the Ralston Buttes district, Jefferson County, Colorado: U.S. Geological Survey Professional Paper 520, 121 p.

Sheridan, D.M., Reed, J.C. Jr., and Bryant, B., 1972, Geologic map of the Evergreen quadrangle, Jefferson County, Colorado: U. S. Geological Survey Miscellaneous Geologic Investigations Map I-786-A, scale 1:24,000, 1 sheet.

Sims, P.K., 1983, Geology of the Central City area, Colorado—a Laramide mining district, *in* Babcock, J.W., ed., The genesis of Rocky Mountain ore deposits—changes with time and tectonics: Denver, Colorado, Proceedings of the Denver Region Exploration Geologists Society Symposium, November 4–5, 1982, p. 95–100.

Sims, P.K., 1988, Ore deposits of the Central City-Idaho Springs area, *in* Holden, G.S., ed., Geological Society of America Field Trip Guidebook 1988: Golden, Colorado School of Mines Professional Contributions no. 12, p. 81–83.

Sims, P.K., Drake, A.A. Jr., and Tooker, E.W., 1963, Economic geology of the Central City district, Gilpin County, Colorado: U.S. Geological Survey Profession Paper 359, 231 p.

Sterns, D.W., 1978, Faulting and forced folding in the Rocky Mountains foreland, *in* Matthews, V., III, ed., Laramide folding associated with basement block faulting in the western United States: Geological Society of America Memoir 151, p. 1–37.

Steven, T.A., Evanoff, E., and Yuhas, R.H., 1997, Middle and Late Cenozoic tectonic and geomorphic development of the Front Range of Colorado, *in* Bolyard, D.W., and Sonnenberg, S.A., eds., Geologic history of the Front Range: Denver, Colorado, Rocky Mountain Association of Geologists, p. 115–123.

Tweto, O., 1975, Laramide (Late Cretaceous–Early Tertiary) orogeny in the Southern Rocky Mountains, *in* Curtis, B.F., ed., Cenozoic history of the Southern Rocky Mountains: Geological Society of America Memoir 144, p. 1–44.

Tweto, O., 1979a, The Rio Grande rift system in Colorado, *in* Riecker, R.E., ed., Rio Grande rift—tectonics and magmatism: Washington, D.C., American Geophysical Union, p. 33–56.

Tweto, O., 1979b, Geologic Map of Colorado: U.S. Geological Survey Special Map, scale 1:500,000.

Tweto, O., 1987, Rock units of the Precambrian basement in Colorado: U.S. Geological Survey Professional Paper 1321-A, p. A1–A54.

Tweto, O., and Lovering, T.S., 1977, Geology of the Minturn 15-minute quadrangle, Eagle and Summit Counties, Colorado: U.S. Geological Survey Professional Paper 956, scale 1:62,500, 96 p.

Tweto, O., and Sims, P.F., 1963, Precambrian ancestry of the Colorado Mineral Belt: Geological Society of America Bulletin, v. 74, p. 991–1014.

Tweto, O., Moench, R.H., and Reed, J.C. Jr., 1978, Geologic map of the Leadville 1° × 2° quadrangle, northeastern Colorado: U.S. Geological Survey Miscellaneous Investigations Series Map I-999, scale 1:250,000, 1 sheet.

Ulrich, G.E., 1963, Petrology and structure of the Porcupine Mountain area, Summit County, Colorado [Ph.D. thesis]: Boulder, University of Colorado, 205 p.

Unruh, D.M., Snee, L.W., and Foord, E.R., 1995, Age and cooling history of the Pikes Peak batholith and associated pegmatites: Geological Society of America Abstracts with Programs, v. 27, no. 6, p. A-468.

Varnes, D.J., Radbruch-Hall, D.H., and Savage, W.Z., 1989, Topographic and structural conditions in areas of gravitational spreading of ridges in the western United States: U.S. Geological Survey Professional Paper 1496, 28 p.

West, M.V., 1978, Quaternary geology and reported surface faulting along east flank of Gore Range, Summit County, Colorado: Quarterly of the Colorado School of Mines, v. 73, no. 2, p. 66.

Weimer, R.J., and Ray, R.R., 1997, Laramide mountain flank deformation and the Golden fault zone, Jefferson County Colorado, *in* Bolyard, D.W., and Sonnenberg, S.A., eds., Geologic history of the Colorado Front Range: Denver, Rocky Mountain Association of Geologists, p. 49–64.

White, W.H., Bookstrom, A.A., Kamilli, R.J., Ganster, M.W., Smith, R.P., and Steininger, R.C., 1981, Character and origin of climax-type molybdenum deposits: Economic Geology, 75th Anniversary Volume, p. 270–316.

Widmann, B.L., Kirkham, R.M., and Beach, S.T., 2000, Geologic map of the Idaho Springs quadrangle, Clear Creek County, Colorado: Colorado Geological Survey Open File Report 00–02, scale 1:24,000, 22 p.

Geological Society of America
Field Guide 5
2004

Continental accretion, Colorado style: Proterozoic island arcs and backarcs of the Central Front Range

Lisa Rae Fisher*
Thomas R. Fisher*

Colorado School of Mines, Department of Geology and Geological Engineering, Golden, Colorado 80401-1887, USA

ABSTRACT

The Central Front Range of the Colorado Rockies is dominated by an early Proterozoic (ca. 1.8–1.7 Ga) metamorphosed volcanic and sedimentary sequence. In terms of plate tectonics, these rocks are interpreted as island arc, backarc, and sedimentary basin-fill units formed during the accretion of Colorado onto the North American craton. Despite good exposures, which we will be able to observe throughout the field trip, and their proximity to a large metropolitan area, these rocks are still not well understood. New research is under way to better understand accretionary processes in this region. The boundaries of the Central Front Range arc sequence are currently undefined. On the east and west, the sequence is terminated by Laramide-age faulting. The Pike's Peak Batholith obscures the southern boundary, and the northern boundary is problematic.

The main units present in the Central Front Range arc sequence are amphibolites, felsic gneisses, calc-silicate gneisses, mica schists and gneisses, iron formations, metagraywackes, quartzites, and metaconglomerates. These units as a whole are often called the "Idaho Springs Formation," which is no longer considered a valid formation name. The degree of metamorphism is generally upper amphibolite grade, high temperature–low pressure. Anatectic conditions were reached in the felsic gneisses and mica schists over much of the area. This field trip examines these units in an area of slightly lower grade, where the character of the rocks is not masked by complications of anatectic melting.

Keywords: Proterozoic, Colorado Front Range, accretionary tectonics, Idaho Springs Formation, Colorado Province.

INTRODUCTION

The Precambrian schists and gneisses of the Colorado Front Range have often been called the Idaho Springs Formation. This is no longer considered a valid formation name, but it remains in popular use as a way to refer to this sequence of lithological units.

The history of geologic study of the Colorado Front Range Precambrian rocks is a fascinating one, begun by the Hayden Survey in 1874 (Marvine, 1874). I (L.R. Fisher) included a detailed review of the various workers and their ideas in my Master's thesis (Finiol, 1992). What I find most interesting is that the early scientists did not merely lump the metamorphic rocks into "basement," but they broke them into units that even today still convey meaning to those of us who work in the area, despite

*e-mails: lfisher@mines.edu, tfisher@mines.edu

Fisher, L.R., and Fisher, T.R., Continental accretion, Colorado style: Proterozoic island arcs and backarcs of the Central Front Range, *in* Nelson, E.P. and Erslev, E.A., eds., Field Trips in the Southern Rocky Mountains, USA: Geological Society of America Field Guide 5, p. 109–129. For permission to copy, contact editing@geosociety.org. © 2004 Geological Society of America the out-of-date interpretations. The geologic quadrangle mapping of the area in the 1950s and 1960s by U.S. Geological Survey workers (e.g., Sheridan et al., 1967; Wells et al., 1964) gave us excellent field maps and descriptions of the units and a good base for further interpretive study.

In the 1980s, new ideas were brought forward by geologists to interpret the regional Precambrian geology of the western United States in terms of modern plate tectonics theory. As is usual in science, the more we learn, the more questions are raised. The findings of the 1980s–1990s that accretionary tectonics were responsible for formation of the region give way today to a desire to better define the details of how this occurred.

Despite the proximity to a large metropolitan area, the metamorphic rocks in the Central Front Range have not been extensively studied or seriously visited by many geologists. Our objective on this field trip is to introduce more geologists to these fascinating rocks and how they fit into the new story of the accre-

tion of the North American craton. Further, we wish to share our plans for how we can improve our understanding of the accretionary process through our current study of the Coal Creek Quartzite and generate discussion and new ideas for future investigation.

TECTONICS AND GEOLOGY

The Proterozoic rocks of Colorado represent the addition of the Colorado Province to the Wyoming craton in a 1.8–1.7 Ga accretionary event (Condie, 1986). Current studies (e.g., CD-ROM Project) are endeavoring to determine details concerning how and when the individual components of the terrane were formed and location of boundaries of accreted arc sequences (geologic elements), and thus gain a better understanding of the tectonic processes involved (Fig. 1). The metamorphosed volcanic and sedimentary arc sequence of the Central Front Range is one component of the accreted terrane in the Colorado Province.

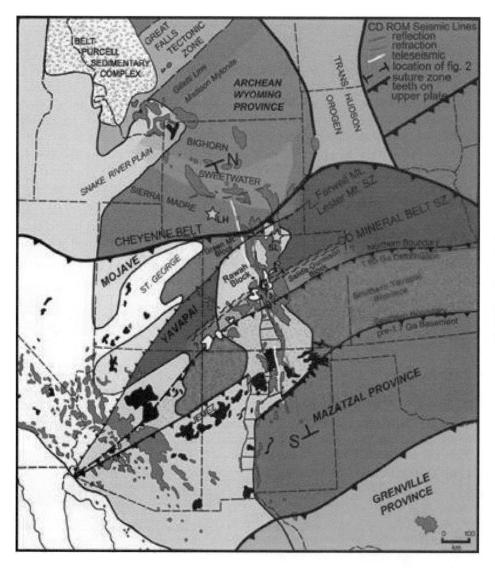

Figure 1. Geologic elements of southwestern North America. The CD-ROM Working Group has suggested several provinces in the accretion of North America. It is not yet clear to which province the Central Front Range should be assigned. Map from CD-ROM Working Group, 2002.

The boundaries of the Central Front Range arc sequence are currently undefined. The eastern and western extents are terminated by Laramide-age faulting. To the south, a boundary separating the sequence from that of the Wet Mountains must exist, but it has yet to be established. This boundary may be obscured by the Pikes Peak Batholith. The northern boundary is also difficult to determine as we have not yet recognized any clear lithological or geochemical breaks. The geology of the Colorado Front Range is shown in Figure 2.

The arc sequence of the Central Front Range has characteristics in common with others of the Colorado Province but still differs in some respects. The metamorphic grade of the central Front Range is upper amphibolite. This is higher than some other Colorado sequences, such as the Gunnison Greenstone Belt, which is greenschist grade. The protolith of metasedimentary rocks in the Central Front Range was pelitic shale, quartzitic sandstones, and conglomerates. Elsewhere in Colorado these may be graywackes, the more common sediment type found in similar sequences in the Gunnison area.

CENTRAL FRONT RANGE ARC SEQUENCE

The main units present in the Central Front Range arc sequence are metamorphosed volcanic and sedimentary units consisting of mica schists and gneisses, iron formations, calc-silicate gneisses, hornblende gneisses, amphibolite gneisses, felsic gneisses, and quartzites (all part of a package unofficially called the "Idaho Springs Formation"), with plutons of Boulder Creek (ca. 1700 Ma), Silver Plume (ca. 1400 Ma), and Pikes Peak (ca. 1000 Ma) ages. Isotopic data restrict the age of the gneisses to between 1700 and 1900 Ma. The geology of the area we will visit on the field trip is shown in Figure 3, where all these units are present at slightly lower pressure-temperature (*P-T*) conditions than elsewhere in the Central Front Range. The units are interpreted as follows (Finiol, 1992):

- **Interlayered Gneiss**: Interlayered metamorphosed intermediate felsic and mafic volcanics, volcaniclastics, and related intrusives, representing a low-K tholeiitic immature bimodal volcanic arc assemblage related to subduction occurring south of the Wyoming craton.
- **Hornblende Gneiss**: Metamorphosed submarine volcanic sequence with related carbonates and cherts, representing backarc generation of submarine tholeiitic basalt flows, with interlayered carbonates, cherts, graywackes, and other minor sediments that accumulated during periods of volcanic quiescence.
- **Transition Zone**: Metamorphosed laterally variable package of chert, sediment, stratabound sulfides, and iron formations, representing exhalative-related deposits related to declining volcanic activity. The cherts and iron formations represent the more distal or lower temperature deposits of hydrothermal vent systems, and the sulfides represent the higher temperature deposits nearer to the vents.
- **Mica Schist**: Metamorphosed pelitic shales containing sandy channels and cherty carbonate pods, representing basin sedimentation in a backarc or continental margin arc.
- **Coal Creek Quartzite**: Metamorphosed sandstone with intercalated conglomerate and shale layers, representing stacked fluvial cycles deposited into the basin.

This arc sequence collided with the growing Wyoming craton to the north, resulting in the deformation and metamorphism of the package. Syntectonic emplacement of the Boulder Creek age plutons occurred as the area became part of the magmatic arc.

METAMORPHISM

The metamorphic rocks of the Central Front Range are of upper amphibolite grade, high *T*–low *P* metamorphism, where anatectic melting reactions were reached. This indicates that heat added to the crust from intrusive bodies, rather than deep burial, was more important to the regional metamorphism.

One small area in the vicinity of White Ranch Park, Jefferson County, is of slightly lower metamorphic grade. *P-T* conditions for anatectic melting were not reached, and sillimanite-muscovite

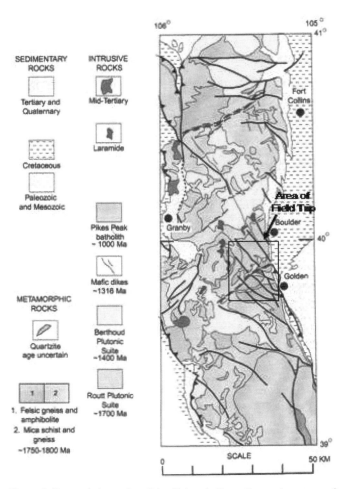

Figure 2. Precambrian units of the Colorado Front Range (courtesy of J.C. Reed).

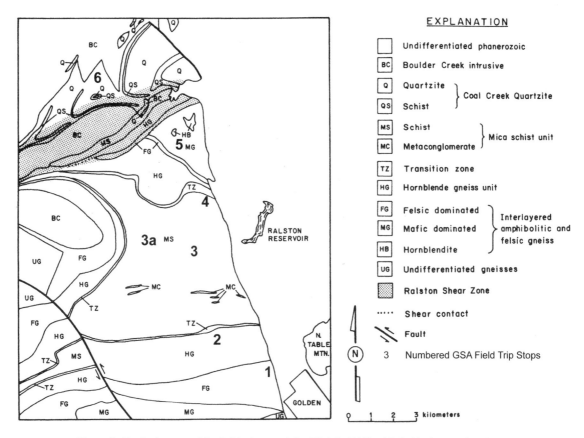

Figure 3. Geologic map of the field trip area (after Finiol, 1992) with field trip stop locations.

and/or andalusite-muscovite were stable. Indicated pressures and temperatures of metamorphism are 525–625 °C at 3–3.75 kbar (~10–12 km depth) (Finiol, 1992).

Outside of the White Ranch Park area, anatectic migmatites are commonly developed and appear to be compositionally controlled. In the mafic and calc-silicate units, *P-T* conditions for anatectic melting were not reached, and the migmatites present were produced by injection, metasomatism, or subsolidus processes, not of anatectic origin. However, *P-T* conditions for anatectic melting are lower for rocks of pelitic and felsic compositions, and anatectic migmatites are common in the pelitic and felsic volcanic units. The degree of migmatization changes across the area. Though not a simple relationship, there is a general increase of metamorphic grade toward the Mount Evans pluton.

NEW DIRECTIONS

We have learned much in the past 20 years about the how the Colorado Province was formed, but we still have much to discover. There is a need to define strategies for determination of arc sequence boundaries, more accurately date the sequences, and determine more about the tectonic processes that formed the Colorado Province.

Through continued study of the Central Front Range, we can work to define and understand the role this sequence plays

in the larger picture. New work is under way to reexamine the Coal Creek Quartzite (Fisher and Fisher, 2004), looking past the metamorphic overprint with attention to sedimentologic and stratigraphic detail. This may help to characterize basin extent, depositional environment, tectonic environment, etc., and may aid in defining arc sequence boundaries.

UNIT DESCRIPTIONS

Interlayered Mafic and Felsic Gneiss Unit

The mafic gneiss and felsic gneiss are interlayered, with the mafic gneiss more prevalent in the lower layers and to the north and the felsic gneiss more prevalent in the upper layers and to the south. The mafic gneiss is a fine to medium-grained, foliated amphibolite, composed of plagioclase and hornblende with possible minor biotite, sphene, clinopyroxene, and microcline (Figs. 4 and 5).

The felsic gneiss is a very fine to fine-grained, foliated metamorphosed quartz latite to dacite, composed of plagioclase, quartz, and microcline with possible minor biotite, hornblende, or muscovite (Figs. 5 and 6). Some layers contain relict phenocrysts of microcline. Both mafic and felsic gneisses contain thin layers of calc-silicates, which are composed of plagioclase,

Figure 4. Amphibolite and hornblendite of the mafic gneiss, Interlayered Gneiss Unit, at the north end of the area, on former Booth Ranch property at Stop 5. (A) Amphibolite, composed of hornblende and plagioclase. (B) Hornblendite, ultramafic intrusive body, now metamorphosed to hornblende. Note the relict phenocrysts, now metamorphosed to tremolite, as "eyes" on the weathered outcrop surface.

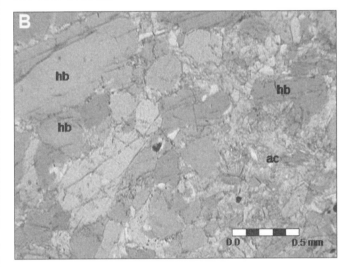

Figure 5. Photomicrographs of the Interlayered Gneiss Unit from Stop 5. (A) Amphibolite, composed of hornblende (hb) and plagioclase (pl); plane polarized light. (B) Hornblendite, composed of hornblende (hb) with relict phenocrysts of tremolite-actinolite (ac); plane polarized light. (C) Felsic gneiss, composed of quartz (q), plagioclase (pl), K-feldspar, and biotite (bt); cross polarized light.

quartz, and epidote with possible calcite, hornblende, and other accessory minerals and indicate submarine deposition as impure carbonates (Fig. 6).

Hornblendites are intrusive mafic to ultramafic bodies in the interlayered gneiss. They are medium-grained, foliated amphibolites composed of hornblende, some with relict phenocrysts that have been metamorphosed to tremolite (Figs. 4 and 5).

The transition to the next unit (the hornblende gneiss unit) is a zone ~1–15 m thick of variable composition, gradational, often containing garnet and sillimanite. In places, this is represented by quartzites, marbles, metaconglomerates, or schists.

On the field trip, we will see the interlayered gneiss unit on the southern end of the area, near the mouth of Golden Gate Canyon. Here it is represented by predominantly felsic gneiss that is a little coarser than on the north. There are several calc-silicate layers (showing some boudinage), indicating submarine deposition of the felsic volcanics. There are some layers of garnet-rich and sillimanite-rich gneiss marking the gradational transition between the felsic gneiss and the overlying hornblende gneiss. The road up the canyon almost follows the contact between the felsic gneiss and the hornblende gneiss and crosses the contact several times. Along the road, the generally pink outcrops are the felsic gneiss and the dark gray-black outcrops are the hornblende gneiss. As we drive up Golden Gate Canyon, there are several outcrops of the felsic gneiss, showing color variations in compositional layering. The pinker layers contain more microcline, the grayer layers more plagioclase. Most layers are 0.5–10 m thick.

If time and weather permit, we will visit the interlayered gneiss unit on the north end of the area, near Coal Creek Canyon at the former Booth Ranch property. At this stop, it is represented by predominantly mafic gneiss. Felsic gneiss units here are generally finer than in the south. Outcrops of hornblendite have "eyes" of relict phenocrysts.

Hornblende Gneiss Unit

The hornblende gneiss unit is complexly interlayered with amphibolite, calc-silicate, and schist. Amphibolite predominates in the lower part of the unit and grades upward into calc-silicates plus or minus schists. The calc-silicates are more prevalent to the north, and the more clastic schist layers are more prevalent to the south.

The hornblende gneiss is a moderately to strongly foliated, fine-grained amphibolite, composed of hornblende and plagioclase with possible minor clinopyroxene (Figs. 7 and 8). Layers can be massive or finely interlayered with calc-silicate. Some pillow and other volcanic textures are preserved (Fig. 9).

The calc-silicate layers contain calcite, plagioclase, hornblende, clinopyroxene, quartz, and epidote, plus accessory minerals. They range from impure marbles to metacherts to true calc-silicate compositions (Figs. 7 and 8). In some places and to the south, there is a more clastic contribution to the layer, which produces schists of variable composition (Figs. 7 and 8). The transition to the next unit (the transition zone) is gradational.

On the field trip, we will see the hornblende gneiss unit at the mouth of Golden Gate Canyon, at the contact with the felsic gneiss. There are several thick layers of amphibolite, with a large, almost Y-shaped pegmatite just around the curve in the outcrop to the west. As we drive up the canyon, the dark outcrops are the hornblende gneiss. We will pass a small, old log cabin next to an old uranium prospect on the northeast side of the road. At this point, there is a thick, black, massive amphibolite layer next to a lighter gray unit of quartz-rich metasediment. We will see these units at the parking lot of the Golden Gate Grange, Stop 2. Also at this stop is a large Silver Plume pegmatite, with quartz, microcline, plagioclase, and muscovite. In several places in the area, the Silver Plume pegmatites also contain tourmaline. We will also be able to see an outcrop of the northern part of the hornblende gneiss near the gate of the Schwartzwalder Mine, where it is a thick amphibolite.

Figure 6. Felsic gneiss, Interlayered Gneiss Unit. (A) South end of the area, Golden Gate Canyon Road, Stop 1. Note the calc-silicate layer in the felsic gneiss. (B) North end of the area, former Booth Ranch property, Stop 5.

Transition Zone

The transition zone is 3–100 m thick and is of laterally variable composition. In most places, the unit is a thin, 3–20 m metachert. In other places, there is a clastic contribution to the unit, producing quartz gneiss. In the area of the Schwartzwalder

Figure 7. Amphibolites and metasediments of the Hornblende Gneiss Unit. (A) Amphibolite from the south end of the area, representing thick volcanic flows. (B) Outcrop of the interlayered amphibolite (dark volcanic flows) and metasediments (lighter gray metagraywackes) from the south end of the area (near Stop 2). (C) Interlayered amphibolite (volcanics) and calc-silicate (impure carbonate layers) from the north end of the area, Schwartzwalder Mine property (near Stop 4); a—amphibolite layers, c—calc-silicate layers.

uranium mine, the unit thickens to 100 m and develops into an iron formation (Figs. 10 and 11), with all four facies of iron formation present. Massive sulfide deposits occur locally. Much of the transition zone is interpreted as exhalative seafloor deposits of chert, iron-rich chert, and massive sulfide deposits in the waning stages of volcanic activity. Locally, there are influxes of clastic sediment, which alter the character of the rocks.

The oxide facies of the iron formation exhibits sedimentary structures, such as cross-bedding and climbing ripples. Garnet and magnetite grains define original bedding layers rich in iron. The transition to the next unit (the mica schist unit) is gradational.

On the field trip, we will see part of the iron formation of this unit at Stop 4, in the canyon cut by Ralston Creek. At this point, we are just outside the Schwartzwalder Uranium Mine property owned by Cotter Corp., and are in the protected area of White Ranch Park of Jefferson County Open Space. There are golden eagles and other raptors nesting in the cliffs above us, and mountain lion and black bear are abundant in the area. Please be respectful of their need for relative quiet and solitude, as well as cautious of their presence.

The iron formation here is oxide facies. There are magnetite-rich schists, iron-rich metacherts, and magnetite-quartz-garnet schists. There are a few thin calc-silicate layers within the iron formation layers, indicating submarine deposition (Fig. 10).

Mica Schist Unit

The mica schist is highly pelitic, composed of quartz, muscovite, biotite, and sillimanite or andalusite, plus or minus garnet or cordierite. The schist varies from very quartz-rich to very andalusite-rich or sillimanite-rich (Figs. 12, 13, and 14). The sillimanite isograd was reached throughout much of the central Front Range, with andalusite stable in the general area of White Ranch Park, where porphyroblasts reach 30+ cm in length. The schist exhibits relict sedimentary features such as trough cross-bedding and climbing ripples, and contains lenses and channels of coarser sand-sized material, metaconglomerates, or calc-silicates. The various lithologies of the schist represent a varied sediment fill of pelitic shales to quartz-rich siltstones in the backarc basin.

The transition to the Coal Creek Quartzite is obscured within the Idaho Springs–Ralston Shear Zone. The deformation within the shear zone is inconsistent, with areas that are relatively undeformed. Within these areas, the transition between the schist and

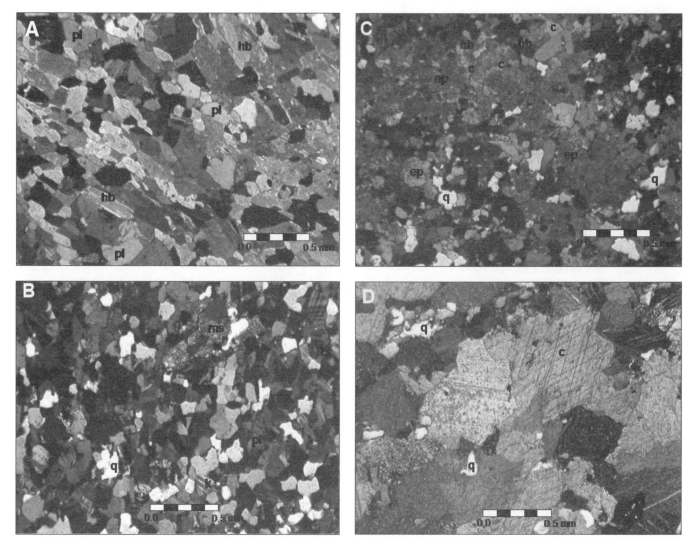

Figure 8. Photomicrographs of the Hornblende Gneiss Unit, all cross polarized light. (A) Amphibolite from the southern part of the unit, composed of hornblende (hb) and plagioclase (pl). (B) Metagraywacke from the southern part of the unit, composed of quartz (q), plagioclase (pl), K-feldspar (k), and biotite (bt). (C) Calc-silicate from the northern part of the unit, composed of epidote (ep), quartz (q), calcite (c), and hornblende (hb). (D) Marble from the northern part of the unit, composed of calcite (c) with a little quartz (q).

the quartzite is gradational. The shear zone is interpreted as a sheared anticline.

On the field trip, we will see the mica schist unit in White Ranch Park. A short walk from the parking lot takes us to an andalusite-rich outcrop of schist, where porphyroblasts of andalusite reach more than 30 cm in length. Laterally, this layer exhibits varying sizes of andalusite porphyroblasts and crosses from the andalusite into the sillimanite zone, where sillimanite occurs in large, ovoid masses. On the drive between the Golden Gate Grange and White Ranch Park, we will pass several outcrops of the schist. It is often silvery in appearance in the road cuts, especially in sunshine. Many Silver Plume pegmatites will be apparent on the drive: one near Van Bibber Creek is rich in tourmaline; one near White Ranch Park is zoned. All are pink with sharp contacts.

Coal Creek Quartzite

The Coal Creek Quartzite (Figs. 15 and 16) is one of several clastic belts of Proterozoic age deposited in syntectonic basins during the accretion of Colorado onto the North American craton. At its type locality in Coal Creek Canyon, the Coal Creek Quartzite consists of at least four quartzite units (A, B, C, and D in ascending order) separated by three major schist units (Wells, 1967). Units A and B are generally white to light gray and pink, fine-grained to conglomeratic, with hematite laminae. Relict sedimentary structures are readily apparent in these units (Fig. 15). Unit C is gray with only occasional pink and white layers. It is generally much more fine-grained and massive in appearance than Units A and B, and bedding is often inconspicu-

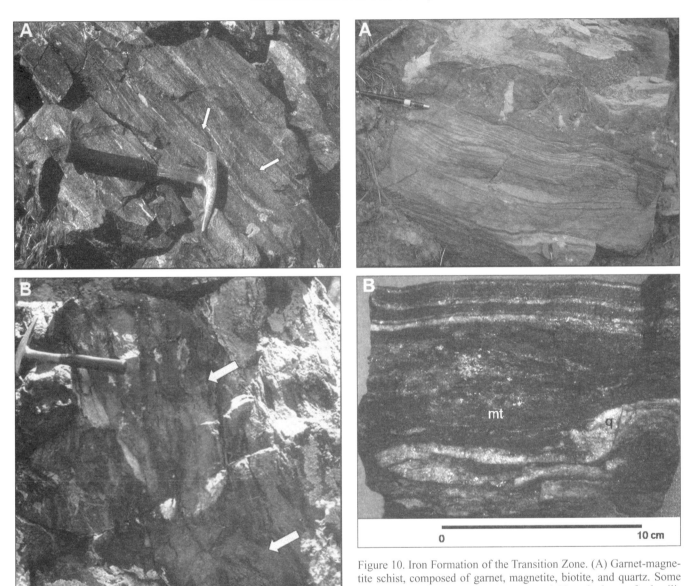

Figure 10. Iron Formation of the Transition Zone. (A) Garnet-magnetite schist, composed of garnet, magnetite, biotite, and quartz. Some relict sedimentary structures are preserved. Note the layer of calc-silicate, indicative of submarine deposition. (B) Banded iron formation of massive magnetite (mt) and quartz (q).

Figure 9. Relict volcanic textures in the northern Hornblende Gneiss Unit, from the Schwartzwalder Mine property (near Stop 4). (A) Fragmental or agglomeratic textures. (B) Pillow lavas.

ous. Unit D, the uppermost unit, is white to pink, generally fine-grained, with abundant lenses of conglomerate, some of which are very arkosic. The separating schist units are in gradational contact with the quartzites. The thickness of the quartzite varies over the main exposure in the Coal Creek Syncline; however, Wells (1967) suggests a total thickness >10,000 ft (~3000 m). The units thin both southward and northward from the axis of the

syncline. The total extent of the Coal Creek is not clear. George and Crawford (1909) reported a Coal Creek–like quartzite present in the Arapahoe Peaks area west of Boulder that may be a time-equivalent unit.

The Coal Creek Quartzite occurs mainly within the Coal Creek Syncline, a northeast-trending and plunging structure that extends into Eldorado Canyon to the north-northeast. This structure was interpreted by Wells (1967) and Wells et al. (1964) as Precambrian in age and associated with emplacement of the Boulder Creek quartz monzanites and granodiorites. The structure probably does not reflect the configuration of the original depositional basin but may mark the original depo-center.

Williams et al. (2003) suggest that the Coal Creek and similar quartzites of Colorado and New Mexico were deposited in syn-tectonic basins developed on a stabilizing crust. They further suggest that the quartzites were deposited during continued thrust convergence in the late stages of the Yavapai orogeny. This interpretation remains somewhat problematic, and it is suggested here that the Coal Creek Quartzite may have been deposited in conditions more similar to those found during the accretion of the Man (Leo) Shield onto the West African Craton. There, similar "bands" of (mineral bearing) quartzites and quartz-pebble con-glomerates accumulated in a series of extensional half-grabens associated with backarc basins. Only further detailed structural, stratigraphic, and sedimentological field work will tell the whole story. It is hoped that this further field work will help reveal the true importance of the Coal Creek and similar units in under-standing the crustal evolution of Colorado.

Recent investigations (Fisher and Fisher, 2004) of the Coal Creek have begun with reevaluation of the lowermost "A" unit, where relict sedimentary structures indicate that the quartzites are fluvial in origin, most likely braided stream deposits. Gen-erally, the lower unit consists of several stacked, fining-upward sequences of a few meters thickness each. Most of these units consist of a conglomeratic unit at the base that transitions upward into small-scale trough cross-bedding, then to more pla-nar-laminar structures with an accompanying decrease in grain size. Some imbrication is apparent in the pebble conglomerates. Pebbles are generally equidimensional to ovoid to flat. Tectonic stretching of the pebbles is not apparent in this lower unit, and individual grains are virtually undeformed (Wells, 1967). Little information has yet been gathered on paleocurrent directions. The intervening schist units are probably derived from pelitic shales. Thin partings of schist within the individual stacking units suggest clay partings within the units. Well-marked lay-ers with hematite alteration suggest the possibility of subareal exposure during deposition, although it is apparent that some remobilization of hematite has occurred. The quartzite units exhibit generally blocky fracturing, with fracture planes tend-ing to exploit depositional bedding planes.

The Coal Creek Quartzite is in contact with the apparently younger Boulder Creek granodiorite and quartz monzanite (Twin Spruce). These igneous units surround and broadly out-line the Coal Creek Syncline, but the contacts are discordant in detail. Where the contact between the Quartzite and the igneous units can be observed, a micaceous selvage exists. This zone may range from absent to a few inches to a few tens of feet thick (Wells, 1967).

On the field trip, we will visit an outcrop of what is believed to be part of the basal-most unit (Unit A of Wells [1967] and Wells et al. [1964]) of the Coal Creek Quartzite, exposed in road cuts along the north side of Coal Creek. We will be able to observe relict sedimentary structures and a sequence of stacked, fining-upward units of a few meters' thickness each. These units have recently been interpreted as fluvial in origin, most likely braided stream–type deposits (Fisher and Fisher, 2004). Most of the units

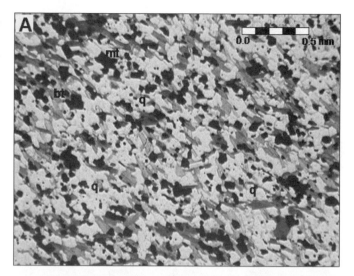

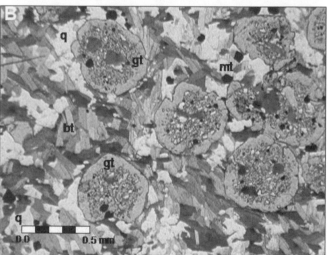

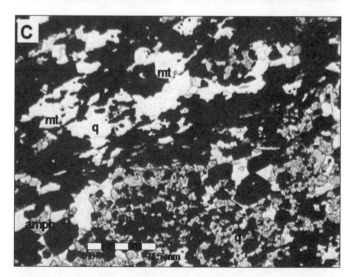

Figure 11. Photomicrographs of the Iron Formation, all in plane polarized light. (A) Magnetite schist, composed of magnetite (mt), quartz (q), and biotite (bt). (B) Garnet-magnetite schist, composed of garnet (gt), mag-netite (mt), biotite (bt), and quartz (q). (C) Banded magnetite (mt) and quartz (q), with smaller amounts of garnet (gt) and amphibole (amph).

Figure 12. Mica Schist Unit in and near White Ranch Park (near Stop 3). (A) Outcrop of aluminous schist (biotite-quartz-muscovite schist). (B) Outcrop of metaconglomerate channel in Mica Schist Unit. This photograph was taken perpendicular to foliation. Parallel to foliation, pebbles are stretched, ranging from 1:2 (quartz clasts) to 1:12 (metavolcanic clasts). (C) Outcrop of quartz-rich schist (biotite-muscovite-quartz schist) exhibiting relict sedimentary trough cross-bedding.

are marked by a basal conglomeratic unit with possible scoured base. The conglomeratic units transition upward to small-scale, trough-type cross-bedding and then to more planar-laminar–like structures. Grain-size appears to decrease upward in relation to the change of sedimentary structures. Quartz pebbles are generally rounded and equidimensional to oval, although some are flat (Wells, 1967). Some indication of imbricate structure is present within the conglomerate beds. More distinct and well-developed channels, up to several tens of meters across and with deep scouring at their base, occur in equivalent units farther to the northeast at Eldorado Canyon. Large-scale trough-type cross-beds also occur at the Eldorado locale.

Thin-section examinations by Wells (1967) show the grains are virtually undeformed at this outcrop. However, stretched-pebble conglomerates have been observed in the upper mica schist in the Golden Gate Canyon–White Ranch area and stratigraphically higher units of the quartzite near the mouth of Coal Creek Canyon. The exposures in Coal Creek Canyon lie in a northeast-trending synclinal structure interpreted by Wells (1967) as Precambrian in age. We will also be able to observe the Boulder Creek quartz monzanite, which apparently intrudes the Coal Creek Quartzite at this locale. Unfortunately, the contact between the two units is not exposed at the level of the roadcut.

Phanerozoic Units

While the focus of this field trip is on Precambrian units, we would like to introduce participants, especially those unfamiliar with Front Range geology, to the Phanerozoic units that we will pass by as we drive to our field trip stops.

Figure 13. Andalusite porphyroblasts, Mica Schist Unit, White Ranch Park, andalusite-rich schist. (A) Large andalusite porphyroblasts at Stop 3. Some porphyroblasts measure >35 cm. (B) Andalusite porphyroblasts near Stop 3. Note the variation of size and abundance in the different compositional layers.

Pennsylvanian Fountain Formation (ca. 300 Ma–280 Ma)

Steeply east-dipping along the eastern flank of the Central Front Range, the Fountain Formation forms prominent "flatirons" and palisades along the Front Range escarpment. Mainly red arkosic sandstones and conglomerates of braided stream/alluvial fan origin, these sediments are derived from erosion of the Ancestral Rockies. They rest unconformably on the ca. 1.8–1.7 Ga "Idaho Springs Formation." Easily weathered interfluvial mudstones, claystones, and siltstones form valleys between the more resistant alluvial fans that form the flatirons. Well developed and notable fans and braided channel complexes occur at Red Rocks Park near Morrison, Ralston Canyon (the "Beartooth"), the Coal Creek Canyon–Booth Ranch (Blue Mountain Subdivision) area south of Colorado

Road 72, and the flatirons between Eldorado Canyon and Boulder, as well as in the Garden of the Gods to the far south near Colorado Springs.

Permian Lyons Sandstone

The Lyons Sandstone consists of braided stream deposits of light gray to tan to reddish arkosic sandstones and conglomerates at its base, grading to and interbedded with interfluvial aeolian deposits. The type locality is at the town of Lyons, Colorado, where it is almost all aeolian in origin. Quarried for dimension stone, it is the source of the majority of the stone facings on the buildings of the University of Colorado at Boulder. Correlation of the Lyons in the Morrison-Golden area is problematic with the type locality sandstones. The Lyons Sandstone

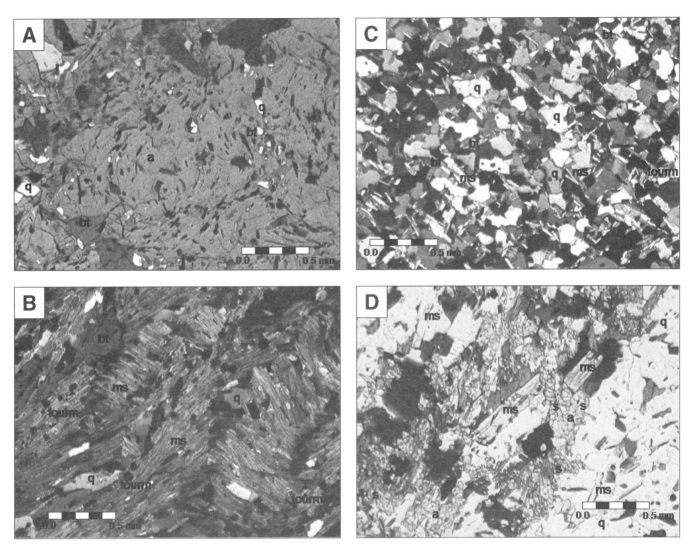

Figure 14. Photomicrographs of various phases of the Mica Schist Unit. These phases represent differences in proportions of clay, silt, and sand in original sedimentary protolith. (A) Large andalusite porphyroblast; cross polarized light. One large andalusite grain (a) with small inclusions of biotite (bt) and quartz (q). (B) Crenulations in schist; cross polarized light. Sample is mostly muscovite (ms), with biotite (bt), quartz (q), and tourmaline (tourm). (C) Quartz-rich schist; cross polarized light. Sample is mostly quartz, with fine muscovite (ms) and biotite (bt). (D) Sillimanite (s) + andalusite (a) in schist; plane polarized light. Sample is composed of quartz (q), biotite (bt), and muscovite (ms), with sillimanite and andalusite. Note the fine prisms of sillimanite growing not only outside of, but also directly within the larger andalusite porphyroblasts.

and Permian-Triassic Lykins Formation are exposed at the east entrance to White Ranch Open Space Park and along the west flank of Ralston Creek to Ralston Reservoir, our route into the Schwartzwalder Mine property.

Permian-Triassic Lykins Formation

This Formation consists mainly of thinly bedded red, sandy, and calcareous mudstones, siltstones, and shales. The Glennon Limestone member near the top of the formation is mainly pinkish to white stromatolitic limestone. The Glennon forms a relatively prominent but low ridge along the west side of the Schwartzwalder Mine road just south of Ralston Reservoir.

Jurassic Morrison Formation

Usually made up of gray, green, to red claystones, thin lacustrine limestones, and tan to reddish alluvial channel sandstones, this unit is not well exposed along the route of our field trip and is generally covered by younger alluvial sediments. The type locality of the Morrison is near the town of Morrison, immediately south of Golden on the extreme southern edge of the field trip area. The town of Morrison is also the site of several of Arthur Lakes' famous saurian quarries during the time of the Cope and Marsh competition for dinosaur fossils. The first Apatosaur (a.k.a., Brontosaur) was found at this locality ca. 1877 (Hunt et al., 2002).

Figure 15. Coal Creek Quartzite. (A) Cyclic deposition as noted by bedsets and channels of Unit A in Eldorado Canyon. (B) Relict cross-bedding in quartzite in Unit A, Stop 6. (C) Metaconglomerate in quartzite from Unit A, Stop 6.

Lower Cretaceous Dakota Sandstone

In the locality of the field trip, the Dakota forms prominent hogback ridges and is well exposed at the I-70 road cut just south of Golden and adjacent to Highway 93 along our route to Coal Creek Canyon. The Dakota Sandstone is the main producer of oil and gas in the Denver Basin immediately east of the field trip area. It is composed of four main units, in ascending order: the Lytle Sandstone, a mainly light-colored, cross-bedded, braided to meandering channel system; the Plainview Sandstone, primarily an estuarine to marine deposit with a few coal zones at its base; and the Skull Creek Shale, which overlies the Plainview and represents a period of moderate to deeper water, open marine deposition along the Cretaceous Seaway. Overlying the Skull Creek in generally unconformable contact is the "J" or Muddy Sandstone, a sequence of valley-fill deposits near the base that fine upward and transition to shallow marine, estuarine, and beach deposits. Bones, tracks, mangrove,

and abundant trace fossils accompany a transgressive cycle at the top of the sequence.

Upper Cretaceous Benton Group

In the field trip area, the Upper Cretaceous Benton group generally forms valleys and is rarely exposed with the exception of outcrops in rail line cuts and quarries near the entrance to Coal Creek Canyon. At the base of the Benton is the silver-gray, siliceous, ammonite-bearing Mowry Shale, followed by the Granerous, a hard, dark gray shale, the Greenhorn Limestone, and the Carlile Shale. The Carlile is made up of the Juana Lopez calcarenite member and the Codell Sandstone (a hydrocarbon producer in some locales of the Denver Basin).

Upper Cretaceous Niobrara Formation

This Formation is composed of the Smoky Hill Shale Member with the Fort Hays Limestone at its base. The Niobrara is a

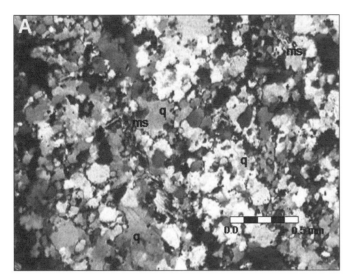 

Figure 16. Comparative photomicrographs of quartzite; cross polarized light. (A) Coal Creek Quartzite (metaclastic) near Stop 6. This and nearby samples are composed of quartz (q), with minor amounts of muscovite (ms). (B) Transition Zone Quartzite (metachert), near the White Ranch Park–Schwartzwalder Mine Property boundary. Sample is composed of quartz (q) and is almost pure SiO_2 with FeO as the only other major oxide present. There are no grains present indicative of relict heavy minerals.

gas producer on the eastern flanks of the Denver Basin. The Fort Hays was locally quarried for cement.

Upper Cretaceous Pierre Shale

This thick (>7000 feet), valley-forming, highly bentonitic shale unit underlies much of our route along Highway 93. Alternating, nearly vertical beds of bentonite pose major geotechnical problems for construction and engineered structures along the Front Range. Three coarsening upward sandy mudstone cycles (the Hygiene Sandstone, the Terry, and the Rocky Ridge Sandstone) occur near the middle of the Pierre. These "tight sand" units produce gas and some oil in parts of the Denver Basin. Several delta-front, fine-grained sandstone and turbidite deposits occur at the top of the Pierre in outcrops near the southern edge of the field trip area along Rooney Road, immediately south of I-70 and the I-70 hogback road cut.

Upper Cretaceous Fox Hills Sandstone

A white to tan, very fine to fine-grained beach to upper shoreface sandstone, the Fox Hills is the youngest and uppermost marine unit present in the sequence deposited in the Upper Cretaceous Seaway. In conjunction with the overlying Laramie Formation, it forms a major aquifer of the Denver Basin.

Upper Cretaceous Laramie Formation

The Laramie Formation consists mainly of delta plain deposits with commercially mineable coals and siliceous clays mined for ceramics. This formation is the main source of coal in the Leyden Gulch coal mine (formerly a Public Service/Xcel Energy gas storage reservoir in the process of conversion to water storage for the City of Arvada). Prominent, nearly vertical to overturned hogbacks of the Laramie Formation, bearing the scars of clay mining from the Coors Ceramics operations, are visible at Leyden Gulch on the east side of Highway 93 at the intersection with 82nd Avenue.

Upper Cretaceous Arapahoe Formation

This formation overlies the Laramie in an unconformable and sharply scoured contact. While not exposed in the immediate vicinity of the field trip, outcrops of the Arapahoe do occur on the Colorado School of Mines campus and bear mentioning here. This unit is one of the most distinctive units in the area and is composed dominantly of conglomerates, arkosic sandstone, and minor layers of claystone and siltstone deposited in a braided channel complex. It is important from the standpoint of tectonics in that the composition of the pebbles contained in the conglomerates and arkoses shows an inverted sequence of younger to older lithologies (including Precambrian metamorphics and igneous rocks), which tells the story of the "unroofing" of the Rockies during the Laramide Orogeny (Weimer, 1996).

Upper Cretaceous to Paleocene Denver Formation

Well exposed in the flanks of North and South Table Mountains, the Denver Formation is capped by 64–66 Ma columnar basalt (shoshonite) flows, which issued from the Ralston Dikes (visible to the west of Highway 93 from the intersection of 56th Street to approximately the Ralston Reservoir Dam). The Denver Formation is composed of sandstones, siltstones, and shales, with local conglomeratic channels. Andesitic material, petrified wood, and some dinosaur bones are present near the base. The K-T

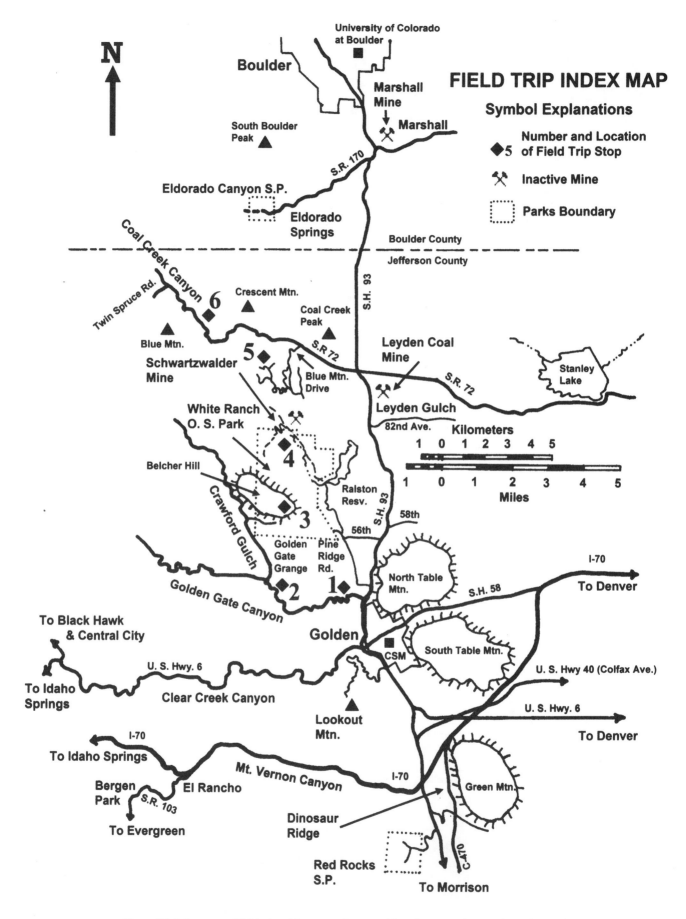

Figure 17. Index map of field trip with stops and geographic reference points. Scale as shown.

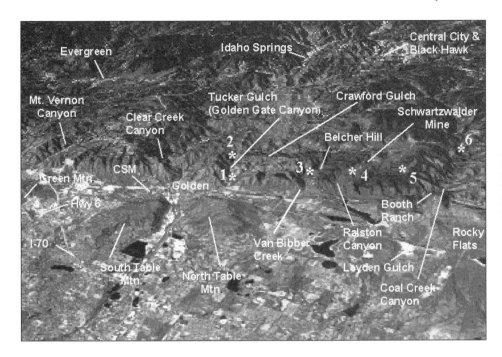

Figure 18. Oblique 30 m three-dimensional satellite image of field trip area with stops. View is to west-southwest into the Central Front Range. Asterisks represent field trip stops 1–6.

boundary is mapped near the center of the Denver Formation at the last occurrence of Triceratops.

Paleocene Green Mountain Conglomerate

This conglomerate dominates the crest of Green Mountain, immediately south of Golden and U.S. Highway 6. It is comprised dominantly of conglomerates, sandstones, and siltstones. Local Precambrian lithologies dominate clasts with andesitic materials at the base of the formation.

Pleistocene Alluvium

Five alluvial terrace sequences occur on the east flank of the Front Range and the Denver Basin. The oldest and highest of these, the Rocky Flats Alluvium, caps the Rocky Flats area along our traverse on Highway 93 to Coal Creek Canyon and northward to Eldorado Canyon. The Rocky Flats is, at this locale, armored with quartzite cobbles and boulders derived from the early Proterozoic Coal Creek Quartzite (discussed elsewhere in this field guide). The Rocky Flats alluvium represents the youngest sedimentary sequence in the field trip area with the exception of Holocene and present day alluvial and fluvial deposits in local stream valleys and terraces.

PROTEROZOIC ORE DEPOSITS OF THE CENTRAL FRONT RANGE

Stratabound deposits of Pb, Zn, Cu, and W are found in the metamorphosed volcanic rocks and their interlayered sediments. Ores are synsedimentary and stratabound. Base metal sulfides of Pb and Zn range from disseminated to localized to true massive sulfide (>50% sulfide) in small pods to large bodies that were originally tabular or lenticular. Deposits contain up to 10^6 tons of ore minerals. In places, the sulfides occur with silicates. Common minerals include sphalerite, chalcopyrite, galena, gahnite, pyrite, pyrrhotite, tetrahedrite, molybdenite, and other secondary and retrograde minerals. The average metal contents from 55 localities in this and similar Colorado Proterozoic terranes are Zn, 5%; Cu, 2.5%; Pb, 0.5%; 1–1.6 ounces Ag/ton; and 0.02–0.1 ounce Au/ton (Raymond et al., 1987). Minor amounts of Mo, W, Ni, Co, Cd, Ti, and Bi occur in some of the deposits (Sheridan and Raymond, 1982) along with occasional graphitic zones (Wallace, 1987, personal commun.). W-Cu deposits tend to be peripheral and spatially related to the Pb-Zn sulfides (Sheridan and Raymond, 1982), with scheelite and malachite the predominant minerals. The average Cu content from nine samples is 1.8% (Sheridan and Raymond, 1984). Deposition of the metals in these sulfide deposits probably resulted from exhalation of hydrothermal fluids from seafloor vents (Raymond et al., 1987). These deposits are most often hosted by more Mg-rich layers (indicating seafloor metasomatic alteration) of calc-silicates, marbles, or anthophyllite-garnet-cordierite schists and gneisses that occur interlayered with submarine volcanics in the hornblende gneiss unit or transition zone.

Iron-rich metacherts and schists or small Algoman-type iron formations occur scattered through the transition zone and occasionally within the hornblende gneiss near its contact with the transition zone. All of the iron formation facies types can be found in the area of the Schwartzwalder Uranium Mine: oxide facies, carbonate facies, sulfide facies, and silicate facies. The main iron-bearing minerals present are magnetite, ilmenite, and hematite (secondary), which occur with various other phases (garnet, grunerite, biotite, muscovite, quartz). The iron is syngenetic and disseminated to massive (~65% FeO + Fe$_2$O$_3$) in the metasediments. These are also probable seafloor exhalites related to backarc submarine volcanism, with differences from the base-metal sulfides due to such factors as fluid temperatures, duration and volume of fluid flow, and/or distance from vents.

Thin, continuous stratiform layers rich in rutile ± corundum ± topaz (locally to 80% topaz) ± gahnite occur in the upper portion of the interlayered amphibolitic and felsic gneisses (originally bimodal arc volcanics) of the Interlayered Gneiss Unit. Their origin is problematic, but exhalative, weathering, and placer processes have been suggested. The weathering horizon origin is favored due to suggestions of a related minor unconformity (Sheridan, 1987, personal commun.; Marsh and Sheridan, 1976; Sheridan and Marsh, 1976).

Laramide age U deposits occur in the transition zone associated with faults that cut the hornblende gneiss. U-Pb isotopic studies indicate that U was remobilized from the metavolcanics of the hornblende gneiss (disseminated in it, not originally U ore deposits) with an original age of 1730 ± 130 Ma (Ludwig et al., 1985). The Schwartzwalder uranium mine has produced 17 million pounds of U$_3$O$_8$, with estimated reserves of at least 16 million pounds (Sheridan et al., 1967).

TRIP LOG

Leave Colorado Convention Center

We will travel west then north to Golden via Highway 6, then proceed north on Highway 93 to the intersection of Hwy 93 and Golden Gate Canyon Road, where we start our mileage log. Locations of stops are shown on Figures 17 and 18.

Cumulative
mi   Description

0.0 Start mileage log at intersection of Hwy 93 and Golden Gate Canyon Road. Turn west on Golden Gate Canyon Road (at traffic light). We are crossing over Quaternary alluvium. As we approach the mouth of the canyon, the red arkoses of the Pennsylvanian Fountain Formation will outcrop on the right. Behind us, you can see North and South Table Mountains east of Golden, with capping lava flows of 64–66 Ma shoshonitic basalt. The K-T boundary lies a little below these flows.

0.5 The Golden Fault and the unconformity between the Fountain (red) and the Precambrian hornblende gneiss (gray and black) can be seen to the right. The hornblende gneiss outcrops around the curve ahead. The Golden Fault is the master fault on which over 14,000 ft of uplift of the Front Range occurred in Laramide time. The Fountain is the oldest Paleozoic unit preserved in this part of the Front Range. It unconformably overlies the Precambrian metamorphics and represents deposition of alluvial fan sediments shed off

the uplift of the Ancestral Rockies in Pennsylvanian time. The Fountain is laterally variable, consisting of conglomerates, arkoses, and siltstones. The unit thickens and thins intermittently north to south, the thicker conglomerates and arkoses representing the alluvial fans and the thinner siltstones and claystones representing the interfluvial areas. The unconformity represents a ca. 1.4 Ga time gap.

0.6 The contact between the hornblende gneiss (black and gray) and the interlayered gneiss (pink) is represented in the outcrop to the right (NW) of the road by a layer rich in sillimanite and garnet. This layer most likely represents a weathering horizon between the units. There is a small pullout on the left side of the road. From here to the first stop, the felsic gneiss of the interlayered gneiss outcrops along the road on the right.

0.9 **Stop 1: Mouth of Golden Gate Canyon.** Pull off at the wide area of the road on the right.

Interlayered gneiss unit: felsic gneiss, calc-silicate layers, hornblende gneiss.

The interlayered gneiss unit consists of interlayered intermediate felsic (dacite to rhyodacite) and mafic (basalt) metavolcanics, interpreted to represent island arc volcanics of the accretionary assemblage. In this area, the felsic volcanics predominate, suggesting proximity to the volcanic center. The K-feldspar–plagioclase feldspar content of the felsic gneiss varies, giving the layers slightly different composition and coloring, pink to gray. Thin layers of calc-silicate indicate submarine deposition of the volcanic layers (Fig. 6).

The contact with the hornblende gneiss unit is visible in the inside curve along the old roadbed. The dark mafic units represent metamorphosed backarc basalts and contain some lighter gray metasedimentary layers.

As we proceed up the canyon (called Tucker Gulch) to the next stop, the road more or less parallels the contact between the felsic gneiss and hornblende gneiss units, crossing the contact several times. We see both units alternate along the road ahead.

3.1 You will see an old log cabin and mine structure on the right side of road. This is an old uranium prospect. The hornblende gneiss here has a massive, dark layer of metabasalt and a lighter layer of metasediments (Fig. 7). We will see examples of these up close at Stop 2.

3.9 **Stop 2: Golden Gate Grange Parking Lot.** Turn right into the parking lot. Use the second driveway entrance. Hornblende gneiss unit: Metamorphosed volcanics and sediments; Silver Plume pegmatite; joints (structural).

The hornblende gneiss at this stop shows good examples of the amphibolites (metabasalts) and metasediments of this unit (Fig. 7). To the north, the metasediments are more commonly calcareous. To the south (here), they are more commonly quartzose. These metamorphosed basalts and sediments represent eruption and sedimentation into a backarc basin formed within the volcanic arc environment. The metabasalts may exhibit relict volcanic textures, such as pillows or fragmental textures (Fig. 9).

The pegmatite in this outcrop is of Silver Plume age, ca. 1.4 Ga. While most of the pegmatites have not been radiometrically dated, it is commonly accepted that the Silver Plume pegmatites contain K-feldspar > plagioclase, muscovite, often tourmaline, have sharp contacts, and are not foliated.

Note the jointing pattern here, related to Laramide uplift of the Rockies.

Exit Golden Gate Grange parking lot and proceed NW along Golden Gate Canyon Road.

4.0   Turn right (north) onto Crawford Gulch Road.

4.5   Note the outcrops of mica schist along the road on the right. We have crossed the contact between the horn-blende gneiss and the mica schist. The schist will appear light gray if the sky is cloudy, but silvery and shiny in bright sunlight.

5.1   Note outcrops of Silver Plume pegmatites along the road on your right. There are several of varying sizes as we drive north. They are similar to the pegmatite at the Grange, but some contain large tourmalines.

6.1   Cross Van Bibber Creek, a small canyon cut through the schist and pegmatites.

6.2   Here are more mica schist and pegmatite outcrops along the road on the right.

7.0   More outcrops of mica schist at road level on the right.

8.1   Turn east onto Belcher Hill Road (at the intersection of Crawford Gulch, Belcher Hill, and Drew Hill Roads).

8.5   Here is a small but zoned Silver Plume Pegmatite on the left along the road.

9.3   Entrance to White Ranch Park, Jefferson County Open Space.

9.7   **Stop 3: White Ranch Park.** Park in the parking lot at the north side of the end of the road (road ends at locked gate).
Mica schist unit: mica schist; andalusite porphyroblasts.

White Ranch Park is a wonderful place to see the mica schist. There are hundreds of outcrops of schist across the park and many hiking, biking, and horse trails. The schist is variable, ranging from quartz-rich to highly aluminous (Fig. 12). We will take a short walk to an outcrop with spectacular andalusite porphy-roblasts (Fig. 13). This particular unit can be traced laterally and remains highly aluminous, with various sizes of andalusites. At Stop 3A, this layer crosses the andalusite-sillimanite isograd, and both are present in the outcrop (Fig. 14). There are several quartz-rich lay-ers of schist; several exhibit relict sedimentary struc-tures, such as cross-bedding and channels (Fig. 12). The metasediments of the mica schist represent pelitic sediments filling the backarc basin. The highly alumi-

nous andalusite-sillimanite–rich layers were sediment layers rich in clay minerals; the more quartz-rich schist layers contained more sand and silt.

Retrace the route back to the intersection of Craw-ford Gulch, Belcher Hill, and Drew Hill Roads.

11.6   Turn north (right) onto Drew Hill Road at the intersection of Crawford Gulch, Belcher Hill, and Drew Hill Roads.

12.0   Pavement ends at Homestead Road; the outcrop of inter-est is between the Y arms of Homestead Road and Drew Hill Road

12.0   **Stop 3A**—Possible extra stop, depending on time and weather.
Mica schist unit: mica schist; andalusite + sillimanite.

This small, scattered outcrop is interesting because of the occurrence of both andalusite and sillimanite in the mica schist. The reaction is apparently prograde, and the grains of andalusite and sillimanite are in direct contact with each other, rather than having the intermediate muscovite present (Fig. 14). The andalu-site-sillimanite masses are large (1–5 cm), and to the west where only sillimanite is present, the sillimanite masses are often 2–5 cm.

12.3   Safe turnaround at Schoolhouse Road.

13.0   Back to the intersection of Crawford Gulch, Belcher Hill, and Drew Hill Roads.

20.8   Retrace route back to Hwy 93.

0.0   At the original mileage 0 point, turn north (left) onto Hwy 93 (at traffic light).

0.4   Turn west onto Pine Ridge Road (Slee Off-Road Co. is on the NW Corner); follow the curve right and proceed northward. The mountainside to the west is composed of the mica schist. The prominent ridge to the east is the Dakota Hogback, a Cretaceous sandstone unit known for dinosaur trackways and other interesting fossils.

2.2   Intersection of Pine Ridge Road and 56th Ave. Turn north through a gate at a sign that reads "Black Forest Mine." We will punch in a code at the gate and proceed north into the Beartooth Ranch gated community. Pri-vate Property and Open Range!

2.4   Turn left at the intersection of the mine service road and Glencoe Valley Road. Proceed north.

2.8   On the west side of the road you will see the restora-tion and reclamation efforts of the old Coors property.

3.0   Pass the intersection of Dakota Ridge Road.

3.4   15 mph! Go left at fork; remain on Glencoe Road.

3.5   Pass the intersection with Bear Point Trail Road.

3.55   Pavement ends. At the yellow iron gate, we will open gate with a key. We will now be on Jefferson County Open Space Protected Area. Please respect the wildlife.

As we drive downhill, we approach Ralston Reser-voir. In this area, we can observe outcrops of the Perm-ian Lyons and Permian-Triassic Lykins Formations on the left. The Lyons Formation overlies the Foun-tain Formation with fluvial and interfluvial aeolian deposits. The Lykins contains the Glennon Member, a limestone with prominent algal stromatolites.

4.0   Curve left through the road cut in the Lyons/Lykins sequence, past the arkoses of the Fountain Formation.

4.3   Cattle guard. Weir and flow station building to right. Ralston Creek is now on left.

4.5   Outcrops of Fountain arkose on the right.

4.9   A faulted contact of Precambrian mica schist and the Pennsylvanian Fountain Formation lies close to the creek. The mountainside to our left is mica schist; the prominent cliffs on the hillside to our right result from the thick, resistant conglomerates and arkoses of the alluvial fan and braided stream complex in the Fountain. Similar Fountain outcrops comprise the Boulder Flatirons, Red Rocks between Golden and Morrison, and the Garden of the Gods in Colorado Springs, among other places less well known along the Front Range.

5.4   The same contact crosses the road here. The knob on the left is schist. Eagle and other raptor nests are on the cliff above. Bear and mountain lion are common.

5.5   **Stop 4: Schwartzwalder Uranium Mine Road.** Outcrops of the iron formation begin here. We are still on Jefferson County Open Space Protected Area land here; please continue to respect the wildlife and the land. Iron formation; hornblende gneiss.

The transition zone between the mica schist and the hornblende gneiss in most places is represented by a thin quartzite (metachert) (Fig. 16). In the vicinity of the Schwartzwalder Mine, the transition zone widens and becomes a small Algoman-type iron formation (Figs. 10 and 11). Farther into the mine property, this unit is thicker and better exposed. At this time of year (November winter/snow conditions), we cannot usually access that area by road. The outcrops in this canyon are not as extensive, but still exhibit many of the features of interest. At the northern end of this set of outcrops, the iron formation is represented by a magnetite-rich schist. It may not look very different from the mica schist, but a magnet will stick to it! As you walk south, the magnetite and quartz contents increase. Relict sedimentary features, such as cross-bedding and troughs, occur in these metasediments.

The hornblende gneiss outcrops at the north end of this part of the canyon. The outcrop here is massive, dark metabasalt.

6.5   Walk/drive up the canyon and turn around; retrace route back to intersection of Pine Ridge Road and 56th Ave.

9.8   Turn east onto 56th Ave and proceed to Hwy 93. There is a view of Ralston dikes to the north. These are the feeder dikes of the 64–66 Ma lava flows that cap North and South Table Mountains east of Golden.

10.2   Turn north onto Hwy 93. From this point to the turnoff to Coal Creek Canyon, we will drive past many outcrops of Cretaceous and Tertiary sedimentary units.

Please refer to the descriptive portion of this field guide for more information about these units.

13.2   The view to left is of Ralston Reservoir dam. The gravel quarry operation to the south is quarrying rock from the Ralston dikes.

14.6   On the right at Leyden Gulch is a hogback of the Cretaceous Laramie Formation, delta-plain sandstones, siltstones, claystones, and coal derived from the Laramide uplift of the Rockies. The Leyden Coal Mine, active in the early part of the twentieth century, lies behind the hogback to the east. The mine was later used by Public Service (now Xcel Energy) for gas storage, and is currently being converted for local municipality water storage.

16.2   Turn west onto Hwy 72 (at traffic light). We are now on the Rocky Flats Alluvium surface.

18.2   *Turn south onto Blue Mountain Drive (we will skip to the next * if the roads are icy). As we drive south, the ridge to the left is Fountain Formation. The mountainside to the right is Precambrian interlayered gneiss.

19.6   Turn west (right) on Ute Drive.

19.8   Turn uphill (left) onto Westridge Drive; proceed uphill.

20.7   Intersection of Westridge Drive and Brumm Trail. Westridge Drive goes to the left; Brumm Trail is straight ahead. Drvie northwest onto Brumm Trail (Private Property!).

21.3   **Stop 5: Old Booth Ranch Property** (weather permitting). Please note that we are on private property here, with permission from the property owners for this visit. Interlayered gneiss: mafic and felsic gneiss; Boulder Creek pegmatites.

At this point, we are back into the interlayered gneiss, at the northern end of the syncline. Here the mafic gneiss predominates (Figs. 4 and 5), with layers of felsic gneiss also present (Fig. 6). There are also several small pegmatites, but they are plagioclase-rich, biotite rather than muscovite, with no tourmaline, and are generally foliated. These are interpreted as Boulder Creek pegmatites. The felsic gneiss here is finer grained than at Stop 1 and in thinner layers. The mafic gneiss is salt-and-pepper in appearance, a fine- to medium-grained amphibolite interpreted as metamorphosed island arc basalts. There are also intrusive hornblendite bodies (metamorphosed ultramafic intrusives) present here (Figs. 4 and 5). Relict phenocrysts now form eyes of tremolite apparent on the weathered surface of an outcrop.

24.6   *Retrace route back to Hwy 72, turn west (left) onto Hwy 72. The steep, rugged mountains here are composed of the Coal Creek Quartzite.

26.2   On the left, we will see a massive outcrop of quartzite. The quartzite at this outcrop is from unit C, and is fairly dark in color (gray to purplish), more massively bedded, and contains more mica and andalusite than units A and B. As we drive up the canyon, we will see more outcrops of the quartzite. Some of these contain stretched pebble conglomerates.

28.3 **Stop 6: Coal Creek Canyon.**
Coal Creek Quartzite; Boulder Creek Batholith; quartz monzonite.

We will pull off to the right onto a wide shoulder past the quartzite outcrop and walk back to it. On the short walk, we will first see the quartz monzonite of the Boulder Creek pluton. It is intrusive into the quartzite. The older granodiorites have been dated at 1714 Ma, the quartz monzonites a little younger. These represent syntectonic intrusions as the area became part of the magmatic arc.

The quartzite outcrop here is the basal unit A (Fig. 15). There are several interesting relict sedimentary features in this outcrop and a sequence of stacked, fining-upward units of a few meters thickness each. Most of the units are marked by a basal conglomeratic unit with usually scoured bases. The conglomeratic units fine upward to small-scale, trough-type cross-bedding then to more planar-laminar–like structures. Grain-size appears to decrease upward in relation to the change of sedimentary structures. Some indication of imbricate structures is present within the conglomerate beds. More distinct and well-developed channels, up to several tens of meters across and with deep scouring at their base, occur in equivalent units farther to the northeast at Eldorado Canyon. Large-scale, trough-type cross-beds also occur at the Eldorado locale.

28.6 Chapel of the Hills turnoff. This is a good place to safely turn around and retrace the route back to the Colorado Convention Center.

ACKNOWLEDGMENTS

We thank the Cotter Corporation for their continued welcome and access to the Schwartzwalder Mine properties; Jefferson County Open Space for access to the environmentally sensitive raptor nesting areas of White Ranch Open Space Park, which are normally closed to the public; the Beartooth Ranch Homeowners Association, the Blue Mountain Homeowners Association, and the Rodgers Family for their courtesy and permission to cross their properties; and the Colorado Scientific Society for their support and for hosting trial runs of this field trip and the helpful comments from participants.

REFERENCES CITED

CD-ROM Working Group, 2002, Structure and evolution of the lithosphere beneath the Rocky Mountains: Initial results from the CD-ROM experiment: GSA Today, v. 12, no. 3, p. 4–10, doi: 10.1130/1052-5173(2002)0122.0.CO;2.

Condie, K.C., 1986, Geochemistry and tectonic setting of early Proterozoic supracrustal rocks in the southwestern U.S.: Journal of Geology, v. 94, p. 845–864.

Finiol, L.R., 1992, Petrology, paleostratigraphy, and paleotectonics of a Proterozoic metasedimentary and metavolcanic sequence in the Colorado Front Range [M.S. Thesis T-3762]: Golden, Colorado School of Mines, 283 p.

Fisher, L.R., and Fisher, T.R., 2004, Preserved sedimentary structures in the Proterozoic Coal Creek Quartzite as indicators of tectonic basin development, Central Front Range, Colorado: A multidisciplinary approach (abs.): Geological Society of America Abstracts with Programs, v. 36, no. 5 (in press).

George, R.D., and Crawford, R.D., 1909, The main tungsten area of Boulder County, Colorado, in George, R.D., Colorado Geological Survey First Report 1908, p. 19–20.

Hunt, A., Lockley, M., and White, S., 2002, Historic dinosaur quarries of the Dinosaur Ridge area: Denver, Dinosaur Ridge Publication Series 4, Friends of Dinosaur Ridge and Dinosaur Trackers Research Group at the University of Colorado at Denver, 44 p.

Ludwig, K.R., Wallace, A.R., and Simmons, K.R., 1985, The Schwartzwalder uranium deposit, II: Age of uranium mineralization and lead isotope constraints on genesis: Economic Geology and the Bulletin of the Society of Economic Geologists, v. 80, p. 1858–1871.

Marsh, S.P., and Sheridan, D.M., 1976, Rutile in Precambrian sillimanite-quartz gneiss and related rocks, ease-central Front Range, Colorado: U.S. Geological Survey Professional Paper 959-G, 17 p.

Marvine, A.R., 1874, Report on the geology of the region traversed by the Northern or Middle Park division during the working season of 1873: U.S. Geological and Geographical Survey of the Territories 7th Annual Report (Hayden), p. 83–192.

Raymond, C.H., Sheridan, D.M., Taylor, R.B., and Hasler, J.W., 1987, Proterozoic stratabound sulfide deposits in Colorado: Geological Society of America Abstracts with Programs, v. 19, no. 7, p. 814.

Sheridan, D.M., and Marsh, S.P., 1976, Geologic map of the Squaw Pass quadrangle, Clear Creek, Jefferson, and Gilpin Counties, Colorado: U.S. Geological Survey Geologic Quadrangle Map GQ-1337, scale 1:24,000, 1 sheet.

Sheridan, D.M., and Raymond, C.H., 1982, Stratabound Precambrian sulfide deposits, Colorado and southern Wyoming: U.S. Geological Survey Open-File Report 82-795, p. 92–99.

Sheridan, D.M., and Raymond, C.H., 1984, Precambrian deposits of zinc-copper-lead sulfides and zinc spinel (gahnite) in Colorado: U.S. Geological Survey Bulletin 1550, 31 p.

Sheridan, D.M., Maxwell, C.H., and Albee, A.L., 1967, Geology and uranium deposits of the Ralston Buttes District Jefferson County, Colorado: U.S. Geological Survey Professional Paper 520, 120 p.

Weimer, R.J., 1996, Guide to the petroleum geology and Laramide Orogeny, Denver Basin and Front Range, Colorado: Denver, Colorado Geological Survey Department of Natural Resources Bulletin 51, 127 p.

Wells, J.D., 1967, Geology of the Eldorado Springs quadrangle Boulder and Jefferson Counties, Colorado: U.S. Geological Survey Bulletin 1221-D, 85 p.

Wells, J.D., Sheridan, D.M., and Albee, A.L., 1964, Relationship of Precambrian Quartzite-Schist sequence along Coal Creek to Idaho Springs Formation, Front Range, Colorado: U.S. Geological Survey Professional Paper 454-O, 25 p.

Williams, M.L., Karlstrom, K.E., Jessup, M., Jones, J., and Connelly, J., 2003, Proterozoic rhyolite-quartzite sequences of the Southwest: syntectonic "cover" and stratigraphic breaks (ca. 1695 and ca. 1660 Ma) between orogenic pulses: Geological Society of America Abstracts with Programs, v. 35, no. 5, p. 42.

Geological Society of America
Field Guide 5
2004

Eco-geo hike along the Dakota hogback, north of Boulder, Colorado

P.W. Birkeland
E.E. Larson
Department of Geological Sciences (both retired), University of Colorado, Boulder, Colorado 80309, USA

C.S.V. Barclay III
U.S. Geological Survey (retired), Denver, Colorado, USA

E. Evanoff
Department of Geological Sciences, University of Colorado, Boulder, Colorado 80309, USA

J. Pitlick
Department of Geography, University of Colorado, Boulder, Colorado 80309, USA

ABSTRACT

Sedimentary rocks, ranging in age from Pennsylvanian to Cretaceous, crop out in the vicinity of the Dakota hogback. The rocks overlie Precambrian igneous and metamorphic rocks. The strata were deformed in late Cretaceous and early Tertiary time and dip to the east. Downcutting occurred in the Tertiary and Quaternary, formed the major canyons, and was interrupted by intervals of alluvial deposition east of the range front. Landslides mantle the east front of the hogback.

Keywords: Dakota hogback, Colorado Front Range, Laramide orogeny, Front Range erosion history, upper Paleozoic and Mesozoic strata.

INTRODUCTION

The style of this trip—public and human-powered transportation—is taken from that of Clyde Wahrhaftig. Clyde was one of the more highly respected geomorphologists in the decades after World War II, and he never drove a car nor had a driver's license. He devised field trips in San Francisco using public transportation. His most well-known trip was titled "A Streetcar to Subduction." In the spirit of his innovative trips, we dedicate this trip to him.

The purpose of this field trip is to examine various aspects of bedrock and surficial geology while hiking along the easternmost hogback of the Rocky Mountain Front Range. Take the Skip bus north on Broadway from downtown Boulder to the northernmost stop, which is Lee Hill Road. Walk west on Lee Hill Road to the City of Boulder Open Space and Mountain Parks (OSMP) parking lot (Fig. 1).

The route for this trip follows OSMP trails. The trip is ~6 mi long with a total elevation gain of 840 ft as it goes up and down the hogback. We rank the trail difficulty as moderate for someone accustomed to the altitude. Long parts of the trail, however, are flat, and therefore are ranked as easy.

STOP 1. OSMP Parking Lot

From here we have a good view of the junction between the Great Plains and the Rocky Mountains. The prominent topographic features of the range front are the first hogback, underlain by the Dakota Group rocks, the second hogback, underlain by the Lyons and Fountain Formations, and Precambrian igneous and

Birkeland, P.W., Larson, E.E., Barclay III, C.S.V., Evanoff, E., and Pitlick, J., Eco-geo hike along the Dakota hogback, north of Boulder, Colorado, *in* Nelson, E.P. and Erslev, E.A., eds., Field trips in the southern Rocky Mountains, USA: Geological Society of America Field Guide 5, p. 131–142. For permission to copy, contact editing@geosociety.org. © 2004 Geological Society of America metamorphic rocks that make up the high country farther west (Figs. 2 and 3). From here, we will discuss the geologic history of the area (Table 1).

Hike southwest, and cross Fourmile Canyon Creek. No recent study of the Quaternary alluvial deposits has been made; however, excavation for new homes on the high fluvial terrace north of Lee Hill Road exposes soils with a red Bt horizon (2.5YR 3/5d; sandy clay, granitic clasts weathered to grus) overlying a stage III to IV K horizon in places. Colton (1978) mapped the alluvium underlying the terrace as Slocum Alluvium, whereas Wrucke and Wilson (1967) mapped it as Verdos Alluvium (Fig. 4). We prefer a Slocum age and, using that as an age datum, estimate an incision rate of Fourmile Canyon Creek of 0.02–0.04 m/k.y., a low rate for the area. Note: the term terrace here refers to a fluvial terrace, most of which have a relatively smooth surface sloping to the east and are underlain by gravel deposits of varying thickness.

Hike southwest, and at the intersection with the main north-south trail, turn to the right (N).

STOP 2. Exposure of the Smoky Hill Member of the Niobrara Formation (Figs. 2 and 3)

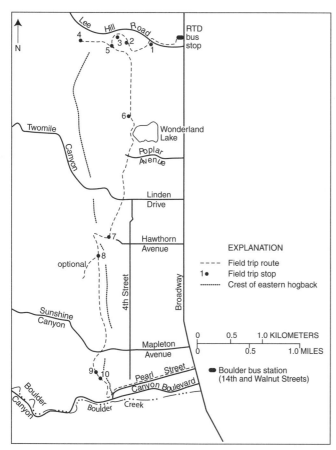

Figure 1. Map showing field trip route.

STOP 3. Exposure of the Ft. Hays Member of the Niobrara Formation (Figs. 2 and 3)

Continue going uphill on the trail, and at the first junction take the right (W) fork. Continue uphill to the west to the OSMP gate.

STOP 4. Section from the Lykins Formation to the Niobrara Formation

We are standing on the Lykins Formation: notice the deep red color of the strata, compared to the white color of the Lyons Formation to the west. Later today we will see outcrops of both red and white Fountain Formation.

The Formation of the Color in Red Beds

The principal red bed units in the Boulder area are the Fountain, Lyons, and Lykins Formations. The red color is imparted by tiny (<0.3 μ to ~2 μ) diagenetically formed hematite crystals and crystal aggregates that are dispersed on the surfaces of and between detrital grains. These tiny pigmenting grains, which generally represent less than 1 wt% of the rock, are translucent to red wavelengths of light. In most cases, red beds are non-marine deposits (e.g., alluvium, dune sand, tidal flat beds) that lacked sufficient organic material to counter post-depositional diagenesis by oxidizing intrastratal solutions.

According to T.R. Walker (a.k.a. Red Bed Ted), red bed sediments when initially deposited were drab in color (Walker et al., 1981). At some time after burial, oxygenated surface waters permeated the sedimentary section, bringing about extensive diagenetic alteration. If Fe-bearing minerals, either oxides (ilmenite, magnetite, hematite) or silicates (notably, hornblende, biotite), were present, their alteration and/or partial solution would release iron ions into the circulating groundwater. In an oxidative environment, the ions would not travel far before forming small crystals of hematite on and between grains—thereby imparting the bright red color.

Alteration undoubtedly proceeded in an inhomogeneous manner determined by local conditions of fluid and grain compositions, porosity, and permeability. Given sufficient time (perhaps as much as several million years), the susceptible sedimentary units would have acquired the red color.

If, at some later time, the intrastratal waters became reducing (as, for example, up-dip from a petroliferous source region), their circulation through the coarser, more permeable beds, would have led to reduction and leaching of much of the pigmenting hematite grains in those units. The end result would be a mottled red bed sequence in which some beds were bleached white or greenish white. There are numerous examples of this occurrence in the Fountain Formation.

As an afterthought, it should be pointed out that the paleomagnetic directions associated with red beds are acquired during the diagenetic stage following burial, when hematite is formed both as pigment and the product of oxidation of deposited magnetite grains. Uncertainties about when and over what time span the

Sedimentary Rocks of the Boulder Valley

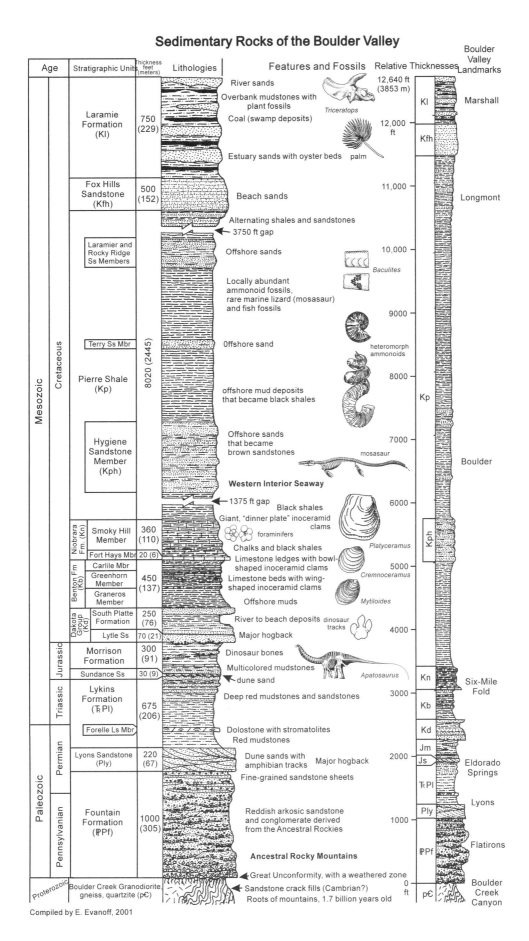

Figure 2. Stratigraphic section for the Boulder area.

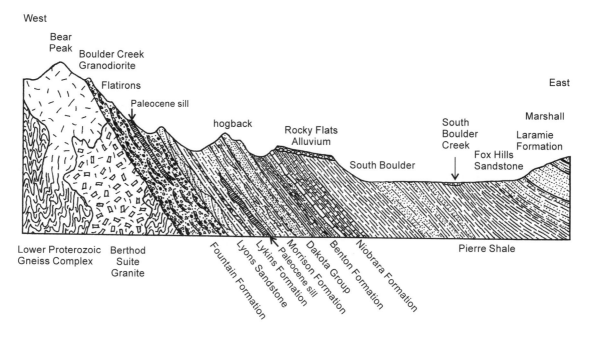

Figure 3. Generalized geologic and topographic cross section for a transect in south Boulder. This cross section is ~4 mi long, and relief is ~3000 ft. Many of the same units and landforms are along the field trip route.

post-depositional diagenesis occurred, therefore, lead to uncertainties in magnetic pole positions acquired from red bed studies.

Walk down the trail, up-section, to view rocks from the Lykins Formation to the Fort Hays Limestone (Figs. 2 and 3).

STOP 5. Previous Trail Junction at the Benton-Niobrara Contact

The downcutting history starts with the Tertiary erosion surface to the west, out of view. In places there are extensive low-relief areas of the surface, and in other places the surface shows up as accordant ridges (Fig. 5). The erosion surface extends from about the Peak-to-Peak Highway (Nederland and Ward) on the west (~2750 m) and slopes gently eastward to ~2500 m just west of the Flatirons. We favor the hypothesis that the surface formed at a relatively low elevation and was subsequently uplifted during regional epeirogenic uplift in the last 5 m.y. (Trimble, 1980; Steven et al., 1997).

Several north-south topographic transects show the canyons with respect to the erosion surface (Fig. 6). The canyons have their greatest relief at the range front, and the relief decreases upstream, so much so that the valley relief is relatively insignificant along the Peak-to-Peak Highway transect. The accordant ridges show up especially well. Taking the initiation of canyon cutting at 5 Ma (Steven et al., 1997), the incision rate is halved when comparing the mountain front rate with the Peak-to-Peak Highway rate.

Alluvial deposits have been mapped and dated only in Boulder Canyon, and these serve to connect the deposits in the

glaciated area with those in the piedmont area (Schildgen, 2000; Schildgen et al., 2002). They identify deposits of Holocene, Pinedale, and Bull Lake ages. The respective ages are <10 ka, 32–10 ka, and >100 ka, based on radiocarbon and cosmogenic dating (compare with ages in the caption to Fig. 7).

Alluvial deposits have been mapped in the piedmont along the range front (Colton, 1978; Trimble and Machette, 1979). Scott (1963) set up the alluvial stratigraphy that has been used throughout the area, but it could be time for a revision considering the numerous climatic changes during the Pleistocene that might have affected stream behavior. Madole (1991) reviewed the deposits recently, and Birkeland et al. (2003) reviewed the soils.

The alluvial deposits underlie a series of river terraces that are progressively younger closer to the drainages (Fig. 7). Close to the mountain front, the deposits are bouldery, relatively thin, and rest on a bedrock surface of low relief, commonly referred to as a pediment. The relative height of the terraces above nearby streams, and soils (Fig. 8), has proven useful in correlating deposits along the range front. The oldest deposit near here is the Rocky Flats Alluvium, and remnants of it form high-level deposits east and south of here. Verdos Alluvium is the next younger deposit, and is dated by association with the Lava Creek B ash (640 ka) derived from the Yellowstone area. The nearest ash localities are south of here at Ralston Reservoir and in Golden. Slocum Alluvium underlies the next younger terrace. Both Verdos Alluvium and Slocum Alluvium have red, clay-rich soils with K horizons that contrast markedly with soils associated with younger units (Fig. 8). Mappers are not always in agreement in

TABLE 1. SUMMARY OF THE GEOLOGIC HISTORY OF THE BOULDER AREA

ca. 2000–1750 Ma	A thick sequence of interlayered sedimentary (conglomerate, sandstone, shale, limestone) and volcanic (basalt, rhyolite) rocks apparently accumulated along the margin of an Archean craton.
ca. 1750–1700 Ma	Two-stage metamorphism of the rocks above formed mostly a high-grade assemblage of schist, gneiss, amphibolite, and migmatite; syntectonic intrusion of granitic plutons occurred toward the end of the metamorphic event (e.g., Boulder Creek Granodiorite).
ca. 1450–1350 Ma	A number of non-tectonic two-mica granitic plutons (e.g., Silver Plume, St. Vrain Granites) were intruded, many of batholithic dimensions.
ca. 1400–1200 Ma	NNW- and NE-trending faults developed and cut deeply into the crust, thereby initiating major, long-lasting zones of weakness along which subsequent deformation has commonly taken place.
ca. l200–525 Ma	No record in the Boulder area of geologic activity exists other than erosion and the intrusion of a kimberlite pipe (ca. 600–550 Ma) in Precambrian rocks at Green Mountain, directly west of Boulder (no diamonds have been found).
ca. Late Cambrian to Mississippian	Colorado was submerged under shallow marine waters transgressing from the west (beginning during Late Cambrian), followed by repeated intervals of minor emergence interspersed with shallow submergence, resulting in deposition of thin sequences of sandstone, limestone, and dolomite.
Middle Pennsylvanian	The Ancestral Front Range crustal block was uplifted (one of several crustal uplifts in the western United States). The range was initially surrounded by marine waters. Erosion of the range was initiated, during which the thin, covering section of Paleozoic strata and some of the underlying Precambrian rocks were eroded. Coarse debris from upland erosion was deposited in the marginal marine basin to the east.
Late Pennsylvanian through Jurassic	As the eastern flank continued to become topographically subdued by erosion, some later basin-filling sediments encroached westward onto the eastern flank of the range. Sedimentary units (oldest to youngest) include coarse arkose and conglomerate (Fountain Formation), coastal dune deposits (Lyons Formation), tidal-flat mudstone (Lykins Formation), a thin dune sand (Sundance Formation), and fluvial and, locally, lacustrine beds (Morrison Formation). The mountain range, except for a few low hills, was completely eroded away by the end of the Jurassic.
Early Cretaceous	Little or no geologic record is preserved; this was probably an interval of erosion. There was late Early Cretaceous deposition of the Lytle Sandstone of the Dakota Group by streams. Clasts consist of reworked sedimentary rocks only, indicating a lack of Precambrian rocks in the source areas.
late Early through Late Cretaceous (ca. 70 Ma)	The North American continent was transected by a N-S–trending marine seaway, with Boulder about midway between the eastern and western uplands. Early deposits reflect complex interplay between transgression and regression of marine waters. These strata include the South Platte Formation of the Dakota Group, near-shore black shale units of the Benton Formation, and offshore limestone units of the Niobrara Formation. The water level attained during limestone deposition of the Greenhorn Member of the Benton Formation and the Niobrara Formation marked the maximum extension of the Western Interior Seaway. The principal marine unit is the thick, black Pierre Shale. A complex regression of the seaway resulted in the shoreline Fox Hills Sandstone and variable estuarine, swamp, and deltaic deposits of the Laramie Formation.
ca. 70–50 Ma	(A) The Laramide Front Range rose as a large, crustal anticline, in places bounded by steep to low-angle thrust faults, in the same general region as the Ancestral Front Range. The Laramide Front Range was one of many, mostly N- to NW-trending ranges that grew throughout the Rocky Mountain Region. Uplift-induced erosion quickly led to removal of the relatively soft Late Pennsylvanian to Cretaceous strata from the range uplands and dissection of the underlying Precambrian rocks. By ca. 50 Ma, the relief of the range had been reduced by a combination of erosion of the uplands and alluvial infilling of the subaerial Denver basin to the east by sediments of the Arapahoe and Denver Formations.
	(B) Mafic, intermediate, and felsic magmas (none larger than a stock) were emplaced, some of which fed a line of stratovolcanoes in a NE-SW trend across the eastern two-thirds of the Front Range. Along the eastern range edge, in the Boulder area, where activity occurred generally ca. 65 Ma, the E-W shoshonitic (potassium-rich basaltic andesite) Valmont dike and small plutons of rhyolitic composition were emplaced, some as sills, in the Fountain and Lykins formations and Cretaceous strata. Near Golden, a shoshonitic sill and associated dike were emplaced; the dike fed at least three shoshonitic lava flows.
ca. 50–5 Ma	Little direct evidence of geologic activity in the Boulder area. Evidence from nearby areas indicates, however, that by the Late Eocene, rivers, which had cut large paleovalleys in the Front Range, flowed east and southeast across the adjacent Great Plains. During the Late Eocene to Early Miocene, these paleovalleys were infilled by volcaniclastic sediments, which also accumulated on the Great Plains to the east. By the Middle Miocene, flow directions of streams from the Front Range switched toward the northeast. By then, large pediments had been cut into the eastern mountain flank, an indication of stable local base levels between the Great Plains and adjacent Front Range. Extensive, overlapping gravels prograded outward from the range, resulting in infilling of the paleovalleys and deposition of sediments on the upper Great Plains.
ca. 5 Ma	Epeirogenic uplift of a large portion of western North America began, including the surface of Colorado. Higher portions (such as the core of the Laramide Front Range) eventually attained elevations of 14,000 feet or more.
ca. 5 Ma to the present	Increased snow precipitation in the Front Range uplands (as well as the rest of the uplifted region) led to renewal of vigorous erosion. In the harder rocks of the Precambrian range core and of some of the well-indurated Late Paleozoic and Cretaceous sandstones (e.g., the Fountain and Lyons Formations and the Dakota Group) along the range flank, the streams and rivers were restricted to cutting narrow canyons. Rivers that emerged from the mountain front, however, encountered more easily erodable rocks (Late Cretaceous and Tertiary) on the adjacent Great Plains, enabling extensive surfaces to be cut. Continued stripping of softer sediments east of the mountain front resulted in exhumation of the preexisting Laramide Front Range, thereby producing the modern Front Range. Remnants of the river-exhumation process younger than ca. 2 Ma are locally preserved east of the mountain front (e.g., the Rocky Flats surface). Climate change probably contributed to the late Cenozoic reactivation of erosion of the area, but the uplift of the land surface beginning ca. 5 Ma has undoubtedly been a major factor.

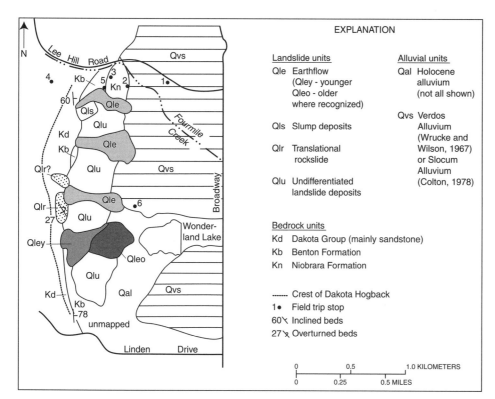

Figure 4. Geologic map of landslides and alluvial units between Linden Drive and Lee Hill Road. Not all bedrock units (Wrucke and Wilson, 1967) are shown.

EXPLANATION

Landslide units

Qle Earthflow
     (Qley - younger
     Qleo - older
     where recognized)

Qls Slump deposits

Qlr Translational
     rockslide

Qlu Undifferentiated
     landslide deposits

Bedrock units

Kd Dakota Group (mainly sandstone)
Kb Benton Formation
Kn Niobrara Formation

------ Crest of Dakota Hogback
1•   Field trip stop
60∖   Inclined beds
27∖   Overturned beds

Alluvial units

Qal Holocene
     alluvium
     (not all shown)

Qvs Verdos
     Alluvium
     (Wrucke and
     Wilson, 1967)
     or Slocum
     Alluvium
     (Colton, 1978)

0   0.5   1.0 KILOMETERS
0   0.25   0.5 MILES

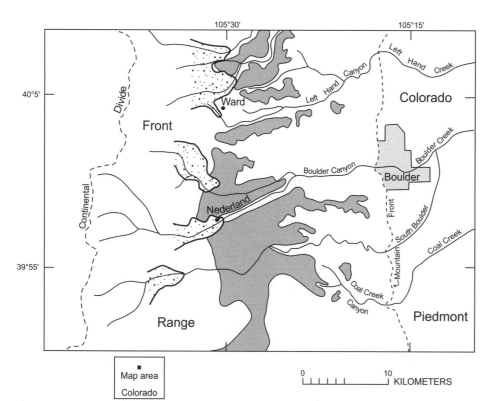

Figure 5. Map of Tertiary erosion surface in the Front Range near Boulder (Fig. 1 of Birkeland et al., 2003). Heavy lines near Ward and Nederland indicate lower limit of Pleistocene glaciers and dotted pattern indicates till and related deposits.

distinguishing between Verdos Alluvium and Slocum Alluvium. For example, Wrucke and Wilson (1967) mapped the deposits along the main north-south trail just below here as Verdos alluvium, but Colton (1978) mapped them as Slocum Alluvium. The two younger pre-Holocene deposits that form prominent terraces are the Louviers Alluvium and Broadway Alluvium. These are mapped in the vicinity of Boulder Creek.

The times of deposition of the alluvial units have been placed in various parts of the glacial-interglacial cycle by different workers. Most workers seem to agree that the Louviers and Broadway Alluviums were deposited during glacial times (see caption to Fig. 7), but there is less agreement on the timing

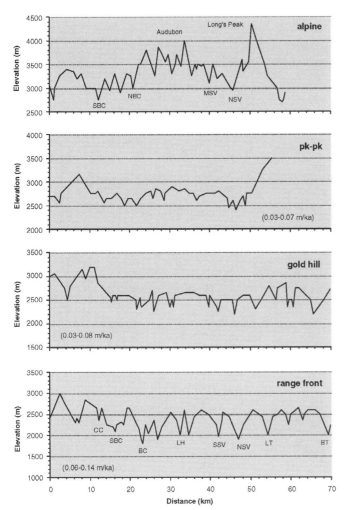

Figure 6. North-south topographic profiles west of Boulder. From top to bottom, the transects are in the alpine zone, along the Peak-to-Peak Highway (pk-pk), near Gold Hill, and just west of the range front. The canyons are Coal Creek (CC), South Boulder Creek (SBC), Boulder Creek (BC), North Boulder Creek (NBC), Left Hand Creek (LH), South St. Vrain Creek (SSV), Middle St. Vrain Creek (MSV), North St. Vrain Creek (NSV), Little Thompson (LT), and Big Thompson River (BT). The numbers in parentheses are estimates of the incision rate for each transect, assuming downcutting was initiated 5 Ma.

of deposition of the older units. One hypothesis that takes into account both the downcutting and glacial-interglacial cycles is the following: Stream downcutting was the dominant mode since initial incision of the Tertiary erosion surface. With the onset of glacial conditions, the changes in sediment load and discharge were such that downcutting was halted and lateral planation by the rivers was dominant, followed by aggradation of various amounts depending on the characteristics of the river. This would help explain both the lateral extent of associated pediments and the relatively thin nature of many deposits. Downcutting would resume during the following interglacial. Given the many glaciations during the Pleistocene, it seems that many glaciations are not represented by alluvial deposits. Still another terrace-forming mechanism is stream capture at the mountain front, as described by Ritter (1987). We have several examples of this in the area. Finally, Leonard (2002) has calculated that about half of the post-Tertiary erosion surface downcutting is due to tectonic uplift and half is due to erosion-induced isostasy.

Proceed southeast down to the main north-south trail and then turn south on the main trail. The E-facing slope of the hogback is mantled with landslides (Fig. 4).

STOP 6. Overlook of Wonderland Lake and the Hogback to the West

The valley west of the lake has been filled with a Quaternary earthflow that contains numerous blocks of Dakota sandstone. The fines responsible for the earthflow were derived from shales of both the Dakota Group and the Benton Formation. The western part of the earthflow has well-preserved landslide features (lateral and transverse ridges with steep slopes and closed depressions) that suggest a relative age younger than the rest of the earthflow. Such morphology suggests a late Quaternary age (Pinedale?). The toe of the earthflow merges with the Quaternary alluvium west of Wonderland Lake, where age relations are not clear. This stop is on alluvium mapped as Verdos Alluvium (Wrucke and Wilson, 1967) or Slocum Alluvium (Colton, 1978), but all this demonstrates is that the earthflow is younger than these ages. We believe it is much younger.

North of the head of the above-mentioned earthflow is a feature mapped as a Quaternary translational rockslide (Fig. 4). As the slide moved, a buckle developed, and the Dakota sandstone was overturned. There will be a detailed discussion of such slides at Stop 7.

All valleys between here and Stop 5 are filled with similar earthflows (Fig. 4). Between the earthflows is topographically higher (some form topographic knobs) and older landslide material of unknown origin. It is old enough that it lacks depositional form.

Continue south along the trail and eventually cross Poplar Avenue. Continue uphill on a concrete path with houses on both sides. Notice the concrete drop-structures along the drainage. We have no idea why they were constructed; they appear to be for flood control, yet this small drainage produces little runoff. The drainage divide is hidden amongst the houses straight ahead.

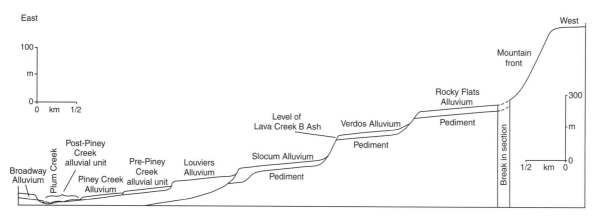

Figure 7. Diagrammatic cross section of alluvial units in the Colorado Piedmont near Denver (from Madole, 1991, his Figure 14). Approximate ages are post-Piney Creek, Piney Creek, and pre-Piney Creek alluvial units, Holocene; Broadway Alluvium, Pinedale glaciation, ca. 14–47 ka (Nelson and Shroba, 1998); Louviers Alluvium, Bull Lake glaciation, ca. 120–160 ka (Nelson and Shroba, 1998); Slocum Alluvium, 240 ka; Verdos Alluvium, similar to age of intercalated Lava Creek B ash, 640 ka; Rocky Flats Alluvium, ca. 2 Ma (see age discussion in Birkeland et al., 2003).

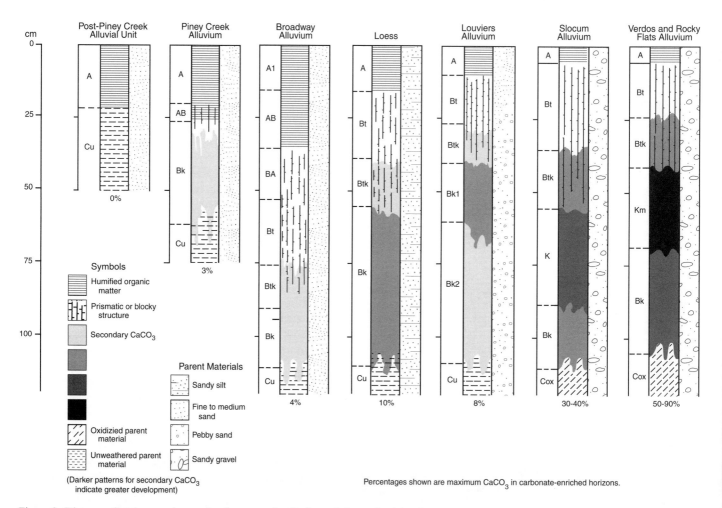

Figure 8. Diagram showing maximum development of soils formed from alluvial units and loess, typical of the Denver-Boulder area (from Madole, 1991, his Figure 15). Letters are standard soil-horizon nomenclature (see Birkeland et al., 2003).

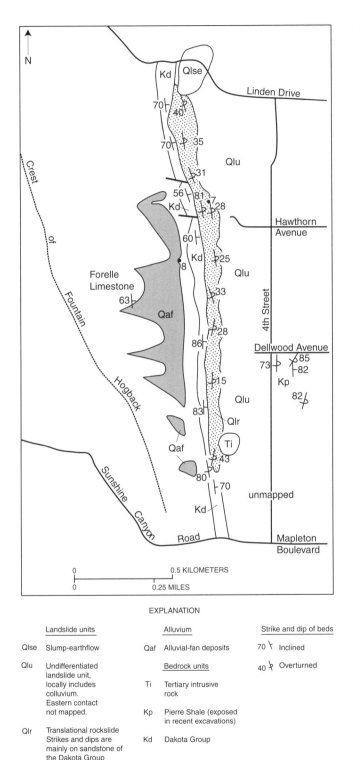

Figure 9. Geologic map of Dakota Group rocks, landslides, and al-luvium between Linden Drive and Sunshine Canyon. Not all bedrock units (Wrucke and Wilson, 1967) are shown.

Take the first path on the right, go left at the road to the end of the cul-de-sac, and exit onto a path going south to a cul-de-sac on Lakebriar Drive. Continue south, eventually onto Spring Valley Drive, cross Linden Drive, and take an informal trail that goes south and diagonally up the lower part of the hogback.

The drainage just crossed is Twomile Canyon Creek. Deposits underlying the highest terrace to the south are mapped as either Verdos Alluvium (Wrucke and Wilson, 1967) or Slocum Alluvium (Colton, 1978). The resulting incision rate for the creek is either 0.04 m/k.y. or 0.1 m/k.y., respectively.

Look at the hogback to the west. The outcrop is sandstone of the Dakota Group, and it is overturned and dips west (Fig. 9). Hunter (1955) mapped the sandstone as the lower limb of a thrust fault. The fault strikes N-S and extends ~0.8 km. Wrucke and Wilson (1967) mapped the overturned bed but offered no structural interpretation. We interpret the overturned beds as the northernmost limit of a large translational rockslide, described at the next stop.

STOP 7. Translational Rockslide Exposed in a Small Canyon Cut in the Dakota Hogback

This is located just west of the intersection of Hawthorn Avenue and 4th Street. Stop at the first trail junction and look south.

Here we will view a large translational rockslide with a buckle-like fold in Dakota Group rocks (Fig. 9). Braddock and Eicher (1962) and Braddock (1978) have described glide blocks (old term) in Dakota Group rocks farther north in the Front Range. There the rocks dip ~20°, and the up-dip sandstone slab broke loose from, or within, the underlying shale and rode along the landslide surface over the same in situ sandstone located down dip. With subsequent erosion, some of these features crop out as ridge-top synclines. This locality differs in that we see what happens when the dip is steep and the overriding slab buckles to form a fold. Bill Braddock, a former structural geologist at the University of Colorado, helped with the interpretation below.

The Dakota Group in this area consists of the basal Lytle Formation (sandstone and conglomerate) and the overlying South Platte Formation. From bottom to top, the South Platte consists of the Plainview Member, the Skull Creek Shale, and the Muddy Sandstone. The Lytle Formation makes up the hogback here and dips 56°–86° west. South of here, across the canyon, is overturned Muddy Sandstone (attitude: N35°W, 30° dip W); ripple marks in beds facing down and to the east indicate it is overturned. The overturned beds north and south of the canyon dip 15°–81° west. Follow the beds to the right across the fold axis, where they eventually dip east (attitude in one place: N5°E, 45° dip E). The prominent tight fold in the sandstone suggests brittle deformation (blocks are broken and rotated) and is strong evidence that this is a surface feature rather than a deep-seated fold. Skull Creek Shale is in the interior of the fold. We interpret the fold as a buckle at the lower end of a large translational rockslide, with movement along incompetent beds such as the Skull Creek Shale (Fig. 10). The slide extends ~0.5 km north of here and 1 km south of here (Fig. 9).

The previous paragraph describes the upper part of the buckle. To view the lower part, go down the trail to the concrete structure in the drainage and look north. There you will see that the near vertical rocks change to an increasingly greater overturned dip to the west as they are traced upslope.

Regional erosion had to be sufficient to allow adequate space for slide movement and for the buckle to form as a surface feature (see A on Fig. 10). We can estimate the time at which regional lowering reached the elevation of the buckle by projecting heights of terraces to the site (Wrucke and Wilson, 1967; Colton, 1978). This suggests formation near Verdos or Rocky Flats time (640 ka to 2 Ma; see Fig. 7). Finally, the reconstructed buckle-fold axis is higher in elevation north of the drainage than it is south of the drainage, suggesting that the former may have moved independently before the latter.

Walk west up the canyon trail. The first outcrops of the Dakota sandstone dip west and are part of the translational rockslide. Higher up, the Dakota sandstone dips east and is in situ bedrock. To the west and beneath the in situ Dakota sandstone are mudstones, sandstones, and limestones of the Morrison Formation. At the top of the canyon are red beds of the Lykins Formation.

At the junction with the Sanitas Valley trail, turn left (S) and go down the valley underlain by Lykins Formation. Dakota sandstone and conglomerate form the hogback to the left (E), and Fountain Formation forms the hogback to the right (W). The Lyons Formation forms the face of the latter hogback, and the flagstone used in many buildings in Boulder was taken from some of the small quarries here.

STOP 8. First Trail Junction, with a Side Trail Heading Southwest from the Sanitas Valley Trail

If time allows, we will take this side trail to see the Forelle limestone of the Lykins Formation, interbedded eolian and fluvial beds at the Lyons-Fountain contact, and exposures of cross-bedded eolian sandstone of the Lyons Formation (Walker and Harms, 1972).

The side trail crosses a dissected alluvial fan. Only the northernmost three zero-order basins that cut into the Fountain hogback are large enough to have generated sufficient runoff to produce debris flows to form the fans. No doubt several ages of fan deposits are present, but a few pitted boulders of Fountain Formation suggest an age for some fan deposits of ca. 100 ka.

Continue south down the main valley. Take the fork to the right just before Mapleton Avenue, and go through the small shelter. Sunshine Canyon Creek is slightly incised here. Using the alluvium on which Mapleton Avenue is located as an age datum (mapped as Verdos Alluvium by Wrucke and Wilson [1967] and Colton [1978]), the incision rate is estimated to be 0.003 m/k.y. The preservation of deep grus in roadcuts along the canyon sides to the west could be explained by such a low incision rate.

Cross Mapleton and follow the wide trail up to a prominent saddle.

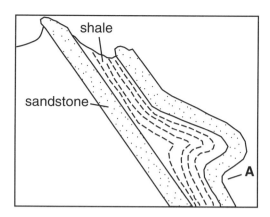

Figure 10. Schematic cross section of translational rockslide in Dakota Group rocks at Stop 7, looking north. This cross section is ~400 ft east-west, with ~120 ft of relief. A is the level of erosion required to allow the buckle to form (from Dethier et al., 2003).

STOP 9. Saddle Marking the Contact between the Fountain Formation and Boulder Creek Granodiorite

This saddle marks the contact between the Fountain Formation to the east and the Boulder Creek Granodiorite to the west (see description to the south in Wahlstrom [1948]). The hill of Fountain Formation to the east forms the easternmost hogback in this area. There is no Dakota hogback here. Many faults were mapped here by Wrucke and Wilson (1967) and to the south by Wells (1967), and the general pattern is a reverse fault or faults extending from Sunshine Canyon southward for ~4 km to the National Center for Atmospheric Research (NCAR). The overall result of the faulting is to cut out the Dakota Group rocks and the associated hogback. South of NCAR, the Dakota Group rocks and associated hogback reappear and trend southward until they are cut out again near Coal Creek, ~11 km south of NCAR. Wrucke and Wilson (1967) also show one fault trending north-northeast from Red Rocks, east of Stop 9, and it probably explains the common occurrence of steeply overturned Pierre Shale observed in recent excavations as far north as Dellwood Avenue and west of 7th Street (Fig. 9).

Boulder Canyon is the canyon north of here. It is one of the higher relief canyons in this part of the Front Range (Fig. 6). Schildgen et al. (2002) calculate an incision rate of 0.15 m/k.y. within the canyon and used the elevation of the creek relative to that of mapped terraces at the range front to calculate an incision rate of 0.04 m/k.y. there. These rates are within the regional range (Dethier, 2001) and help explain the presence of steep side slopes, cliffs, and young soils along the canyon walls (Birkeland et al., 2003). In contrast, Coal Creek, ~15 km to the south, has the lowest mountain-front incision rate (0.006 m/k.y.), and old, red, clay-rich soils are present on the slopes (Birkeland et al., 2003).

The main trail goes south from the saddle. Take the first side trail to the left, and follow it downhill along the hillside to outcrops of igneous rocks, just above the ditch.

STOP 10. Laramide Magmatism along the Eastern Margin of the Front Range

During the Laramide pulse of magmatism in Colorado (ca. 72–45 Ma), several sills, dikes, and flows were emplaced along the eastern edge of the Front Range (Mutschler et al., 1987). Within the precision of the dating methods (whole-rock K/Ar, K/Ar on biotite, fission track on zircon), all of the magmatism occurred from ca. 66 to 64 Ma.

About 7 km north of Golden, the shoshonitic (potassium rich, mafic) Ralston sill and associated dike were intruded into the Pierre Shale. This dike is thought to have fed lava flows, three and two of which are preserved on North and South Table Mountains in Golden, respectively. Comparison of the paleomagnetic directions of the sill with those of the three nearly flat-lying flows indicates that the sill, after intrusion into nearly flat-lying sediments, was tilted eastward ~70°. It is not known whether the deformation occurred during one event or multiple events. The east-west, 65 Ma shoshonitic Valmont dike, best exposed at Valmont Butte (Butte Mill Road and Valmont Drive in Boulder), extends eastward from the foothills in west Boulder about five miles.

Along the eastern Front Range, from Boulder to Lyons, there are several sills that have been intruded into the east-flanking strata. All of these, which are rhyolitic in composition (~70 wt% silica), were intruded between ca. 65 and 64 Ma. On Flagstaff Mountain, a 15-ft-thick rhyolitic sill in the Fountain Formation dips ~50° eastward. This same sill can be identified northward, but is broken by two cross-cutting faults. At Red Rocks, ~0.75 mi north, one of the sill segments is vertical, whereas in Sunshine Canyon, ~0.25 mi farther north, the segment dips eastward at ~70°. About six miles north of Boulder, in the Left Hand Canyon area, a 20-ft-thick sill in the Lykins Formation dips eastward at ~25°, and just southwest of Lyons, a 75-ft-thick sill that intruded the Fountain Formation is nearly flat-lying.

Several small, irregular pods of microcrystalline rhyolite have also intruded Cretaceous strata near Boulder. Most, however, are poorly exposed.

Paleomagnetic directions (Hoblitt and Larson, 1975) strongly suggest that the Left Hand Canyon sill was intruded after the Lykins Formation was already tilted ~25° eastward. No further deformation occurred subsequent to intrusion. In contrast, paleomagnetic data from the sill segments at Flagstaff Mountain, Red Rocks, and Sunshine Canyon lead to the following sequence of events: first, tilting of the sediments eastward 25°–30°, then intrusion of the sill, followed by further tilting of the sedimentary beds to their present attitudes. It is noteworthy that sediments of the Arapahoe and Denver Formations also suggest two pulses of Laramide uplift and subsequent erosion of the Front Range, one occurring between ca. 70 and 64 Ma, and a later one subsequent to ca. 55 Ma (Raynolds and Johnson, 2003).

Continue eastward on this trail. At the next trail junction, go south and downhill, crossing good exposures of Fountain Formation. At Pearl Street, go left (E) to downtown Boulder.

If time allows, we will walk downtown along Boulder Creek and discuss its long-term incision rate, the history and extent of flooding, and what the city has done to help avoid flood damage to bridges and buildings.

ACKNOWLEDGMENTS

Mary Ann Berger and Ralph Shroba of the U.S. Geological Survey prepared many of the illustrations, and made helpful comments, respectively.

REFERENCES CITED

Birkeland, P.W., Shroba, R.R., Burns, S.F., Price, A.B., and Tonkin, P.J., 2003, Integrating soils and geomorphology in mountains—an example from the Front Range of Colorado: Geomorphology, v. 55, p. 329–344, doi: 10.1016/S0169-555X(03)00148-X.

Braddock, W.A., 1978, Dakota Group rockslides, northern Front Range, Colorado, *in* Voight, B., ed., Rockslides and avalanches, 1: New York, Elsevier, p. 439–479. (Note: some of the rockslides are shown on the U.S. Geological Survey quadrangle map GQ-1805).

Braddock, W.A., and Eicher, D.L., 1962, Block-glide landslides in the Dakota Group of the Front Range foothills, Colorado: Geological Society of America Bulletin, v. 73, p. 317–324.

Colton, R.B., 1978, Geologic map of the Boulder-Fort Collins-Greeley area, Colorado: U.S. Geological Survey Map I-855-G, scale 1:100,000, 1 sheet.

Dethier, D.P., 2001, Pleistocene incision rates in the western United States calibrated using Lava B tephra: Geology, v. 29, p. 783–786, doi: 10.1130/0091-7613(2001)0292.0.CO;2.

Dethier, D.P., Benedict, J.B., Birkeland, P.W., Caine, N., Davis, P.T., Madole, R.F., Patterson, P.E., Price, A.B., Schildgen, T.F., and Shroba, R.R., 2003, Quaternary stratigraphy, geomorphology, soils, and alpine archaeology in an alpine-to-plains transect, Colorado Front Range, *in* Easterbrook, D.J., ed., Quaternary Geology of the United States: INQUA 2003 Field Guide Volume, Reno, Nevada, Desert Research Institute, p. 81–104.

Hoblitt, R., and Larson, E.E., 1975, Paleomagnetic and geochronologic data bearing on the structural evolution of the northeastern margin of the Front Range, Colorado: Geological Society of America Bulletin, v. 86, p. 237–242.

Hunter, Z.M., 1955, Geology of the foothills of the Front Range in northern Colorado: Denver, Rocky Mountain Association of Geologists, two maps, scale 1:48,000.

Leonard, E.M., 2002, Geomorphic and tectonic forcing of late Cenozoic warping of the Colorado Piedmont: Geology, v. 30, p. 595–598, doi: 10.1130/0091-7613(2002)0302.0.CO;2.

Madole, R.F., 1991, Colorado Piedmont section, *in* Morrison, R.B., ed., Quaternary nonglacial geology, conterminous United States: Boulder, Colorado, Geological Society of America, The Geology of North America, v. K-2, p. 456–462.

Mutschler, F.E., Larson, E.E., and Bruce, R.M., 1987, Laramide and younger magmatism in Colorado—New petrologic and magmatic variations on old themes, *in* Drexier, J.W., and Larson, E.E., eds., Cenozoic volcanism in the southern Rocky Mountains revisited: Colorado School of Mines Quarterly, v. 82, p. 1–47.

Nelson, A.R., and Shroba, R.R., 1998, Soil relative dating of moraine and outwash—terrace sequences in the northern part of the upper Arkansas Valley, U.S.A.: Arctic Alpine Research, v. 30, p. 349–361.

Raynolds, R.G., and Johnson, K.R., 2003, Synopsis of the stratigraphy and paleontology of the uppermost Cretaceous and Lower Tertiary strata in the Denver Basin, Colorado: Rocky Mountain Geology, v. 38, p. 171–178.

Ritter, D.F., 1987, Fluvial processes in the mountains and intermontane basins, *in* Graf, W.L., ed., Geomorphic Systems of North America: Boulder, Colorado, Geological Society of America Centennial Special Volume 2, p. 220–228.

Schildgen, T.F., 2000, Fire and ice: The geomorphic history of middle Boulder Creek as determined by isotopic dating techniques, Colorado Front Range [Honors Thesis]: Williamstown, Massachusetts, Williams College, 103 p.

Schildgen, T., Dethier, D.P., Bierman, P., and Caffee, M., 2002, [26]Al and [10]Be dating of late Pleistocene and Holocene fill terraces: a record of fluvial deposition and incision, Colorado Front Range: Earth Surface Processes and Landforms, v. 27, p. 773–787, doi: 10.1002/ESP.352.

Scott, G.R., 1963, Quaternary geology and geomorphic history of the Kassler quadrangle, Colorado: U.S. Geological Survey Professional Paper 421-A, 70 p.

Steven, T.A., Evanoff, E., and Yuhas, R.H., 1997, Middle and late Cenozoic tectonic and geomorphic development of the Front Range of Colorado, *in* Bolyard, D.W., Sonnenberg, S.A., eds., Geologic history of the Colorado Front Range: Denver, Rocky Mountain Association of Geologists Guidebook, p. 115–124.

Trimble, D.E., 1980, Cenozoic tectonic history of the Great Plains contrasted with that of the Southern Rocky Mountains: a synthesis: The Mountain Geologist, v. 17, p. 59–69.

Trimble, D.E., and Machette, M.N., 1979, Geologic map of the greater Denver area, Front Range urban corridor, Colorado: U.S. Geological Survey Map I-856-H, scale 1:100,000, 1 sheet.

Wahlstrom, E.E., 1948, Pre-Fountain and recent weathering on Flagstaff Mountain near Boulder, Colorado: Geological Society of America Bulletin, v. 59, p. 1173–1189.

Walker, T.R., and Harms, J.C., 1972, Eolian origin of flagstone beds, Lyons Sandstone (Permian), type area, Boulder County, Colorado: The Mountain Geologist, v. 9, p. 279–288.

Walker, T.R., Larson, E.E., and Hoblitt, R.P., 1981, Nature and origin of hematite in the Moenkopi Formation (Triassic), Colorado Plateau: A contribution to the origin of magnetism in red beds: Journal of Geophysical Research, v. 86, p. 317–333.

Wells, J.D., 1967, Geology of the Eldorado Springs Quadrangle, Boulder and Jefferson Counties, Colorado: U.S. Geological Survey Bulletin 1221, 85 p.

Wrucke, C.T., and Wilson, R.F., 1967, Preliminary geologic map of the Boulder quadrangle: U.S. Geological Survey Open-File Report 67-281.

Geological Society of America
Field Guide 5
2004

The South Cañon Number 1 Coal Mine fire: Glenwood Springs, Colorado

Glenn B. Stracher

Division of Science and Mathematics, East Georgia College, Swainsboro, Georgia 30401, USA

Steven Renner

Colorado Division of Minerals and Geology, Inactive Mines Program, 101 South 3rd Street, Grand Junction, Colorado 81501, USA

Gary Colaizzi

Goodson and Associates, Inc., 12200 West 50th Place, Wheat Ridge, Colorado 80033, USA

Tammy P. Taylor

Chemical Division, C-SIC, Mail Stop J514, Los Alamos National Laboratory, Los Alamos, New Mexico 87545, USA

ABSTRACT

The South Cañon Number 1 Coal Mine fire, in South Canyon west of Glenwood Springs, Colorado, is a subsurface fire of unknown origin, burning since 1910. Subsidence features, gas vents, ash, condensates, and red oxidized shales are surface manifestations of the fire. The likely success of conventional fire-containment methodologies in South Canyon is questionable, although drilling data may eventually suggest a useful control procedure. Drill casings in voids in the D coal seam on the western slope trail are useful for collecting gas samples, monitoring the temperature of subsurface burning, and measuring the concentration of gases such as carbon monoxide and carbon dioxide in the field. Coal fire gas and mineral condensates may contribute to the destruction of floral and faunal habitats and be responsible for a variety of human diseases; hence, the study of coal gas and its condensation products may prove useful in understanding environmental pollution created by coal mine fires. The 2002 Coal Seam Fire, which burned over 12,000 acres and destroyed numerous buildings in and around Glenwood Springs, exemplifies the potential danger an underground coal fire poses for igniting a surface fire.

Keywords: South Cañon Number 1 Coal Mine, coal fires, coal fire gas, mineral condensates, coal pollution.

INTRODUCTION

The South Cañon Number 1 mine is located ~6.4 km (4 mi) west of Glenwood Springs, in Garfield County, Colorado (Fig. 1). The room and pillar bituminous mine, burning since 1910, was ignited by an undetermined source.

On May 18, 2004, the authors visited the South Cañon mine fire in preparation for Geological Society of America Field Trip No. 412, scheduled for November 6, 2004. This report presents a synopsis of coal mining and fire in the canyon. In addition, recent subsidence features associated with the burning D coal seam are illustrated as are coal gas condensates and the instrumentation

Stracher, G.B., Renner, S., Colaizzi, G., and Taylor, T.P., 2004, The South Cañon Number 1 Coal Mine fire: Glenwood Springs, Colorado, *in* Nelson, E.P. and Erslev, E.A., eds., Field trips in the southern Rocky Mountains, USA: Geological Society of America Field Guide 5, p. 143–150. For permission to copy, contact editing@geosociety.org. © 2004 Geological Society of America

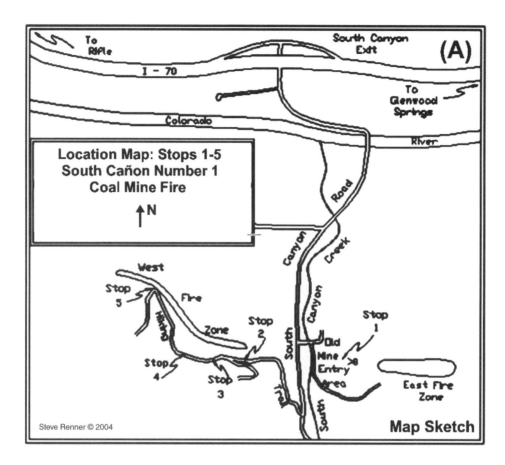

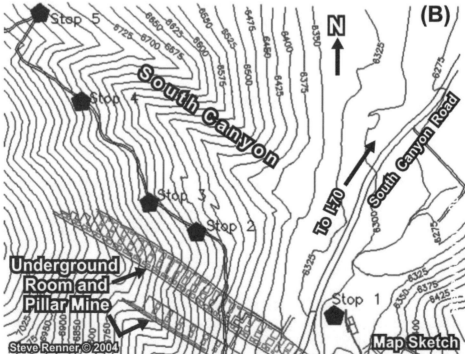

Figure 1. (A) Location map of the South Cañon Number 1 underground coal mine fire (west and east fire zones) and stops 1–5, relative to South Canyon Road and I-70, west of Glenwood Springs, Colorado. (B) Map of South Canyon, stops 1–5, and surface projection of the room and pillar coal mine on fire beneath the western slope of the canyon.

used to collect coal fire gas and thermocouple-temperature data. Techniques for using the instrumentation, a part of the field trip agenda, will provide participants with the skills necessary for collecting valuable data for laboratory analysis.

GEOLOGIC SETTING

The Upper Cretaceous sedimentary rocks in South Canyon are exposed in the Grand Hogback monocline. The monocline, formed during the later part of the Laramide orogeny (Middle to Late Eocene), extends northward from McClure Pass in Pitkin County to the town of Meeker in Rio Blanco County. It separates the Colorado Plateau to the west from the southern Rocky Mountains to the east (Rushworth et al., 1989; Kirkham and Matthews, 2000). Bedding plane attitudes in South Canyon consistently cluster around N35°W, 55°SW.

The Williams Fork Formation, an important coal-bearing unit in western Colorado, consists of alternating coal, sandstone, and shale beds of variable thickness. Numerous surface mines in the Williams Fork Formation are present in northwest Colorado. In South Canyon, and along the Grand Hogback in Garfield County, there are five coal seams that were mined in the Williams Fork Formation. These are, from the stratigraphically lowest to highest, the U, E, Wheeler, D, and Allen seams. The Wheeler, D, and Allen seams were the most important commercially; the Allen is reported to be up to 4.3 m (14 ft) thick near New Castle, west of South Canyon (Colorado Geological Survey, 2004). Drilling data indicate that in South Canyon, the D seam is 1.8–2.4 m (~6–8 ft) thick and the Wheeler seam is 6.1 m (20 ft) thick.

MINING METHODOLOGY

The South Cañon Number 1 mine, one of eleven burning coal mines in Garfield County, lies beneath the western and eastern slopes of South Canyon. Because of steeply dipping beds there, the mining techniques employed were analogous to those used in the anthracite fields of eastern Pennsylvania. Main entries were driven into a coal seam parallel to strike, near canyon-bottom (creek) level. Stope-like raises, or rooms, were then developed as much as 91 m (300 ft) up dip and parallel to strike, above the mains. Coal pillars, 12–18 m (~40–60 ft) thick, separated the rooms. Man-ways, or small passages, in the pillars were excavated between rooms, providing access to various levels of the active mine at different elevations above the mains. Typically, the coal was drilled and shot to bring it down dip. Doors were affixed immediately above coal passes in the mains so that the coal could be easily loaded into carts that traveled on rails within the main entries. In this manner, coal extraction proceeded along strike and up dip within the outcrop in a steep hillside.

In South Canyon, the D seam was mined more aggressively than others. It was initially mined at near-creek elevation, and subsequently, at below-creek elevation. Maps indicate that the same mining method was employed down dip of the initial

workings, with 23-m-thick (75 ft) coal pillars separating adjacent stopes (Colorado Geological Survey, 2004).

MINING HISTORY

Published reports about the history of mining in South Canyon are essentially non-existent. Anecdotal information and early mine maps indicate that two stratigraphically lower seams, the E and U, were mined in the late 1880s. At some later date, mines in the Wheeler and D seams were developed. A small town was constructed at the site, and the area was inhabited until the mid-1950s. According to Rushworth et al. (1989), coal production totaled 925,000 tons during the life of the operation, from 1887 to 1953.

MINE FIRE

The U.S. Geological Survey first reported an underground mine fire in South Canyon in 1910. It was located in the Wheeler seam, on the west side of South Canyon (Rushworth et al., 1989). Anecdotal information, supplied by a retired miner, indicates that the fire spread from the west side of the South Cañon Number 1 mine to the east side in 1953, resulting in closure of the mine. Subsidence, red oxidized shales, gas vents, white ash, and condensates are surface manifestations of the fire today (Figs. 2, 6, 7, and 8).

The Colorado Division of Minerals and Geology has drilled into the east side of South Canyon and is in the process of drilling into the west side. The purpose of this exploratory drilling is to determine the locations of underground stopes and to determine which seams are burning at particular locations. The fire is currently thought to be burning in the upper stopes of the E, D, and Wheeler seams on the east side of the canyon and the Wheeler and D seams on the west side.

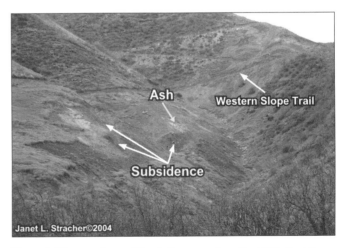

Figure 2. Western slope of South Canyon, where the 2002 Coal Seam Fire is thought to have started (see text). Red oxidized shale (area devoid of vegetation), subsidence, and white ash are present along strike.

It is likely that a second drilling program will be undertaken to more closely evaluate specific subsurface locations on fire. The drilling data will be reduced and evaluated, and will assist future decisions regarding actions necessary to slow the progress of fire on either side of South Canyon.

COAL SEAM SURFACE FIRE

On June 8, 2002, a brush and forest fire referred to as the "Coal Seam Fire" (Piper and Farer, 2002; Rocky Mountain News, 2004) erupted on the western slope of South Canyon. U.S. Forest Service investigators attributed the fire to a surficial gas vent from the underground coal fire in South Canyon. A spark or cinders from the vent is thought to have ignited foliage on the western slope (McKibbin, 2003). The fire spread down-canyon and into West Glenwood Springs after jumping the Colorado River and the Interstate 70 (I-70) highway, burning over 12,000 acres (Piper and Farer, 2002). In addition, 29 homes and 14 other buildings were destroyed, for an estimated loss of $7.3 million dollars (Rocky Mountain News, 2004). The surface fire was eventually extinguished, thanks to the work of nearly 700 fire fighters who created firebreaks and dropped slurry on the fire with air tankers (Piper and Farer, 2002).

ACCESS TO SOUTH CANYON

The base of the western and eastern slopes of South Canyon is readily accessible by automobile. A steep, winding dirt trail cut into the western slope, herein referred to as the "western slope trail," is accessible by foot or off-road vehicle. Surface manifestations of the underground fire in the D coal seam, in addition to burnt trees from the 2002 Coal Seam Fire, are visible from the canyon floor and along the trail. The trail provides a superb overview of the underground fire and the 2002 Coal Seam Fire, and from it, gas vents and drill casings used by the Colorado Division of Minerals and Geology to monitor the mine fire can be seen. Extreme caution must be used along the trail. Its maximum width is 3.4 m (11 ft), and an off road vehicle could easily tumble down slope. Because of subsidence along the slope, hiking off the trail is inadvisable.

GSA Field Trip Route

The trip departs from the Colorado Convention Center in Denver at 6:30 a.m. on November 6th, 2004. The route from Denver is as follows:

Cumulative		Description
mi	km	
160	~257	Follow I-70 west for ~257 km (160 mi) and get off at "South Canyon Exit 111," illustrated in Figure 1A.
160.2	~257.3	Turn left at the bottom of the off ramp and drive under I-70, ~0.3 km (0.2 mi) from the start of the exit.
160.5	~257.8	Drive ~0.5 km (0.3 mi) from the underpass along South Canyon Road (Fig. 1A) to a bridge crossing the Colorado River.
161.7	~259.7	After crossing the river, drive ~1.9 km (1.2 mi) along South Canyon Road (asphalt). At a fork in the road, there is a landfill on the right. Bear to the left where the asphalt ends and South Canyon Road becomes a dirt road.
163.3	~262.3	Drive ~2.6 km (1.6 mi) on this road to the base of South Canyon. This is stop 1, discussed below. From stop 1, it's about another 0.3 km (0.2 mi) to the western slope trail where stops 2–5 are located.

The Western Slope Trail

The hike along the steep western slope trail is strenuous, and sturdy walking shoes are mandatory. Several stops along the trail provide key locations where collection techniques may be readily demonstrated and spectacular views of surficial features associated with the underground mine fire and 2002 Coal Seam Fire are visible. In addition, Storm King Mountain, where 14 firefighters died in 1994, can be seen. The stops in South Canyon (Fig. 1), and related field trip events associated with them, are discussed below.

Stop 1. The first stop of the trip is at the collapsed and overgrown entrances to the South Cañon Number 1 Coal Mine, near the base of the eastern slope of the canyon. The elevation here is ~1928 m (6325 ft). Smoke billowing from these entrances forced closure of the mine in 1953.

Stop 2. Hiking up the western slope trail to the second stop, we can observe the location at which the 2002 Coal Seam Fire is thought to have begun. Surface manifestations of the fire consist of white ash in addition to subsidence and red oxidized shale along strike (Fig. 2).

Stop 3. Continuing up the trail, we see drill sites FA 1 and FA 2, which are used by the Colorado Division of Minerals and Geology to monitor subsurface combustion gas and temperatures (Fig. 3). The steel casings inserted into the drill holes have a nominal inside diameter of ~5.08 cm (2 inches), are ~3.2 m (10 ft) apart along dip, and penetrate a void in the bituminous D coal seam at 1.5 m (5 ft) below the surface, to a depth of ~33 m (108 ft). At this location, a gas collection technique using a LaMotte hand pump, Teflon intake and exhaust lines, and Tedlar gas collection bags is demonstrated. This technique is the same as that used by G.B. Stracher, T.P. Taylor, and J.T. Nolter in June of 2003 (Stracher, 2003) while filming for National Geographic at the Centralia mine fire in Pennsylvania (Fig. 4). A Pasco Scientific thermocouple probe for collecting temperature data and Drager tube field analysis of carbon monoxide and carbon dioxide are also demonstrated at Stop 3 (Fig. 5).

Approximately 30.5 m (100 ft) down slope from Stop 3 is a gas vent encrusted by minerals condensing from exhaled gas. Figure 6 illustrates samples collected for analysis on May 18, 2004, by G.B. Stracher and S. Renner. The temperature measured

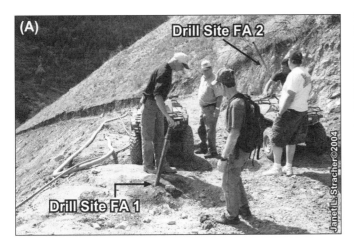

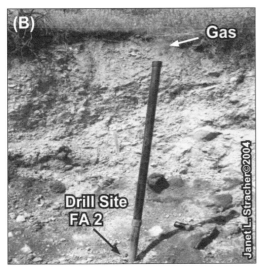

Figure 3. (A) Drill sites FA 1 and FA 2 (steel casings) used to monitor gas and temperatures associated with the burning D coal seam. Left to right: S. Renner, G. Colaizzi, G.B. Stracher, W. Duncan, and G. Griggs. (B) Exhaled coal fire gas from drill casing FA 2 (cap removed).

Figure 4. J.T. Nolter (left) and G.B. Stracher using a coal fire gas-collecting apparatus at the Centralia mine fire in Pennsylvania. This same apparatus is used to demonstrate the collection procedure to field trip participants at drill sites FA 1 and FA 2 in South Canyon. A LaMotte hand pump is used to pump gas from a coal fire vent through Teflon intake and exhaust lines into a Tedlar gas collection bag.

Stop 5. The final stop on the trip along the western slope trail is a switchback at 2115 m (6939 ft). Here, field trip participants can observe (1) overgrown mine entrances at the base of the eastern slope of South Canyon; (2) subsidence features, gas vents, and white ash (at eye level) on the eastern slope, due to subsurface burning of the D coal seam (Fig. 8); (3) subsidence and red oxidized shale along strike of the D coal seam burning beneath the western slope; (4) the devastation in South Canyon caused by the 2002 Coal Seam Fire; and (5) Storm King Mountain, where 14 trapped firefighters died in the "South Canyon Fire" (unrelated to the coal mine fire) on July 6, 1994 (Butler et al., 1998). This fire began on July 2, 1994, when a lightning strike ignited piñon juniper and other foliage. It was contained by July 11, 1994, after burning more than 2,000 acres and triggering debris flows across I-70 as torrential rains poured down on the burn area (Cannon et al., 1995).

DISCUSSION

Subsidence at mine entries, subsequent to closure of the South Cañon Number 1 mine, has not impeded oxygen from reaching the underground fire that continues to burn there. Surface sealing and conventional fire control methods are not likely to work in South Canyon because of former mining practices that left numerous pillars of coal exposed to oxygen circulating through rock fissures. Fire containment is further complicated by the rugged terrain in South Canyon, steeply dipping coal beds there, and the questionable subsurface extent of the fire.

Methods for controlling and extinguishing underground coal mine fires in mountainous terrain include the use of liquid nitrogen and the injection of foamed grout into surface vents and fissures (Feiler and Colaizzi, 1996). However, such techniques

immediately inside the vent with the Pasco probe on May 18, 2004, was 89.4 °C (193 °F). Sample collecting techniques using plastic vials and metal spatulas are discussed at Stop 3, with an emphasis on avoiding sample contamination from soil or altered rock. If conditions permit, G.B. Stracher and one of the trip co-leaders will climb down slope and demonstrate the actual collection process. For safety and liability reasons, field trip participants will not be permitted to accompany them.

Stop 4. Drill site FE, for monitoring subsurface combustion gas and temperatures, is seen at stop 4, further up the western slope trail from stop 3. Those field trip participants who volunteer to do so will take a gas sample and thermocouple temperature measurement here. A section of iron rail from a track, adjacent to a closed depression ~61 m (200 ft) down slope, suggests the presence of a mine entrance on the western slope (Fig. 7).

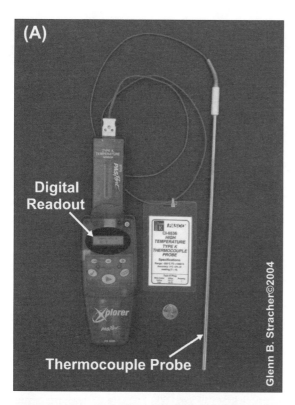

Figure 6. (A) G.B. Stracher (left) collecting condensates encrusting a coal fire gas vent, while S. Renner records GPS coordinates. The temperature ~5 cm (2 inches) inside the vent on May 18, 2004, was 89.4 °C (193 °F). Location is 39°32′09.737″ N, 107°25′14.803″ W, at an elevation of 2028 m (6654 ft) and ~35 m (115 ft) down slope from drill sites FA 1 and FA 2 (Fig. 3). (B) Close-up of condensates and gas vent.

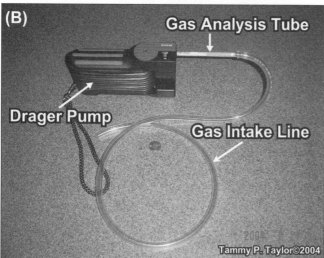

Figure 5. (A) Pasco Scientific thermocouple probe used to measure coal gas temperatures at drill sites FA 1, FA 2, and FE. The metal probe is inserted into the gas, and the temperature is digitally recorded. (B) Drager hand pump and tube apparatus used to extract coal fire gas and measure carbon monoxide and carbon dioxide concentrations in the field.

Figure 7. Iron track rail adjacent to a closed depression, ~61 m (200 ft) down slope from stop 4, suggesting the presence of a mine entrance on the western slope. Left to right: S. Renner, G.B. Stracher.

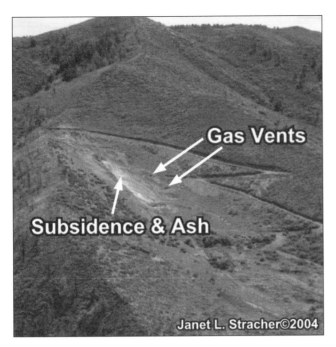

Figure 8. Subsidence, ash, and gas vents (small black spots) on the eastern slope of South Canyon due to subsurface burning of the D coal seam.

may be cost prohibitive in South Canyon, and the potential for these alternate methods to contain the underground fire there is questionable, for the same reasons conventional methods are problematic. Regardless, additional test-drilling data, acquired by the Colorado Division of Minerals and Geology, may suggest a method to contain the fire.

The study of coal fire gas and its condensation products can be critically useful in understanding the pollution caused by coal mine fires. The chemistry of exhaled gas and condensates is reflective of elements and compounds that may be released as pollutants into the atmosphere, soil, or the hydrosphere. Such pollutants may be responsible for a variety of environmental and human health problems, including the destruction of floral and faunal habitats, bronchitis, lung cancer, stroke, pulmonary heart disease, chronic obstructive pulmonary disease, arsenosis, and fluorosis (Finkelman et al., 1999, 2001, 2002; Johnson et al., 1997, p. 19; World Resources Institute, 1999, p. 63–67; Stracher and Taylor, 2004). Consequently, the study of coal mine fire gas and mineral condensates, such as those in South Canyon, can be critically useful in understanding environmental pollution.

The authors sincerely hope that some of their readers, including the field trip participants, will pursue research in the health-related and numerous additional areas of study associated with coal fires science.

ACKNOWLEDGMENTS

The authors thank the Colorado Division of Minerals and Geology, William Duncan, a contractor employed by the Division, and Garrison Griggs for assisting them in preparation for Geological Society of America Field Trip 412. We also thank Chris Carroll, Director of the Coal Division, Colorado Geological Survey, for assisting us with reference material. In addition, we are grateful to Eric Nelson of the Colorado School of Mines and Eric Erslev of Colorado State University for their review of the manuscript.

DEDICATION

This manuscript and the work contained herein are dedicated to Dr. Nancy Lindsley-Griffin and Dr. John R. Griffin of the University of Nebraska–Lincoln. Without their guidance and rigorous training of the senior author in the elegant techniques of complex field geology, this project and others undertaken by him, both in America and abroad, could not have been completed. They have and will always be for him a source of inspiration, two people to whom he can turn, even when they are not beside him, and who, with their "spiritual guidance" in his mind's eye, assist him in determining the most straightforward method by which to proceed while working in the field.

REFERENCES CITED

Butler, B.W., Bartlette, R.A., Bradshaw, L.S., Cohen, J.D., Andrews, P.L., Putnam, T., and Mangan, R.J., 1998, Fire behavior associated with the 1994 South Canyon Fire on Storm King Mountain, Colorado: Ogden, Utah Rocky Mountain Research Station, U.S. Department of Agriculture (Forest Service) Research Paper RMRS-RP-9, 82 p.

Cannon, S.H., Powers, P.S., Pihl, R.A., and Rogers, W.P., 1995, 1995 preliminary evaluation of the fire-related debris flows on Storm King Mountain, Glenwood Springs, Colorado: U.S. Geological Survey Open-File Report 95-908, http://landslides.usgs.gov/html_files/ofr95-508/skrep.html (accessed July 2004).

Colorado Geological Survey, 2004, Unpublished data from the South Cañon No. 1 Mine map, Subsidence Information Library: Historic coal mine maps and maps of mined-out areas, 1 sheet.

Feiler, J.J., and Colaizzi, G.J., 1996, IHI mine fire control project utilizing foamed grout technology, Rifle, Colorado: U.S. Bureau of Mines Report Contract number 14320395H0002, prepared by Goodson and Associates, Inc., 12200 West 50th Place, Wheat Ridge, Colorado 80033, 111 p.

Finkelman, R.B., Belkin, H.E., and Zheng, B., 1999, Health impacts of domestic coal use in China: Proceedings of the National Academy of Sciences of the United States of America, v. 46, p. 3427–3431.

Finkelman, R.B., Skinner, H.C., Plumlee, G.S., and Bunnell, J.E., 2001, Medical geology: Geotimes, v. 46, no. 11, p. 21–23.

Finkelman, R.B., Orem, W., Castranova, V., Tatu, C.A., Belkin, H.E., Zheng, B., Lerch, H.E., Marharaj, S.V., and Bates, A.L., 2002, Health impacts of coal and coal use: possible solutions: International Journal of Coal Geology, v. 50, p. 425–443, doi: 10.1016/S0166-5162(02)00125-8.

Johnson, T.M., Liu, F., and Newfarmer, R., 1997, Clear water, blue skies: China's environment in the new century: Washington, D.C., World Bank, 124 p.

Kirkham, R.M., and Matthews, V., 2000, Guide to the geology of the Glenwood Springs Area, Garfield County, Colorado: Denver, Colorado Geological Survey, 41 p.

McKibbin, M., 2003, South canyon checked for underground fire: The Daily Sentinel (11 Nov. 2003), http://www.gjsentinel.com/news (accessed June 2004).

Piper, J., and Farer, P., 2002, Life in Glenwood Springs getting back to normal: 9News KUSA-TV (10 June 2002), http://www.9news.com/storyfull.asp?id=3636 (accessed June 2004).

Rocky Mountain News, 2004, Colorado Wildfires, RockyMountainNews.Com, http://cfapp.rockymountainnews.com/wildfires/moredata.cfm?fire=1755 (accessed June 2004).

G.B. Stracher et al.

Rushworth, P., Haefner, B.D., Hynes, J.L., and Streufert, R.K., 1989, Reconnaissance study of coal fires in inactive Colorado coal mines, Information Series Number 26: Denver, Colorado Geological Survey, 60 p.

Stracher, G.B., 2003, Coal mine fire—gas and condensation products: Collection techniques for laboratory analysis: Center for Applied Energy Research, University of Kentucky, Lexington, Energeia, v. 14, no. 5, p. 1, 4–5.

Stracher, G.B., and Taylor, T.P., 2004, Coal fires burning out of control around the world: Thermodynamic recipe for environmental catastrophe: International Journal of Coal Geology, v. 59, p. 7–17, doi: 10.1016/J.COAL.2003.03.002.

World Resources Institute, 1999, 1998–1999 World Resources: A guide to the global environment, environmental change and human health: Oxford, United Kingdom, Oxford University Press, 384 p.

Printed in the USA

Geological Society of America
Field Guide 5
2004

Field guide to the paleontology and volcanic setting of the Florissant fossil beds, Colorado

Herbert W. Meyer*
National Park Service, Florissant Fossil Beds National Monument, Florissant, Colorado 80816, USA

Steven W. Veatch
Department of Earth Science, Emporia State University, Emporia, Kansas 66801, USA

Amanda Cook
Boulder Open Space and Mountain Parks, Boulder, Colorado 80306, USA

ABSTRACT

This field trip in the vicinity of the Florissant fossil beds includes five stops that examine the Precambrian Cripple Creek Granite and Pikes Peak Granite, and the late Eocene Wall Mountain Tuff, Thirtynine Mile Andesite lahars, and Florissant Formation. The Cripple Creek Granite and Pikes Peak Granite formed in batholiths ca. 1.46 and 1.08 Ga, respectively. Uplifted during the Laramide Orogeny of the Late Cretaceous and early Tertiary, the Precambrian rocks were exposed along a widespread erosion surface of moderate relief by the late Eocene. The late Eocene volcanic history of the Florissant area is dominated by two separate events: (1) a caldera eruption of a pyroclastic flow that resulted in the emplacement of the Wall Mountain Tuff, a welded tuff dated at 36.73 Ma; and (2) stratovolcanic eruptions of tephra and associated lahars from the Guffey volcanic center of the Thirtynine Mile volcanic field. This volcanic activity from the Guffey volcanic center had a major influence on the development of local landforms and on sedimentation in the Florissant Formation, which was deposited in a fluvial and lacustrine setting and is dated as 34.07 Ma. The Florissant Formation contains a diverse flora and insect fauna consisting of more than 1700 described species. Most of these fossils are preserved as impressions and compressions in a diatomaceous tuffaceous paper shale and as huge petrified trees that were entombed in a lahar deposit.

Keywords: paleontology, Florissant, Colorado.

INTRODUCTION

Florissant represents one of the most taxonomically diverse fossil sites in the world, with ~1700 described species of mostly plants, insects, and spiders, and also a few vertebrates. These

fossils range from tiny impressions in paper shale to enormous petrified tree stumps. The Florissant fossil beds have been collected and researched for more than 130 years, and the important collections that early paleontologists made from this area have ended up in at least 25 museums. Following years of exploitation while in private ownership, a large portion of the fossil beds was designated as Florissant Fossil Beds National Monument in 1969.

---

*Herb_Meyer@nps.gov

Meyer, H.W., Veatch, S.W., and Cook, A., 2004, Field guide to the paleontology and volcanic setting of the Florissant Fossil Beds, Colorado, *in* Nelson, E.P. and Erslev, E.A., eds., Field trips in the southern Rocky Mountains, USA: Geological Society of America Field Guide 5, p. 151–166. For permission to copy, contact editing@geosociety.org. © 2004 Geological Society of America

The upper Eocene Florissant Formation was deposited in a paleovalley on a widespread erosion surface of moderate relief. The formation consists predominantly of lacustrine sediments but also includes fluvial and lahar deposits. Much of the geologic history of the Florissant area is related to the nearby Thirtynine Mile volcanic field. This volcanic field includes the Guffey volcanic center, which consisted of a large coalesced stratovolcano during the late Eocene and early Oligocene. Ash and pumice falls as well as lahars from the Guffey volcano had a major influence on the deposition of the Florissant Formation.

The town of Florissant, Colorado, is located on U.S. Highway 24 ~35 mi (56 km) west of Colorado Springs. The area of this field trip is located to the south of the town. The field trip will make five stops in this area (Fig. 1): The first will examine an outcrop of the Wall Mountain Tuff, which is a welded tuff that was erupted from a caldera ~50 mi (80 km) to the west and predates the Florissant Formation. Next, we will proceed to an overlook of the Thirtynine Mile volcanic field, followed by a stop to see one of the lahars. We will then have a long stop at the visitor center at Florissant Fossil Beds National Monument to see the exhibits and the petrified forest area. Finally, we will stop at a privately owned quarry where you will have a chance to look for plant and insect fossils.

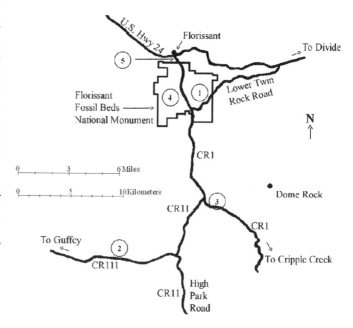

Figure 1. Map of the Florissant area indicating Stops 1–5 of this field trip. Stop 1—Barksdale picnic area; Stop 2—overlook of the Thirtynine Mile volcanic field; Stop 3—Evergreen Station; Stop 4—Florissant Fossil Beds National Monument visitor center; Stop 5—Florissant Fossil Quarry.

HISTORY OF SCIENTIFIC STUDY AT FLORISSANT

The fossils from Florissant have been known and studied for more than 130 years. By the late 1860s, regional newspapers had mentioned Florissant's fossils. The *Daily Central City Register* (April 18, 1871) reported "...about a mile from Castello's ranch in Florissant Valley, is a petrified forest near which are found, between sedimentary layers, the most beautiful imprints of leaves differing entirely from any that grow in the valley now-a-days."

Theodore L. Mead (1852–1936), a 19-yr-old college student at Cornell, went on a butterfly collecting trip to South and Middle Park in 1871. While at the Hartsel Ranch in South Park, Colorado, Mead heard about fossil insects and fossil tree stumps near Castello's ranch at Florissant and decided to investigate the site (Brown, 1996). In a letter to George M. Dodge, Mead described his Florissant outing: "Visited some petrified stumps thirty miles from our stopping place and beside the wood secured about 20 fossil insects from the shale nearby. They were mostly Diptera and Hymenoptera" (Brown, 1996, p. 66). It was Theodore Mead who made the first scientific collection at Florissant. The Hayden Survey, one of several scientific surveys sponsored by the federal government, made the fossil beds known to science. Albert Charles Peale (1849–1914), as part of the Hayden Survey, was exploring South Park in Colorado in 1873 when he rode into the Florissant valley. In his 1874 report he mentioned the fossil beds,

... around the settlement of Florissant is an irregular basin filled with modern [sic] lake deposits. The entire basin is not more than 5 miles in diameter ... just below Florissant, on the north side of the road, are

bluffs not over 50 feet in height, in which are good exposures of the beds.... About one mile south of Florissant, at the base of a small hill of sandstone, capped with conglomerate, are 20 or 30 stumps of silicified wood. This locality has been called 'Petrified Stumps' by the people in the vicinity. (Peale 1874, p. 210)

Three students from Princeton (then the College of New Jersey), William Scott, Henry Fairfield Osborn, and Frank Speir Jr., organized a student expedition in 1877 to the American West to search for vertebrate fossils. On Wednesday, July 11, 1877, 18 students, two professors, and two Princeton employees arrived at the fossil beds (Journal of the Paleontologists of the Princeton Scientific Expedition, 1877; used by permission of the Princeton University Library). The students spent two days collecting some of the most important fossils to come from Florissant. The students also made arrangements with Charlotte Hill, the landowner, to send additional fossils to Princeton (Scott, 1939). For description, the fossil insects were sent to Samuel Scudder and the fossil plants were sent to Leo Lesquereux. At least 180 of the plant and insect fossils became type specimens (Veatch, 2003).

Samuel H. Scudder (1837–1911), a pioneering paleoentomologist, became very interested in fossils from Florissant. Based on specimens obtained by other collectors, Scudder began publishing papers on Florissant in 1876. Scudder arrived at the fossil beds in 1877, just after the Princeton students had left. Arriving with F.C. Bowditch of Boston and Arthur Lakes from Golden, Scudder collected fossils for five days. In 1878, Scudder described *Prodryas persephone* (Scudder, 1878), regarded as one of the finest fossil butterfly specimens known. *The Tertiary Insects of North America*, published in 1890, is regarded as Scudder's greatest work and includes many descriptions of new species of Florissant's fossil insects. Later publications (Scudder, 1893, 1900) described numerous fossil beetles from Florissant.

Leo Lesquereux (1806–1889) emigrated from Switzerland to Ohio in 1847, and he soon became one of America's early paleobotanists. Lesquereux never obtained permanent scientific employment and supported himself as a watchmaker while working as a part-time paleobotanist describing specimens collected by state and federal government scientific surveys. Although he never collected at Florissant, Lesquereux was the first to name Florissant's fossil plants, and he described more than 100 new species (Meyer, 2003). In fact, he produced the first scientific paper about Florissant (Lesquereux, 1873). Later, he published lengthy monographs on the fossil plants (Lesquereux, 1878, 1883).

One of the most influential contributors to the scientific study of Florissant was T.D.A. Cockerell (1866–1948), a highly regarded naturalist and entomologist and a professor at the University of Colorado for most of his career (Weber, 2000). Cockerell organized expeditions to Florissant from 1906 through 1908 in cooperation with the University of Colorado, the American Museum of Natural History, Yale University, and the British Museum. Most of the fossils from the Florissant shales consist of two corresponding halves, and one of Cockerell's common practices was to retain one of these halves for the University of Colorado and send the counterparts of these same specimens to other institutions. Between 1906 and 1941, Cockerell published 140 papers on Florissant fossils (e.g., Cockerell, 1906, 1908a, 1908b, 1908c)—more than any other paleontologist has written about the site.

Harry D. MacGinitie (1867–1987), encouraged by Cockerell, came to Florissant and excavated three new sites in 1936 and 1937 using a horse pulling a plow (Meyer, 2003). MacGinitie published *Fossil Plants of the Florissant Beds, Colorado* in 1953, the most comprehensive work on the systematic paleontology of Florissant's plants. MacGinitie completely revised and updated the fossil plants, adding new species and placing some of the older taxonomic assignments into synonymy. He was the first to consider some of the broader implications of Florissant's fossils, such as the stratigraphic context, the basis for age determination, paleoecology, paleoclimate, and paleoelevation.

Other early scientists who worked on Florissant's fossils include E.D. Cope, who, between 1874 and 1883, published on fossil fish. W. Kirchner published on fossil plants in 1898. H.F. Wickham worked on fossil beetles between 1907 and 1908. C.T. Brues worked on fossil bees from 1906 through 1910. F.H. Knowlton published on fossil plants in 1916. M.T. James published on fossil flies between 1937 and 1941, and A.L. Melander published on fossil flies in 1949. F.M. Carpenter published a monograph on the fossil ants in 1930. Paul R. Stewart, between 1940 and 1971, made large collections of fossils through his summer geology camp for Waynesburg College.

There has been a considerable amount of recent scientific work. Manchester (2001) has worked on further revising the taxonomic assignments of the fossil plants and has recognized new extinct genera. Evanoff (1992) revised the stratigraphy of the Florissant Formation and produced a detailed (unpublished) geologic map. O'Brien et al. (2002) used scanning electron microscopy to examine the influence of diatoms and microbial biofilms on the taphonomic processes of fossilization. Studies on paleoelevation have been made by Meyer (1992), Gregory (1994), and Wolfe et al. (1998) (a more complete history of studies is summarized in Meyer, 2001), all suggesting that the fossils were deposited at a relatively high elevation. Fossil pollen has been studied by Wingate and Nichols (2001) and Leopold and Clay-Poole (2001).

A paleontology research program was initiated by Florissant Fossil Beds National Monument in 1994 when the first position for a paleontologist (held by the first author) was established. A database Web site (http://planning.nps.gov/flfo/) has been developed recently to document all of the published Florissant specimens from various museums and the publications in which these specimens were first described and later revised. Excavations have resumed, yielding new specimens that now include stratigraphic data.

TOURISM AND THE ESTABLISHMENT OF THE NATIONAL MONUMENT

The Centennial from Georgetown, Colorado (February 1876), reported, "Florissant …has become a noted resort for tourists passing through that portion of the Territory. Until about 3 years ago the locality went by the name of The Petrified Stumps, there being within a circuit of about a mile in diameter 15 or 20 of, probably, the largest petrified stumps of trees in the world....The land on which these mammoth stumps stand is now principally occupied by settlers...." In 1887, the Colorado Midland Railway came to Florissant with a special excursion train that allowed passengers to view wildflowers and collect fossils.

Commercial operations developed the petrified forest as tourist attractions. By 1922, the Coplen Petrified Forest was in operation around Big Stump. New stumps were excavated and visitors were allowed to collect fossils. The old Colorado Midland Railroad station was moved from the town of Florissant to a site near Big Stump, where it was used as a lodge. This site was later operated by the Singer family as the Colorado Petrified Forest from 1927 until the national monument was established.

Another private operation, the Pike Petrified Forest, was located adjacent to the Singer's venture. Situated near the Redwood Trio and at the same location as the current national monument headquarters, the Pike Petrified Forest was developed by the Henderson family as a tourist attraction by 1922. The operation was later owned by H.D. Miller and then by John Baker, who operated the Pike Petrified Forest until it closed in 1961. During the summer of 1956, Walt Disney visited the Pike Petrified Forest and purchased a large stump, which was removed with a crane and shipped to California, where it remains on exhibit in Frontierland at Disneyland Park (Meyer, 2003).

The Florissant Fossil Quarry, opened in the 1950s by the Clare family, is a private fossil site that is still operating today. The quarry provides a site that allows public collecting, and it will be visited as Stop 5 of this field trip.

The fossil beds remained in private ownership through the 1960s. During the late 1960s, real estate developers planned a subdivision of A-frame cabins at the fossil beds. The Defenders of Florissant was soon established by scientists and other citizens, and this organization became a strong advocate for the protection of the valuable fossil resources. While members of the Defenders of Florissant were ready to defiantly stand at the face of bulldozers, a court injunction was issued that temporarily halted development. Soon after, Congress passed a bill to establish the national monument and purchase 6,000 acres of private land. President Nixon signed the act into law on August 20, 1969.

GEOLOGIC HISTORY OF THE FLORISSANT AREA

The stratigraphic column for the Florissant area (Fig. 2) consists of Precambrian and Cenozoic rock formations. Although rocks of Paleozoic and Mesozoic age comprise much of the landscape in nearby areas to the south (near Cañon City) and east (near Colorado Springs), they are absent in the area of this field trip, apparently due to erosion concomitant with uplift during the Laramide Orogeny in the Late Cretaceous to early Tertiary. By the late Eocene, a widespread erosion surface of low to moderate relief had developed in the region, and it was on this erosion surface that the volcanic rocks of the Tertiary were deposited (Epis and Chapin, 1975). Paleobotanical estimates suggest that this erosion surface existed at a high paleoelevation (Gregory, 1994; Meyer, 2001; Wolfe et al., 1998). Geologic maps of the field trip area include Wobus and Epis (1978) and Scott et al. (1978). An overview of the region's geology is also provided in the guidebooks by Henry et al. (1996) and Henry et al. (2004).

Precambrian Rocks

Various Precambrian granites and metamorphic rock units make up the basement rock of the region, although only two of these will be seen during the field trip. Both of these are granites that form large batholiths. The Cripple Creek Granite is dated ca. 1.46 Ga and is distributed primarily to the south and southwest of Florissant (Wobus, 2001). The Pikes Peak Granite is dated as 1.08 Ga (Wobus, 2001), occurs in the immediate Florissant vicinity and to the east, and was the bedrock unit on which the Florissant Formation was deposited. Both the Pikes Peak Granite and Cripple Creek Granite are rich in reddish-orange potassium feldspar, but the Cripple Creek Granite is a fine to medium-grained muscovite-biotite granite, whereas the Pikes Peak Granite is a coarser-grained biotite and hornblende granite lacking muscovite (Wobus, 2001). The outcrop patterns also help to distinguish the two, with the Cripple Creek Granite forming more angular, blocky exposures and the Pikes Peak Granite forming more rounded outcrops including exfoliation domes. These

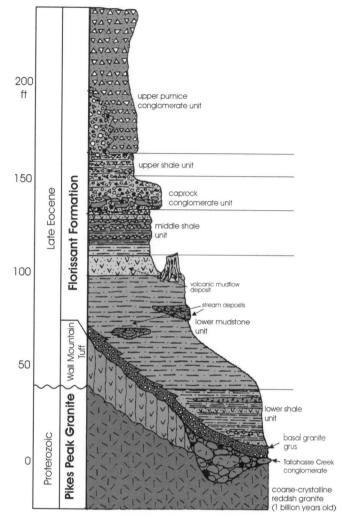

Figure 2. Stratigraphic column showing the rock units in the vicinity of Florissant Fossil Beds National Monument. Stratigraphy based on Evanoff et al. (2001). Diagram provided courtesy of Emmett Evanoff.

batholiths were uplifted during the late Paleozoic development of the Ancestral Rocky Mountains and again during the Late Cretaceous to early Tertiary Laramide Orogeny. By the late Eocene, they formed a widespread erosion surface with incised drainages, and it was on this surface that the Wall Mountain Tuff and Florissant Formation were deposited.

Wall Mountain Tuff

The Wall Mountain Tuff is a rhyolitic welded tuff (ignimbrite) that is $^{40}Ar/^{39}Ar$ dated as 36.73 Ma (Evanoff et al., 2001). Regionally, it covers a wide area in central Colorado and apparently originated from a caldera eruption from the vicinity of modern Mount Princeton in the Sawatch Range ~50 mi (80 km) west of Florissant, and it extends eastward another 45 mi (70 km) to the area near Castle Rock between Colorado Springs and Denver. Locally, it has a patchy, remnant distribution around the Florissant area and lies unconformably on the Pikes Peak Granite. It was emplaced by a pyroclastic flow that draped the landscape but tended to follow paleodrainages such as the Florissant paleovalley. The thickness varies greatly from one location to another. The devitrified welded tuff consists dominantly of sanidine and plagioclase, with lesser amounts of biotite, clinopyroxene, and minor hornblende (Shannon et al., 1987). Flattened and stretched pumice lapilli are present, but quartz is absent as phenocrysts (Epis and Chapin, 1974; Wobus and Epis, 1978).

Thirtynine Mile Andesite

The Thirtynine Mile volcanic field is located in a large area beginning ~10 mi (16 km) west and southwest of Florissant and extending as much as 20 mi (32 km) beyond that. The Guffey volcanic center was a late Eocene stratovolcanic complex within the Thirtynine Mile volcanic field, post-dating the Wall Mountain Tuff. It produced eruptions such as ashfalls, lava flows, domes, and lahars. The stratified lava flows and laharic breccia from this volcanic field have been mapped as the Thirtynine Mile Andesite (Wobus and Epis, 1978; Scott et al., 1978). A series of lahars originating from the slopes of the Guffey volcanic center flowed into the surrounding paleovalleys. Some of these lahars flowed down the ancestral paleodrainage of Four Mile Creek to its confluence with the Florissant paleovalley, where they formed a natural dam in the vicinity of Evergreen Station (Stop 3). This impoundment of the drainage created the late Eocene "Lake Florissant" to the north, and it was in this basin that the Florissant Formation accumulated.

Florissant Formation

The Florissant Formation is correlative in time with at least part of the Thirtynine Mile Andesite. The formation has been subdivided into six informal units (Fig. 2) by Evanoff et al. (2001). This formation was deposited in close association with the volcanic eruptions from the Guffey volcanic center, which is located ~18 mi (29 km) southwest of the Florissant area. Pyroclastic eruptions and lahars from the Guffey volcanic center influenced the deposition and paleogeography of the Florissant area over a long interval of time, beginning with the lahars of the Thirtynine Mile Andesite that created Lake Florissant. At least two lacustrine episodes are evident in the Florissant Formation (Evanoff et al., 2001), and these developed because of impoundment of the Florissant paleodrainage by the lahars to the south. During the intervening period between these two lacustrine events, the earlier lake apparently was infilled by sediments and/or drained by erosional breaching of the lahar dam, forming an alluvial valley during this time. The lower shale unit was deposited in the first generation of Lake Florissant, and the middle and upper shale units, as well as the caprock conglomerate, were deposited in the later lake (Fig. 2). The mudstones and lahar of the lower mudstone unit were deposited during the intervening period (Fig. 2).

This lahar of the lower mudstone unit buried large redwoods and other trees in the Florissant valley.

Pyroclastic eruptions from the Guffey volcanic center deposited layers of ash and pumice in and around the lake basin. Other eruptive centers that may have provided sources for some of this tephra include pre-ignimbrite eruptions of the 33.8 Ma Mount Aetna caldera or post-ignimbrite eruptions from the 34.3 Ma Grizzly Peak caldera (Evanoff et al., 2001). Some of the ash was partially weathered to clay and was deposited in the lake as thin laminae that alternated with thin laminae consisting of diatoms that settled following diatom blooms in the lake. These alternating laminae of ash-clay and diatoms formed 0.1–1.0-mm-thick couplets (O'Brien et al., 2002) of diatomaceous tuffaceous paper shales that contain the extraordinarily well-preserved plant and insect fossils of the lower, middle, and upper shale units. Deposition of these paper shales is thought to have involved mucilaginous biofilms that developed by mucus secretions of diatoms, and these biofilms may have been important contributors to the exceptional fossil preservation at Florissant (Harding and Chant, 2000; O'Brien et al., 2002). Occasionally, more voluminous pyroclastic eruptions produced thicker, interbedded tuffaceous layers of ash and pumice between the paper shale layers within the three shale units.

A later lahar or debris flow produced the caprock conglomerate, which separates the middle and upper shale units in the Florissant Formation (Fig. 2). This lahar was deposited subaqueously in Lake Florissant. The final infilling of the lake corresponds to the deposition of the upper pumice conglomerate unit, which makes up the uppermost layer of the Florissant Formation. This layer is composed mostly of pumice and ash. The upper part of the Florissant Formation has been ^{40}Ar/^{39}Ar dated at 34.07 Ma (Evanoff et al., 2001).

Quaternary Deposits

During the Quaternary, glaciers existed on the higher elevations of nearby Pikes Peak, although the area around Florissant was unglaciated. Murphey (1992) and Evanoff and Murphey (1994) described the Quaternary deposits at Florissant as Pleistocene granular, grussic terrace gravels derived from the Pikes Peak Granite, and Pleistocene/Holocene colluvium and alluvium, composed predominantly of granite grus with shale fragments, silicified wood fragments, andesitic clasts, and mud. Three units consisting of Pleistocene gravels, Quaternary colluvium, and Holocene alluvium were mapped by Evanoff (1992). More detailed mapping and definition of Quaternary units is needed in order to more completely understand the ages and depositional settings in which these sediments accumulated.

PALEONTOLOGY

Late Eocene Fossil Plants

The fossil plants from Florissant include huge petrified tree stumps, impressions and compressions of leaves, fruits, seeds, and flowers, and pollen. These various plant organs have contributed significant information about understanding the ecology, climate, and elevation of this upland late Eocene community. They also provide a broader perspective relating to paleogeographic dispersal patterns, plant community dynamics, and the nature of biotic and climatic events associated with the major climatic cooling of the Eocene-Oligocene transition.

The tree stumps in the petrified forest were preserved in situ by a lahar deposit that makes up the upper part of the lower mudstone unit of the Florissant Formation (Fig. 2). Most of the stumps represent a species of *Sequoia*, which also has been referred to as the fossil wood genus *Sequoiaoxylon* (Andrews, 1936) and is related to the modern coast redwood of California. The largest of these at Florissant measures more than 4 m in diameter at breast height. Although less common, five species of hardwoods have also been described based on wood anatomy (Wheeler, 2001). Growth rings from some of the *Sequoia* stumps have provided information for interpreting growing conditions of the trees, indicating that the Florissant trees grew under more favorable conditions than the modern coast redwoods of California (Gregory-Wodzicki, 2001).

About 120 species of plants are known from leaves, fruits, seeds, and flowers. These are preserved as impression and compression fossils in the lower, middle, and upper shale units of the Florissant Formation. The flora was comprehensively revised in the classic monograph by MacGinitie (1953), and more complete, updated taxonomic lists are provided by Manchester (2001) and Meyer (2003). The flora is dominated by numerous species of broad-leaved hardwoods, but also includes eleven species of conifers. These plants probably represent a mosaic of habitats from the late Eocene landscape, ranging from a mesic environment near the lakeshore, to drier habitats along the surrounding ridges, to coniferous forests in the nearby volcanic and granitic uplands.

The conifers include *Torreya*, *Chamaecyparis* (white cedar or false cypress, Fig. 3), *Sequoia* (redwood), *Abies* (fir; only one known specimen), *Picea* (spruce), and six species of *Pinus* (pine). At the collecting site for this field trip (Stop 5), *Chamaecyparis* (Fig. 3) and *Sequoia* are common. Both were probably very tall, long-lived trees in the late Eocene forest near the lake. Some of the species of *Pinus* probably lived near the lake as well, whereas others inhabited the drier habitats nearby. *Abies* and *Picea* may have been transported into the depositional basin from more distant sources in the surrounding uplands.

The broad-leaved hardwoods are very diverse, with more than 100 species. Some of the more diverse hardwood families include Fagaceae (beech family), Juglandaceae (walnut family), Salicaceae (willow family), Ulmaceae (elm family), Rosaceae (rose family), Fabaceae (legume family), Sapindaceae (soapberry family), and Anacardiaceae (cashew family). Work in recent years (e.g., Manchester, 1989, 1992; Manchester and Crane, 1983, 1987) has recognized that some of the Florissant genera are extinct, and in fact, this includes the two most common plants, *Fagopsis* (Fig. 4) and *Cedrelospermum* (Fig. 5), both of which are found at Stop 5. The largest number of hardwood

Figure 3. Foliage of *Chamaecyparis linguaefolia* (false cypress or white cedar in the family Cupressaceae). Specimen UCMP-3780, courtesy of the University of California Museum of Paleontology. Scale bar is 1 cm.

species were deciduous, although a number of evergreens were also present. Leaf size tends to be small for most of these plants.

The palynoflora (pollen and spores) from Florissant has been examined by Leopold and Clay-Poole (2001) and Wingate and Nichols (2001), and it provides additional insight into the composition and ecology of the Florissant forest community. Pollen was incorporated into the shale during deposition and can be extracted chemically by dissolving the shale detritus in various acids. Because of various aspects of taphonomy, certain plants that are known as leaves and fruits are not known in the palynofloral record, and vice versa. About 25 genera are known exclusively from pollen and spores (Leopold and Clay-Poole, 2001). Although not definitive, pollen evidence suggests that Florissant may hold the earliest record for the Asteraceae (sunflower family), which is today one of the world's largest plant families.

There is no plant community in the modern world in which all members of the late Eocene paleocommunity still coexist, and many of the Florissant genera now inhabit widely different regions

Figure 4. Leaves and attached flowering head of *Fagopsis longifolia* (an extinct genus in the family Fagaceae). Specimen YPM-30121, courtesy of Peabody Museum of Natural History, Yale University. Scale bar is 1 cm.

Figure 5. Leaf of *Cedrelospermum lineatum* (an extinct genus in the family Ulmaceae). Specimen UF-7279, courtesy of the Florida Museum of Natural History, University of Florida. Scale bar is 1 cm.

of the world. For example, some of the plants are now endemic to eastern Asia, such as *Koelreuteria*; others are restricted to the west coast of the United States, such as *Sequoia*; and still others have a modern distribution limited to Mexico. Florissant provides important information for understanding patterns of evolution and paleobiogeographical dispersal. For example, the earliest known fossil record for *Carya* (hickory) is at Florissant; hickories spread into Europe by way of the north Atlantic land connection and later into Asia, but today they live only in eastern North America (including Mexico) and eastern Asia (Manchester, 1999).

Fossil plants provide the best evidence for interpreting Florissant's paleoclimate and paleoelevation, although estimates by different workers vary widely (see summary provided in Meyer, 2001). MacGinitie (1953) was the first to make such estimates, and he concluded that the Florissant mean annual temperature was >18 °C and that the paleoelevation was ~300–900 m. However, more recent estimates by workers such as Meyer, Gregory, and Wolfe (data summarized in Meyer, 2001) have all proposed cooler temperatures (ranging from ~11–14 °C) and relatively high paleoelevation (ranging from 1900 to 4100 m).

Florissant's position in time near the close of the Eocene, shortly preceding the climatic cooling of the Eocene-Oligocene transition, makes it an important point of reference for under-

standing the nature of this climatic change and the importance of upland biotic communities in understanding the nature of biotic community change during this time. For example, many of the plant genera known from the late Eocene upland community at Florissant appear in the early Oligocene lowland of the Pacific Northwest following the climatic cooling, suggesting that upland floras such as Florissant served as important source areas for plant dispersal (Meyer, 2002).

Late Eocene Fossil Insects

The insects and spiders of Florissant have been of special interest to the paleontologic and entomologic community for over a century. In the past, researchers were concerned primarily with the systematic placement of the fossil insects and their diversity in the Florissant Formation. However, recent research has begun to use the fossil insect record to ask questions about the paleoecology of the lake and surrounding ecosystems as well as the paleoclimate at that time (Moe, 2003).

The insects are preserved in the Florissant Formation shales as impression and/or compression fossils. The paper shales represent layers of volcanic ash that weathered into the lake, along with dead insects and other organic debris from the lake shore. The paper shales have a very fine-grained matrix which allows for detailed preservation of fossil organisms. In some cases, you can see the individual lenses of the compound eye on a fly! Many of the insects are preserved in their entirety; however, it is also common to find individual parts, such as wings and elytra. Color patterns are also commonly preserved on insect wings. (This is because color patterns on insect wings are structural, not pigmented.) However, the actual color is not preserved—just a "black and white" version of the pattern.

Evidence of fossil insects is also preserved as feeding traces on plants. As insects feed on plant tissues, the plants react by producing "scar tissue" where the feeding has occurred. This reaction tissue is much more robust than the normal plant tissue and is preserved as darker areas on a plant fossil. Insects leave distinct patterns of feeding traces depending on how they feed on the plant and what part of the plant they feed on. These feeding traces are of special interest to researchers studying the coevolution of plants and insects (Labandeira, 1997; Smith, 2000). By studying the damage left on plant fossils from insects, researchers are able to broaden their knowledge of how these organisms have interacted and how the patterns of their interactions have changed over time.

Over 1500 species of insects and spiders have been described in the Florissant Formation. This is unusual for a lake deposit and seems to be much more similar to amber deposits, which are known for their high diversity. Samuel Scudder's 1890 monograph, *The Tertiary Insects of North America*, was the first comprehensive publication of the Florissant fossil insects (Scudder, 1890). Many of the insects described from Florissant were placed in modern genera; however, all were described as new, extinct species. Many people expect that the world would have looked drastically different 34 million years ago, but in reality, some organisms such as insects looked very similar to the way they look today. However, many of the insects found in the Florissant Formation no longer inhabit Colorado. One of the most striking examples is the tsetse fly, which has its only known fossil occurrence at Florissant, but today lives only in equatorial Africa.

The climate and ecology of Colorado has changed significantly since the late Eocene. Although many paleobotanists and paleoentomologists have shown that Florissant's paleoclimate was probably warm-temperate (Meyer, 2001), we commonly find tropical to sub-tropical taxa that indicate relicts of a warmer climate. Ancient Lake Florissant preserved a dynamic late Eocene ecosystem that soon was to become subject to major climatic changes during the Eocene-Oligocene transition. Overall global climate cooled during this time due to a variety of factors. One of the major reasons for climate change involved the development of the Circum-Antarctic oceanic current. As this formed, it brought cold ocean water to other parts of the globe, which cooled terrestrial climates.

Some of the more common insects are the winged reproductive ants (Family Formicidae; Fig. 6). These are commonly

Figure 6. Winged reproductive phase of *Protazteca elongata* (Formicidae), one of the common ants at Florissant. Specimen YPM-10004, courtesy of Peabody Museum of Natural History, Yale University. Scale bar is 1 cm.

confused with bees or other members of the Order Hymenoptera because of their large, segmented abdomen. Beetles (Order Coleoptera) are also common at Florissant. Many of the beetles are very small (<5 mm) and can be difficult to see on a piece of shale. Some of the more commonly preserved beetles are the weevils (Family Curculionidae; Fig. 7) and the scarabs (Family Scarabaeidae). Flies (Order Diptera) are also commonly preserved. They can be easily distinguished from other orders of insects because they only have two wings. The Florissant Formation is also known for its high diversity of the Lepidoptera, or moths and butterflies (Emmel et al., 1992; Fig. 8). Although these are not commonly found in the fossil record, they are easily distinguished by their wing shape and venation as well as their antennae.

Late Eocene Fossil Vertebrates

The most common vertebrate fossils from Florissant are fish, and four genera are known: *Amia* (bowfin), *Amyzon* (sucker), *Ictalurus* (catfish), and *Trichophanes* (pirate perch). All of these were originally described by E.D. Cope during the 1870s, when he obtained material from Florissant during the time that he was studying the dinosaurs from the area near Cañon City. About five kinds of birds are known, all of them very rare, including *Eocuculus* (cuckoo) and *Palaeospiza* (roller). Only one mammal has been collected from the shale units (*Herpetotherium*, a small opossum), although several others have been found rarely from the lower mudstone unit. These include *Mesohippus* (small horse), *Merycoidodon* (oreodont), and a brontothere. More recent work by Marie Worley indicates the presence of small mammals as well, including rodents and lagomorphs.

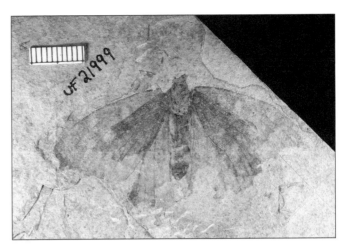

Figure 8. *Vanessa amerindica*, a butterfly in the family Nymphalidae. Specimen UF-21999/FLFO-108. Scale bar is 1 cm.

Figure 7. "*Crptorhynchus*" *fallii* (family Curculionidae), a small weevil that measures only 5 mm in length. Specimen USNM-90639, courtesy of the National Museum of Natural History. Increments of scale bar are 1 mm.

Quaternary Fossil Vertebrates

Fossils from the overlying Quaternary sediments are rare at Florissant. New fragmentary material from a mandible and molar tooth of *Mammuthus* has been recovered along the road leading into the visitor center for the Florissant Fossil Beds National Monument. The mammoth occurs in Pleistocene gravels, which form a thick alluvial fill. The tooth has been radiocarbon dated on purified collagen at 49,830 ± 3290 yr (CAMS-22182). Even though this date exceeds the reliable range for radiocarbon dating, it indicates that mammoths were at high elevations before the last glacial maximum (ca. 18 ka). The thickness of the enamel and lamellar frequency of the tooth suggest that the specimen should be referred to *M. columbi* rather than *M. primigenius*, which was adapted to tundra conditions farther north.

FIELD TRIP ROAD LOG

Stop 1: Florissant Fossil Beds National Monument—Barksdale Picnic Area

From the intersection of the road to the Florissant Fossil Beds visitor center and Teller County Road 1, proceed south 0.9 mi (1.5 km) to the junction with Lower Twin Rock Road and turn left. Continue for 1.6 mi (2.5 km) and turn left into the Barksdale picnic area.

The rock outcrop around the parking area (Fig. 9) consists of the Wall Mountain Tuff. This welded tuff is dated as 36.73 Ma, predating the fossil-rich Florissant Formation by 2.6 m.y. The Wall Mountain Tuff represents a simple cooling unit, and it originated

from a caldera eruption located in the vicinity of the modern Sawatch Range near Mount Princeton, more than 50 mi (80 km) west of here. This was an instantaneous, catastrophic eruption that extended as far as Castle Rock, 40 mi (65 km) northeast of Florissant. Florissant was about midway between the source and the most distant extent of this eruption. The total extent of the Wall Mountain Tuff covered at least 4000 square miles (10,400 km^2).

Depressurization during the caldera eruption resulted in a rapid and explosive eruption in which pieces of pumice, ash, volcanic glass, rock fragments, and mineral crystals were suspended and buoyed in gases as a superheated (500–1000 °C) pyroclastic flow that probably moved at speeds of 50–100 mph (80–160 kph). Within about an hour or two, this flow traveled nearly 100 mi (160 km) from its source near Mount Princeton to the area near Castle Rock. It generally tended to follow the courses of the stream valleys, as it did here along the Florissant paleovalley, where it rests unconformably on the late Eocene erosional surface cut into the Pikes Peak Granite. As it came to rest, the individual components of the gaseous flow were densely welded. The tuff is chemically a low silica, calc-alkali rhyolite, and mineralogically it is a phenotrachyte (Shannon et al., 1987). On close examination, individual mineral crystals (such as sanidine, plagioclase, biotite, clinopyroxene, and hornblende) and volcanic rock fragments occur within a finer groundmass. Accessory minerals include orthopyroxene, magnetite, sphene, allanite, apatite, and zircon. Much of the Wall Mountain Tuff was eroded and only remnants exist along the valley margins.

Stop 2: Overlook of the Thirtynine Mile Volcanic Field

From the intersection of the road to the Florissant Fossil Beds visitor center and Teller County Road 1, proceed 6.7 mi (10.8 km) south (or, if returning to Teller County Road 1 from Stop 1 above, proceed south for 5.8 mi or 9.4 km) and turn right

Figure 9. Outcrop of Wall Mountain Tuff at the Barksdale picnic area in Florissant Fossil Beds National Monument (Stop 1).

on County Road 11 just before Evergreen Station. Follow that road ~4 mi (6.5 km), until you reach a "Y" in the road. Turn right onto Guffey Rd. and follow it for 2.7 mi (4.4 km) until you see a large turnoff on the left side of the road (just past mile marker 14) at the crest of a hill. Park here, and cross the road. Walk out ~150 yards (140 m) perpendicular to the road (heading north) to the overlook of Thirtynine Mile volcanic field (Fig. 10).

You are standing on the Cripple Creek Granite, a potassium feldspar–rich, fine to medium-grained muscovite-biotite granite that is dated at 1.46 Ga. The finer-grained, bimodal texture and

more angular outcrop patterns distinguish the Cripple Creek Granite from the younger Pikes Peak Granite to the east. Weathered cobbles and boulders of the Cripple Creek Granite formed some of the larger clasts in the lahars from the Guffey volcanic center, and examples will be seen at Stop 3.

By the late Eocene, a widespread erosional surface with low to moderate relief had developed in this region. Although originally thought to have been at a relatively low elevation of ~300–900 m (MacGinitie, 1953), various paleobotanical interpretations based on the fossil flora at Florissant now suggest that this region

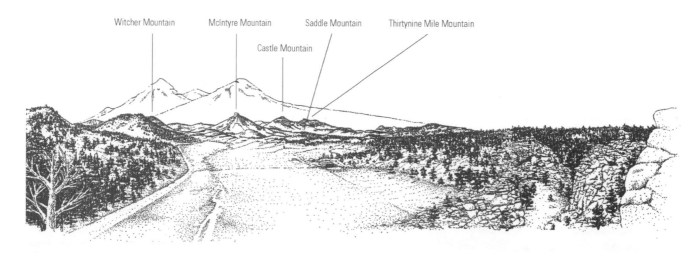

Figure 10. Sketch of the Thirtynine Mile volcanic field as seen from Stop 2. The mountains of the modern landscape are identified by name, and the coalesced stratovolcanoes of the late Eocene Guffey volcanic center are reconstructed in the unshaded profile along the horizon. Original drawing by Peter Gaede. From Henry et al. 2004, used by permission of the Gold Belt Tour Scenic and Historic Byway Association.

may have been at an elevation of 1900–4100 m (Gregory, 1994; Meyer, 1992, 2001; Wolfe et al., 1998). In this area, the erosional surface was developed on the Cripple Creek Granite on which you are standing, and the paleodrainage of Four Mile Creek was incised into this surface. Much of the late Eocene geomorphic surface has been exhumed to form the modern landscape in this area. Scattered outcrops of the Wall Mountain Tuff occur on the top of this surface, and the volcanic field to the west later developed on the surface as well.

Looking to the west from here, you can see the Thirtynine Mile volcanic field (Fig. 10). This includes the Guffey volcanic center, which consists of the deeply eroded remains of the domes and lava flows of a large stratovolcanic complex of late Eocene age. Other mountains of the Thirtynine Mile volcanic field (including Witcher, McIntyre, Castle, and Thirtynine Mile; see Fig. 10) are composed of the stratified lava flows and lahars that formed the flanks of the Guffey volcanic center. The rocks of Witcher Mountain dip south, those of McIntyre Mountain dip east, and those of Thirtynine Mile Mountain dip north; all of these modern mountains form the erosional remnants of the flanks of the late Eocene Guffey volcanic center. The Guffey volcanic center probably consisted of two or more closely coalesced stratovolcanoes, as reconstructed in Figure 10. Evidence for this comes from the presence of multiple vents in the area around Guffey and because the surrounding flanks of the volcanoes are not circular but show an east–west elongation suggesting multiple cones (E. Evanoff, 2004, personal commun.).

The lahars filled the paleovalley that was to the left (south) of the modern Four Mile Creek that you see below. These lahars filled the valley before you to a level that would have extended far above the point on which you are standing. Most of these lahars were later removed by erosion, although remnants can be seen as outcrops along the Four Mile Creek drainage (several of which can be seen between here and Stop 3), and also in the roadcut at Stop 3. When the new drainage was reestablished and eroded through these lahars, it generally followed the course of the original paleodrainage. However, at this point, it encountered the underlying, harder Cripple Creek Granite, formerly an interfluvial ridge that was later covered by the lahars. "Trapped" in the valley that it had already begun carving through the lahars, the stream was forced to continue eroding its channel through this short section of the granite, forming the notch canyon that you see below immediately north of this viewpoint. This paleogeomorphic feature is an excellent example of a *superposed stream* (i.e., one that has a present course which was established on young rocks burying an old surface, with this course being maintained as the stream cut down).

Stop 3: Evergreen Station Roadcut

Return along the route from Stop 2 for 6.7 mi (10.8 km) until you reach the intersection of Teller County Road 1 and High Park Road. Turn right on Teller County Road 1 and park at the Evergreen Station General Store (0.1 mi on the left). Walk ~100 yards (90 m) south to the roadcut. This is a dangerous area for pedestrians. Be very cautious and watch for traffic!

This roadcut exposes a lahar (Fig. 11) that is mapped as the lower member of the Thirtynine Mile Andesite (Wobus and Epis, 1978; Scott et al., 1978). This lahar originated from the Guffey volcanic center in the Thirtynine Mile volcanic field, flowed rapidly eastward down the paleodrainage of Four Mile Creek (along the route that we have just driven from Stop 2), and in this area

Figure 11. Outcrop of lahar at Evergreen Station (Stop 3).

dammed the southward-flowing Florissant paleodrainage, forming Lake Florissant. This lake probably averaged about a mile wide (1.6 km) and extended ~12 mi (19 km) to the north. It is likely that multiple lahars influenced the characteristics of this lake through time and that the lake actually may have had at least two separate generations in its existence, with an alluvial valley developing in the interim. The existence of the lakes ended as they were infilled with volcanic tephra and other sediments and/ or by erosional breaching of the lahar dam.

This outcrop is an erosional remnant of what were once much more voluminous lahar deposits in this vicinity and to the west of here. Notice the heterolithic composition and the extremely poor sorting of clasts in this outcrop of the lahar. The predominant clasts are andesite from the Thirtynine Mile volcanic field and cobbles and boulders of Cripple Creek Granite (coarse-grained with reddish-orange potassium feldspar). Most of the volcanic clasts are angular, but some of the more rounded cobbles (such as those of the Cripple Creek Granite) may have been picked up along the stream channel in the paleovalley that the lahar followed.

The location of this stop is near the contact between the Pikes Peak Granite (1.08 Ga), which lies to the east of this site, and the much older Cripple Creek Granite (1.46 Ga), which lies to the west. By looking to the north, you can see the difference in the outcrop patterns of these two rocks. The Cripple Creek Granite forms more angular, blocky exposures, whereas the Pikes Peak Granite forms more rounded outcrops including exfoliation domes (such as Dome Rock and Sheep Rock just to the east of this stop). Notice that the lahar exposed in the roadcut lacks clasts of Pikes Peak Granite. This is because the source of the lahar was to the west of here, and the Pikes Peak Granite crops out to the east.

Stop 4: Florissant Fossil Beds National Monument— Visitor Center and Petrified Forest

The entrance road to the Florissant Fossil Beds National Monument visitor center is located 2.5 mi (4.0 km) south of the town of Florissant. As you drive along the entry road to the visitor center, the road passes through a roadcut in Pleistocene alluvial terrace gravels from which a fossil mammoth tooth and bone fragments were collected in 1994. This specimen was radiocarbon dated at 49,830 ± 3290 (CAMS-22182), the upper limit of radiocarbon dating.

From the parking area, Pikes Peak can be seen to the east. Much of the area to the east of here, including Pikes Peak, consists of a large batholith of the 1.08 Ga Pikes Peak Granite. Many of the rounded, rocky hills around the Florissant vicinity are composed of Pikes Peak Granite. To the north, conical Crystal Peak is part of a smaller pluton associated with the emplacement of the Pikes Peak Granite, and this area has produced many of the world-class specimens of the feldspar amazonite.

The primary purpose of this stop is to examine the outcrops and fossils of the late Eocene Florissant Formation, isotopically dated at 34.07 Ma. This field guide will highlight the major points; additional publications elaborating on Florissant's paleontology can be purchased in the visitor center. Most of the fossils from Florissant are plants and insects in shale, and examples of these are on exhibit in the visitor center. Large petrified trees can be seen along the trails. All fossils within the boundaries of the National Monument are protected, and collecting is strictly prohibited. Following this stop, you will be able to collect at the privately owned Florissant Fossil Quarry located just north of the monument.

The tremendous diversity of fossil plants and insects, including ~1700 described species, has come from the three shale units within the Florissant Formation (Fig. 2). The vast majority of these fossils are in the research collections at ~20 major museums from Berkeley to London, comprising a total of more than 40,000 specimens. The display in the visitor center includes typical examples of these impression and compression fossils. Most of the specimens currently on exhibit are on loan from Waynesburg College, and despite the size of museum collections worldwide, this is the largest public exhibit of Florissant fossils. In addition to these delicate shale fossils, the Florissant area is also well known for its petrified forest. Large in situ stumps of these petrified trees, mostly of *Sequoia* (redwood), can be seen around the visitor center. Several of the most striking stumps are located immediately behind the visitor center, and others can be seen along trails that lead from here around two different loop hikes, including the Petrified Forest Loop trail.

The recommended walk proceeds from the visitor center to the stumps that are covered by the large shelters and then along the Petrified Forest Loop to the Big Stump. From the Big Stump, you can either follow the loop around, or shorten the walk by backtracking to the visitor center. We will spend two hours here. Along this walk, you will first come to three large petrified stumps behind the visitor center. These were redwood trees similar to those that are now endemic to coastal California. The Redwood Trio (under the larger of the two shelters) is the most unique of these, and represents a rare example of three trunks interconnected at their base (Fig. 12). The trio developed by a form of reproduction in which three trees sprouted around an original parent plant and later grew to maturity after the parent plant died. This type of growth form is sometimes referred to as a "family circle," and similar examples can be seen in the modern redwood forests. The process of fossilization began when the trees were buried by a tuffaceous lahar from the Guffey volcanic center. This lahar flowed off the flanks of the Guffey volcano, which by this time had probably buried much of the older landscape to the west of here under lahars and lava flows, and into the paleovalley where the redwoods grew. This lahar is ~15 ft (5 m) thick, and it forms the upper part of the lower mudstone unit of the Florissant Formation (Fig. 2). The trees were permineralized when groundwater, rich in dissolved minerals, permeated the wood and precipitated silica within the cell walls of the wood. Only the lower portions of the trees were buried by the lahars, and thus only the bases (stumps) were preserved.

Figure 12. The *Sequoia* Redwood Trio near the visitor center at Florissant Fossil Beds National Monument (Stop 4).

Proceed out of the second shelter and along the concrete path, which becomes the Petrified Forest Loop trail, and proceed north for ~200 ft (60 m) to a trail junction. Turn left and walk along this trail for ~60 ft (20 m) to the rock outcrops behind the rail on the right. The lower part of this outcrop exposes the middle shale unit of the Florissant Formation (Fig. 2), which was deposited in a lacustrine setting and is rich in plant and insect impression fossils. This lake formed as the ancient stream drainage was dammed by a lahar ~5 mi south of here (as we saw at Stop 3), and the basin north of this dam filled with water. This created a lake that covered the mudflow lahars containing the petrified forest, burying the remains of the forest at the bottom of the newly formed lake. The Guffey volcanic center erupted episodically throughout the existence of the lake, providing sediments either as tephra that fell directly into the lake, or on the surrounding landscape from which epiclastic sediments washed into the lake. The shales split into paper-like sheets (paper shale) consisting of thin layers of clay (derived from weathered volcanic ash) alternating with thin layers of diatoms. These alternating clay-diatom couplets represent slow deposition in the lake, and it is these layers of paper shale that contain the fossils. Interbedded with the shales are thicker, coarser layers of volcanic ash and pumice that were rapidly deposited into the lake during more voluminous eruptions from the Guffey volcano. The 6-ft-thick (2 m) upper layer in this outcrop is the caprock conglomerate (Fig. 2), a hard rock that is resistant to erosion and forms a ledge. The caprock conglomerate is composed of pebbles of andesite and granite, and was deposited as a debris flow that flowed into the lake and rapidly settled on the lake bottom. If you look carefully along the outcrop to the left, you will see vertically aligned structures that have been interpreted as water-escape tubes through which the

water from the debris flow was squeezed out as the debris flow came to rest and was compacted on the lake bottom.

Return to the trail junction and continue straight ahead along the Petrified Forest Loop trail for ~0.3 mi (0.5 km) to the Big Stump, another large petrified redwood tree. This is the type specimen of *Sequoiaoxylon pearsallii* (Andrews, 1936), which is an organ generic name for the wood. MacGinitie (1953) later placed this into synonymy with *Sequoia affinis*, which is the name based on foliage and cones from the shale deposits. The wood has not been found in attachment to these other organs, however. The rock outcrop behind this stump is a good example of the lahar (in the lower mudstone unit) that buried this forest, with the middle shale unit and caprock conglomerate above (Fig. 2). The flared base of the Big Stump is typical of redwoods, and is an evolutionary adaptation to support a very tall tree. In the late 1800s, an attempt was made to remove the Big Stump, piece by piece, and ship it back east by train for exhibit. The deeply embedded and broken saw blades in the stump attest to the failure of that effort.

Although most of the petrified trees at Florissant are redwoods, five different hardwoods also have been identified by Wheeler (2001). The total number of species of wood is small compared to the 140 species of fossil leaves and pollen. This illustrates the fact that there were taphonomic biases, and in this case, only a small portion of the total flora was preserved as wood. Those plants that were preserved as petrified wood grew along the lower elevations of the stream drainage and were buried by the lahar, whereas leaves, fruits, and pollen from plants in the surrounding basin were more easily transported by wind to become deposited in the lake basin. Thus, growth habits, mechanisms of dispersal, and processes of sedimentation are all components of taphonomy.

Figure 13. Outcrop of the lower shale unit of the Florissant Formation exposed at the Florissant Fossil Quarry (Stop 5).

You can continue around the loop trail to see more petrified trees buried at ground level, or to shorten the return distance, backtrack from here to the visitor center. A short side trail just south of the Big Stump leads to one of the historic excavation sites in the lake shales.

Stop 5: Florissant Fossil Quarry

The Florissant Fossil Quarry (Fig. 13) is located ~0.2 mi (0.3 km) south of the town of Florissant on the west side of Teller County Road 1. The site has been operated by the Clare family for ~50 yr. It is the only site from which Florissant fossils can be legally collected by the general public, educational groups, and scientific field trips, although exceptional or unusual fossil specimens are often retained and donated to institutions that assure scientific access. The Clare family, under a formal partnership agreement with Florissant Fossil Beds National Monument, the Denver Museum of Nature & Science, and the Woodland Park schools, has allowed one section of the quarry face to be systematically excavated for scientific and educational purposes. This collection is housed at the Denver Museum of Nature & Science.

According to the stratigraphy of Evanoff et al. (2001), the site represents the lower shale unit (Fig. 2), which is stratigraphically lower than any unit of the Florissant Formation within Florissant Fossil Beds National Monument. The lithology of the lower shale unit is very similar to the middle and upper shale units. Fossil plants and insects can be found by splitting the delicate paper shale using a tool such as a single-sided razor blade or a microspatula. These are paper shales in the classic sense, and each "sheet of paper" represents a couplet (0.1–1.0 mm in thickness) consisting of a microlayer of diatoms and a microlayer of volcanic ash that is weathering to clay. Interbedded between the layers of paper shale are thicker, coarser layers of volcanic mudstones and tuffs.

Fossils of leaves, fruits, insects, and even spiders are common at this site. Common fossils include the plant genera *Cedrelospermum* (an extinct genus in the elm family), *Fagopsis* (an extinct genus in the beech family), *Chamaecyparis* (false cedar), and *Sequoia* (redwood). Common insects include ants, beetles, and craneflies. Although extremely rare, some of the best bird fossils have come from this site.

REFERENCES CITED

Andrews, H.N., 1936, A new *Sequoioxylon* from Florissant, Colorado: Annals of the Missouri Botanical Garden, v. 23, no. 3, p. 439–446, plates 20–21.

Brown, G.H., editor, 1996, Chasing butterflies in the Colorado Rockies with Theodore Mead in 1871: As told through his letters: Florissant, Colorado, Pikes Peak Research Station, Colorado Outdoor Education Center, 73 p.

Cockerell, T.D.A., 1906, Fossil plants from Florissant, Colorado: Bulletin of the Torrey Botanical Club, v. 33, p. 307–312.

Cockerell, T.D.A., 1908a, Fossil insects from Florissant, Colorado: Bulletin of the American Museum of Natural History, v. 24, art. 3, p. 59–69.

Cockerell, T.D.A., 1908b, The fossil flora of Florissant, Colorado: Bulletin American Museum of Natural History, v. 24, art. 4, p. 71–110.

Cockerell, T.D.A., 1908c, Florissant: A Miocene Pompeii: The Popular Science Monthly, v. 74, no. 2, p. 112–126.

Emmel, T.C., Minno, M.C., and Drummond, B.A., 1992, Fossil butterflies: A guide to the fossil and present-day species of central Colorado: Stanford, California, Stanford University Press, 118 p.

Epis, R.C., and Chapin, C.E., 1974, Stratigraphic nomenclature of the Thirty-nine mile volcanic field, central Colorado: U.S. Geological Survey Bulletin 1395-C, p. C1–C23.

Epis, R.C., and Chapin, C.E., 1975, Geomorphic and tectonic implications of the post-Laramide, late Eocene erosion surface in the southern Rocky Mountains, *in* Curtis, B.F., ed., Cenozoic history of the southern Rocky Mountains: Geological Society of America Memoir 144, p. 45–74.

Evanoff, E., 1992, Surficial geology of Florissant Fossil Beds National Monument, Colorado: Unpublished geologic map prepared for the National Park Service, scale 1:10,000, 1 sheet.

Evanoff, E., and Murphey, P.C., 1994, Rock units at Florissant Fossil Beds National Monument, *in* Evanoff, E., ed., Guidebook for the field trip: Late Paleogene geology and paleoenvironments of central Colorado with emphasis on the geology and paleontology of Florissant Fossil Beds National Monument (1994): Unpublished guidebook for Geological Society of America field trip, p. 40–44.

Evanoff, E., McIntosh, W.C., and Murphey, P.C., 2001, Stratigraphic summary and ^{40}Ar/^{39}Ar geochronology of the Florissant Formation, Colorado, *in* Evanoff, E., Gregory-Wodzicki, K.M., and Johnson K.R., eds., Fossil flora and stratigraphy of the Florissant Formation, Colorado: Proceedings of the Denver Museum of Nature & Science, ser. 4, no. 1, p. 1–16.

Gregory, K.M., 1994, Paleoclimate and paleoelevation of the 35 Ma Florissant flora, Front Range, Colorado: Palaeoclimates, v. 1, p. 23–57.

Gregory-Wodzicki, K.M., 2001, Paleoclimatic implications of tree-ring growth characteristics of 34.1 Ma Sequoiaoxylon pearsallii from Florissant, Colorado, *in* Evanoff, E., Gregory-Wodzicki, K.M., and Johnson, K.R., eds., Fossil flora and stratigraphy of the Florissant Formation, Colorado: Proceedings of the Denver Museum of Nature & Science, ser. 4, no. 1, p. 163–186.

Harding, I.C., and Chant, L.S., 2000, Self-sedimented diatom mats as agents of exceptional fossil preservation in the Oligocene Florissant lake beds, Colorado, United States: Geology, v. 28, p. 195–198, doi: 10.1130/0091-7613(2000)0282.3.CO;2.

Henry, T.W., Evanoff, E., Grenard, D., Meyer, H.W., and Pontius, J.A., 1996, Geology of the Gold Belt Back Country Byway, south-central Colorado, *in* Thompson, R.A., Hudson, M.R., and Pillmore, C.L., eds., Geologic excursions to the Rocky Mountains and beyond: Field Trip Guidebook for the 1996 Annual Meeting of the Geological Society of America, Denver, Colorado, October 28–31: Colorado Geological Survey Special Publication 44, 48 p. on CD-ROM.

Henry, T.W., Evanoff, E., Grenard, D.G., Meyer, H.W., and Vardiman, D.M., 2004, Geologic guidebook to the Gold Belt Byway, Colorado: Cañon City, Colorado, Gold Belt Tour Scenic and Historic Byway Association, 112 p.

Journal of the Paleontologists of the Princeton Scientific Expedition, 1877, Princeton Scientific Expeditions Collection, box 1, folder 7, Princeton University Archives. Used by permission of the Princeton University Library.

Labandeira, C.C., 1997, Insect mouthparts: Ascertaining the paleobiology of insect feeding strategies: Annual Review of Ecology and Systematics, v. 28, p. 153–193, doi: 10.1146/ANNUREV.ECOLSYS.28.1.153.

Leopold, E.B., and Clay-Poole, S.T., 2001, Fossil leaf and pollen floras of Colorado compared: climatic implications, *in* Evanoff, E., Gregory-Wodzicki, K.M., and Johnson, K.R., eds., Fossil flora and stratigraphy of the Florissant Formation, Colorado: Proceedings of the Denver Museum of Nature & Science, ser. 4, no. 1, p. 17–69.

Lesquereux, L., 1873, Lignitic formation and fossil flora: Annual report of the U.S: Geological Survey of the Territories, v. 6, p. 317–427.

Lesquereux, L., 1878, Contributions to the fossil flora of the Western Territories, pt. II, The Tertiary flora: Report of the U.S. Geological Survey of the Territories, v. 7, p. 1–366, 65 plates.

Lesquereux, L., 1883, Contribution to the fossil flora of the Western Territories, pt. III: The Cretaceous and Tertiary floras: U.S. Geological Survey of the Territories Report 8, p. 1–283, 60 plates.

MacGinitie, H.D., 1953, Fossil plants of the Florissant Beds, Colorado: Carnegie Institution of Washington Publication 599, p. 1–198, plates 1–75.

Manchester, S.R., 1989, Attached reproductive and vegetative remains of the extinct American-European genus *Cedrelospermum* (Ulmaceae) from the early Tertiary of Utah and Colorado: American Journal of Botany, v. 76, no. 2, p. 256–276.

Manchester, S.R., 1992, Flowers, fruits, and pollen of *Florissantia*, an extinct malvaean genus from the Eocene and Oligocene of western North America: American Journal of Botany, v. 79, no. 9, p. 996–1008.

Manchester, S.R., 1999, Biogeographical relationships of North American Tertiary floras: Annals of the Missouri Botanical Garden, v. 86, p. 472–522.

Manchester, S.R., 2001, Update on the megafossil flora of Florissant, Colorado, *in* Evanoff, E., Gregory-Wodzicki, K.M., and Johnson, K.R., eds., Fossil flora and stratigraphy of the Florissant Formation, Colorado: Proceedings of the Denver Museum of Nature & Science, ser. 4, no. 1, p. 137–161.

Manchester, S.R., and Crane, P.R., 1983, Attached leaves, inflorescences, and fruits of *Fagopsis*, an extinct genus of fagaceous affinity from the Oligocene Florissant flora of Colorado, U.S.: American Journal of Botany, v. 70, no. 8, p. 1147–1164.

Manchester, S.R., and Crane, P.R., 1987, A new genus of Betulaceae from the Oligocene of western North America: Botanical Gazette (Chicago, Ill.), v. 148, no. 2, p. 263–273, doi: 10.1086/337654.

Meyer, H.W., 1992, Lapse rates and other variables applied to estimating paleoaltitudes from fossil floras: Paleogeography, Paleoclimatology, Paleoecology, v. 99, p. 71–99, doi: 10.1016/0031-0182(92)90008-S.

Meyer, H.W., 2001, A review of the paleoelevation estimates from the Florissant flora, Colorado, *in* Evanoff, E., Gregory-Wodzicki, K.M., and Johnson, K.R., eds., Fossil flora and stratigraphy of the Florissant Formation, Colorado: Proceedings of the Denver Museum of Nature & Science, ser. 4, no. 1., p. 205–216.

Meyer, H.W., 2002, Significance of Eocene-Oligocene floras of the Pacific Coast to models of paleoelevation and plant community dynamics: Geological Society of America Abstracts with Programs, v. 34, no. 5, p. A9.

Meyer, H.W., 2003, The fossils of Florissant: Washington, D.C., Smithsonian Books, 258 p.

Moe, A.P., 2003, Using fossil Diptera from the Florissant Formation as paleoenvironmental indicators [Master's thesis]: Boulder, Colorado, University of Colorado at Boulder, 45 p.

Murphey, P.C., 1992, Stratigraphy of Florissant Fossil Beds National Monument, *in* Evanoff, E., and Doi, K., eds., The stratigraphy and paleontology of Florissant Fossil Beds National Monument, A progress report: Unpublished report submitted to Florissant Fossil Beds National Monument.

O'Brien, N.R., Meyer, H.W., Reilly, K., Ross, A.M., and Maguire, S., 2002, Microbial taphonomic processes in the fossilization of insects and plants in the late Eocene Florissant Formation, Colorado: Rocky Mountain Geology, v. 17, p. 1–11.

Peale, A.C., 1874, Report of A.C. Peale, M.D., geologist of the South Park division: Annual report of the U.S. Geological and Geographical Survey of the Territories, v. 7, p.193–273, 20 plates.

Scott, G.R., Taylor, R.B., and Epis, R.C., and Wobus, R.A., 1978, Geologic map of the Pueblo 1° × 2° quadrangle, south-central Colorado: U.S. Geological Survey Miscellaneous Investigation Series Map I-1022, scale 1: 250,000, 2 sheets.

Scott, W.B., 1939, Some memories of a paleontologist: Princeton, New Jersey, Princeton University Press, 336 p.

Scudder, S.H., 1878, An account of some insects from the Tertiary rocks of Colorado and Wyoming: Bulletin of the U.S. Geological and Geographical Survey of the Territories, v. 4, p. 519–543.

Scudder, S.H., 1890, The Tertiary insects of North America: Report of the U.S. Geological Survey of the Territories, v. 13, p. 1–734, 28 plates.

Scudder, S.H., 1893, Tertiary Rhynchophorus Coleoptera of the United States: Monographs of the U.S. Geological Survey, v. 21, p. 1–206.

Scudder, S.H., 1900, Adephagous and clavicorn Coleoptera from the Tertiary deposits at Florissant, Colorado, with descriptions of a few other forms and a systematic list of the non-rhynchophorous Tertiary Coleoptera of North America: Monographs of the U.S. Geological Survey, v. 40, p. 1–148.

Shannon, J.R., Epis, R.C., Naeser, C.W., and Obradovich, J.D., 1987, Correlation of intracauldron and outflow and an intrusive tuff dike related to the Oligocene Mount Aetna Cauldron, central Colorado, *in* Drexler, J.W., and Larson, E.E., eds., Cenozoic volcanism in the southern Rocky Mountains updated, a tribute to Rudy Epis, part 1: Colorado School of Mines Quarterly, v. 82, no. 4, p. 65–80.

Smith, D.M., 2000, The evolution of plant-insect interactions: Insights from the fossil record [Ph.D. dissertation]: Tucson, Arizona, University of Arizona, 316 p.

Veatch, S.W., 2003, Princeton scientific expedition of 1877: Florissant Colorado segment: Geological Society of America Abstracts with Programs, v. 35, no. 5, p. 16.

Weber, W.A., editor, 2000, The American Cockerell: A naturalist's life, 1886–1948: Boulder, Colorado, University Press of Colorado, 352 p.

Wheeler, E.A., 2001, Fossil dicotyledonous woods from Florissant Fossil Beds National Monument, Colorado, *in* Evanoff, E., Gregory-Wodzicki, K.M., and Johnson, K.R., eds., Fossil flora and stratigraphy of the Florissant Formation, Colorado: Proceedings of the Denver Museum of Nature & Science, ser. 4, no. 1, p. 187–203.

Wingate, F.H., and Nichols, D.J., 2001, Palynology of the uppermost Eocene lacustrine deposits at Florissant Fossil Beds National Monument, Colorado, *in* Evanoff, E., Gregory-Wodzicki, K.M., and Johnson, K.R., eds., Fossil flora and stratigraphy of the Florissant Formation, Colorado: Proceedings of the Denver Museum of Nature & Science, ser. 4, no. 1, p. 71–135.

Wobus, R.A., 2001, Precambrian igneous and metamorphic rocks in the Florissant region, central Colorado: Their topographic influence, past and present: Pikes Peak Research Station Bulletin, v. 5, p. 1–9.

Wobus, R.A., and Epis, R.C., 1978, Geologic map of the Florissant 15-minute quadrangle, Park and Teller Counties, Colorado: U.S. Geological Survey Miscellaneous Investigations Series Map I-1044. scale 1:62,500, 1 sheet.

Wolfe, J.A., Forest, C.E., and Molnar, P., 1998, Paleobotanical evidence of Eocene and Oligocene paleoaltitudes in midlatitude western North America: Geological Society of America Bulletin, v. 110, p. 664–678, doi: 10.1130/0016-7606(1998)1102.3.CO;2.

Geological Society of America
Field Guide 5
2004

Paleoceanographic events and faunal crises recorded in the Upper Cambrian and Lower Ordovician of west Texas and southern New Mexico

John F. Taylor

Geoscience Department, Indiana University of Pennsylvania, Indiana, Pennsylvania 15705, USA

Paul M. Myrow

Department of Geology, The Colorado College, Colorado Springs, Colorado 80903, USA

Robert L. Ripperdan

Department of Earth & Atmospheric Sciences, Saint Louis University, St. Louis, Missouri 63103, USA

James D. Loch

Department of Biology & Earth Science, Central Missouri State, Warrensburg, Missouri 64093, USA

Raymond L. Ethington

Department of Geological Sciences, University of Missouri, Columbia, Missouri 65211, USA

ABSTRACT

A revised lithostratigraphy for Lower Paleozoic strata in New Mexico and west Texas was developed through detailed sedimentological study of the Bliss and Hitt Canyon Formations within a refined temporal framework assembled from precise biostratigraphic (trilobite and conodont) and chemostratigraphic (carbon isotope) data. Member boundaries within the Hitt Canyon now correspond with mappable and essentially isochronous horizons that represent major depositional events that affected sedimentation in basins throughout Laurentian North America. This trip is designed to examine these and other important intervals, such as the extinction horizons at the base and top of the Skullrockian Stage, and to demonstrate the utility of associated faunas and isotopic excursions for correlation within and beyond the region.

Keywords: Cambrian, Ordovician, stratigraphy, New Mexico, Texas.

INTRODUCTION

The lowermost Paleozoic rocks in southern New Mexico and west Texas were deposited when a rugged Precambrian topography was inundated by the Sauk Transgression. They comprise a basal mixed clastic-carbonate package of highly vari-

able thickness and lithology, the Bliss Formation, overlain by a thick succession of carbonates known collectively as the El Paso Group. Over the past 50 years, a bewildering number of names has been assigned to the numerous formations and members recognized within the El Paso Group, as shown in Figure 1. For a more complete and detailed summary of the nomenclatural

Taylor, J.F., Myrow, P.M., Ripperdan, R.L., Loch, J.D., and Ethington, R.L., Paleoceanographic events and faunal crises recorded in the Upper Cambrian and Lower Ordovician of west Texas and southern New Mexico, *in* Nelson, E.P. and Erslev, E.A., eds., Field Trips in the Southern Rocky Mountains, USA: Geological Society of America Field Guide 5, p. 167–183. For permission to copy, contact editing@geosociety.org. © 2004 Geological Society of America history of formations and members, see Clemons (1991). As noted by LeMone (1996), the large number of names assigned to subunits of the El Paso Group belies a fairly consistent physical stratigraphy within the Lower Ordovician carbonates across the region. This situation is comparable to that recently described in western Colorado where deposition of the mixed clastics and carbonates of the Sawatch Formation leveled the submarine surface, allowing for subsequent deposition of laterally persistent packages within the overlying Dotsero Formation (Myrow et al., 2003). As in Colorado, minor miscorrelation of key intervals within the El Paso Group has led to confusion and misinterpretation of the depositional history, with significant implications for the reconstruction of regional paleogeography.

On this field trip, we will visit measured sections in three ranges to examine the considerable lithologic variability of the Bliss Formation and, in contrast, the relatively consistent succession of lithofacies through the Hitt Canyon Formation, the lowest formation within the El Paso Group. Like Clemons (1991, 1998), we have adopted with minor revision the lithostratigraphy proposed by Hayes (1975), who defined the Hitt Canyon Formation; however, we reject Clemons' argument that the Lower Ordovician formations defined by Hayes are too thin to be mapped at a scale of 1:24000, and treat the El Paso as a group with three formations. Unlike previous investigators, we had the benefit of working within a highly refined temporal framework constructed through integration of abundant biostratigraphic and chemostratigraphic data. Numerous new macro- and microfossil collections and detailed carbon isotope ratio (δ^{13}C) profiles from 11 measured sections (Fig. 2), all measured and sampled with

centimeter-scale precision, greatly improved the precision and accuracy of correlation possible across southern New Mexico into west Texas. The expanded data set revealed inconsistencies in the recognition of units in Texas, particularly near the top of the Hitt Canyon Formation. The refined temporal framework more tightly constrains the age of several horizons that represent significant events in the depositional history of the area, such as deepening, shallowing, and the initiation or suppression of microbial reef growth. Several of the formation and member boundaries in our revised lithostratigraphy (Fig. 1, left column) have been repositioned slightly to coincide with such horizons, rendering the formations no less suitable for mapping but considerably more reliable for derivation of the depositional history and for correlation within and beyond the region. We will examine these sedimentological event horizons and compare their position with horizons of faunal turnover to evaluate hypotheses that link the extinction of invertebrate faunas at those biozonal/stadial boundaries to Late Cambrian and Early Ordovician eustatic events. In particular, we will examine the extinction horizons that define the base and top of the Skullrockian Stage: the base of the *Eurekia apopsis* trilobite Subzone (= base of the *Cordylodus proavus* conodont Zone) and the base of the *Leiostegium-Kainella* trilobite Zone, respectively (Fig. 3). Intensive sampling has constrained the position of those stadial boundaries in some sections to less than a meter, making it possible to establish whether there is associated physical evidence of a change in sea level (lithofacies boundary) or ocean chemistry (δ^{13}C excursion).

The δ^{13}C profiles generated from the El Paso Group (Fig. 4) are the most detailed Stairsian-aged results obtained from

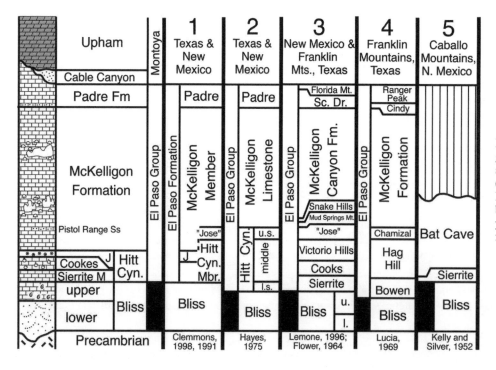

Figure 1. Lower Paleozoic lithostratigraphic units recognized in the present study (left column) and previous studies (columns 1–5) in west Texas and New Mexico. True Jose Member represented by J in the figure; "Jose" refers to younger interval previously misidentified as its equivalent in Texas.

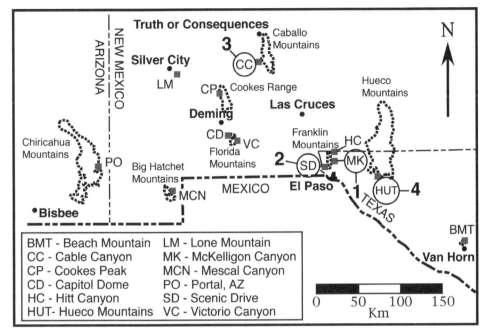

Figure 2. Location map showing sections included in the present study. Field trip stops are circled and labeled with stop number (1–4).

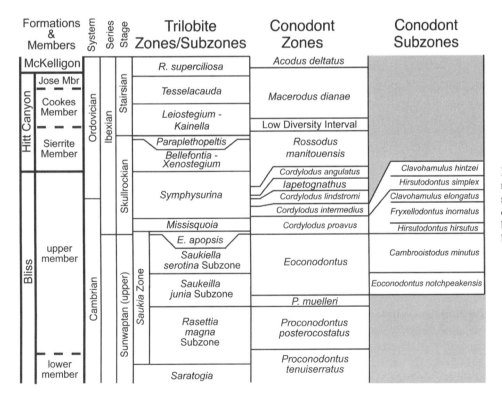

Figure 3. Biostratigraphic and chronostratigraphic units represented in the study interval. Columns with bold lines on the left show position/age of member boundaries where best constrained by biostratigraphic data in New Mexico.

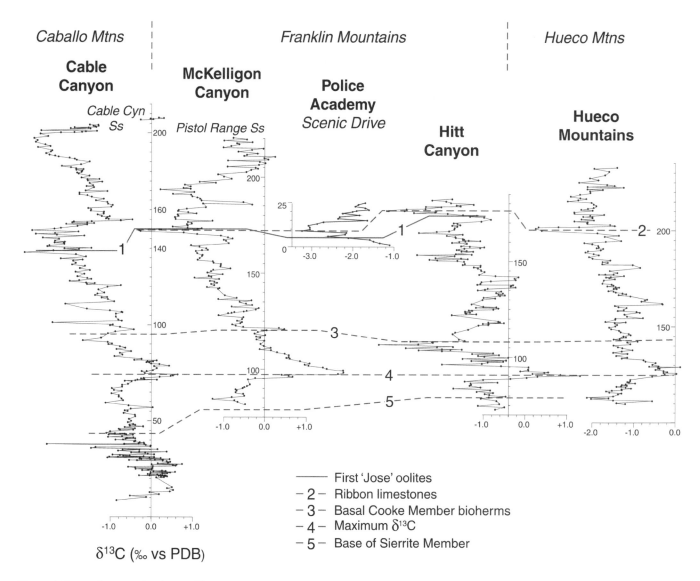

Figure 4. Fence diagram of carbonate δ¹³C profiles across southern New Mexico into west Texas. Datum is the basal Stairsian positive excursion.

Laurentian rocks. Profiles from the four major sections are bracketed by two prominent excursions of >2‰, with ~1‰ variations occurring at several intervening levels. A thin 2‰ positive excursion spans the base of the Stairsian Stage. Smaller-scale variation below this prominent feature provides extremely precise chronostratigraphic correlation of the base of the Sierrite Member in all 4 sections. A second, strongly negative excursion coincides precisely with the base of the Jose Member at the top of the Hitt Canyon Formation—a horizon that we interpret as representing a submergence event. This event caps a long-term trend toward more negative values that is especially prominent in the Cable Canyon section. δ¹³C profiles from McKelligon Canyon, Texas, and Cable Canyon, New Mexico, indicate additional δ¹³C variation above the basal Jose event and suggest that high-resolution

geochemical correlation of the regressive Pistol Range Member of the McKelligon Formation to areas outside the Franklin Mountains might eventually be possible. These δ¹³C excursions already have proven extremely useful for determining the position of the base of the Stairsian Stage and the level representing the Jose submergence event in west Texas, where the scarcity of macrofossils and pervasive dolomitization of critical intervals makes precise correlation within that interval very difficult.

DAY 1—SOUTHERN FRANKLIN MOUNTAINS, TEXAS

Follow McKelligon Canyon Drive west from its intersection with Alabama Street into McKelligon Canyon, and park at the westernmost picnic tables at the head of the canyon. Hike a short distance southwest to a gully where the contact between granitic basement and overlying sandstone is well exposed. This is the base of the section, which follows the gully to the left and ascends the spur at the head of the gully.

Note: Stops on Day 1 are within Franklin Mountains State Park, and collecting without a Texas State Scientific Collecting Permit is prohibited.

Stop 1. McKelligon Canyon

Bliss Formation

The pink granite, nonconformably overlain by quartzitic sandstone at the base of the Bliss, is the Red Bluff Granite, the youngest Proterozoic formation in the Franklin Mountains. For a thorough and useful summary of the units within the Precambrian in the Franklin Mountains, see LeMone (1988). In a few places, presumably the highest points on the inherited Precambrian topography, Lower Ordovician carbonates rest directly on basement, but in most sections, basal Bliss strata consist of sandstone and conglomerate. We divide the Bliss Formation into two informal members: a lower one dominated by quartz-rich sandstone and strongly hematitic lithologies and an upper member with abundant glauconite and significantly more carbonate. The hematitic portions of the lower member in some areas of New Mexico include ore-grade oolitic hematite. The boundary between the members is sharp and likely represents a significant transgression in each section, although perhaps not the same event in all sections. The age of the lower Bliss is poorly established in most sections, but the sparse biostratigraphic data available suggest that the base of the upper Bliss in New Mexico is significantly older than the base of the upper member in Texas.

Hayes (1975) designated this locality the type section for the Bliss Formation. A lithologic column is provided as Figure 5. Here at McKelligon Canyon, the basal 4.67 m comprises a coarse-grained sandstone facies that formed in very shallow water during the initial transgression of the eroded Precambrian surface. It includes massive beds that formed by rapid deposition of coarse sediment from suspension and horizontally laminated beds that record deposition at velocities associated with upper plane bed conditions. The appearance slightly higher of glauconitic, bioturbated sandstone records deepening to where the reduced wave and flow energy allowed the infauna to more completely churn the sediment after deposition. At 42.31 m, a trough cross-bedded sandstone facies, created by deposition from ripples and small dunes that probably formed under unidirectional currents, appears and continues up to the contact with the overlying Sierrite Limestone at 77.64 m. A zone of red-stained (hematitic) beds representing all lithofacies of the lower Bliss occurs from 56.39 to 61.04 m. This interval might correlate with a hematite-rich interval from 84.7 to 92.35 m in the Hueco Mountain section. The top of the hematitic interval, marked by a pebble lag at McKelligon Canyon and by appearance of dense concentrations of glauconite in the Hueco Mountains, is used to define the base of the upper Bliss in both sections.

El Paso Group

Hitt Canyon Formation—Sierrite Member. We will visit the type section of the Sierrite Limestone of Kelly and Silver (1952) at Cable Canyon in the Caballo Mountains, New Mexico (Stop 3). There the base of the unit is marked by the near disappearance of glauconite and other noncarbonate components whose abundance in the underlying Bliss Formation makes it less resistant to erosion than the cliff-forming purer limestone of the Sierrite. We retain the original definition of the Bliss-Sierrite contact as corresponding with the shift to pure carbonate (and the base of the cliffs in western sections) and place the Bliss–Hitt Canyon contact in McKelligon Canyon somewhat arbitrarily at a sharp contact between very sandy dolomite with prominent trough cross-bedding (below) and purer dolomite (above). This is probably the horizon used by LeMone (1969) as the top of his lower member of the Sierrite Formation. Although the base of the Sierrite in McKelligon Canyon lacks the physiographic expression that characterizes the base of the Sierrite in New Mexico, the contact is easily recognized by a shift to more pure carbonate.

The Sierrite Limestone was originally defined as a formation at the base of the El Paso Group. We amend its definition by considering the Sierrite a member of the Hitt Canyon Formation and by moving the top of the member upward to coincide with the appearance of meter-scale microbial biostromes, a horizon that appears to have greater correlation potential than the top of the cliff-forming limestone selected by Kelly and Silver (1952). Unlike the base of the Sierrite, which becomes younger to the east, this new upper contact displays no evidence of diachroneity. To the contrary, conodont and $\delta^{13}C$ data both suggest that the initiation of microbial reef growth recorded by that horizon was effectively synchronous across and perhaps even beyond the study area. The basal bed of our Sierrite Member in McKelligon Canyon yielded conodonts of the Lower Ordovician *Rossodus manitouensis* Zone. Repetski (1988) reported diagnostic elements of this same zone only 20 m above the base of the Bliss at Scenic Drive (Stop 2), where the formation is over 80 m thick, leaving little room for Cambrian or even lower Skullrockian strata.

Hitt Canyon Formation—Cookes Member. This middle member of the Hitt Canyon Formation contains at least two intervals of microbial reefs and associated grainstone to rudstone. It is roughly equivalent to the Cooks and Victorio Mountains Formations of Flower (1964, 1969, 1968) and LeMone (1969, 1996), which have not been adopted by subsequent workers due to the largely biostratigraphic basis of their definition. The name was selected to acknowledge the considerable contributions of Rousseau Flower and David LeMone to the study of Lower Paleozoic strata in the southwest. It is changed slightly in spelling to avoid confusion with the more restricted interval identified by those authors as the Cooks Formation and to match the spelling used on most modern maps for the Cookes Range north of Deming. The base of the Cookes at McKelligon Canyon is slightly less than 120 m above the base of the section (Fig. 5). Many of the characteristic microbial reefs, particularly those low in the member, have been thoroughly dolomitized at this locality. Some better-

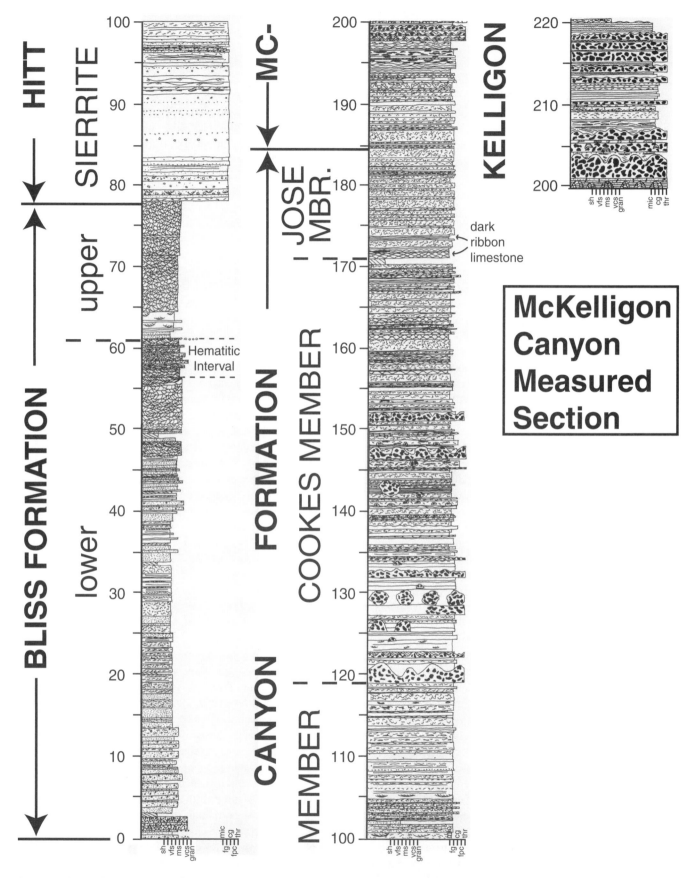

Figure 5. Lithologic column for McKelligon Canyon measured section (Stop 1). Siliciclastic lithologies (inset to left) include shale (sh); very fine, medium, and very coarse sandstone (vfs, ms, vcs); and granules (gr). Carbonate lithologies drawn to far right of column include micrite (mic); fine and coarse grainstone (fg, cg); flat pebble conglomerate (fpc), and thrombolitic boundstone (thr).

preserved reefs do occur higher in the member, but more impressive reefs will be seen in the Cookes Member in the Caballo and Hueco Mountains sections. The highest 15–20 m of the Cookes Member here in the Franklin Mountains consists largely of conspicuously burrowed lime mudstone to fine grainstone with some thin, nodular, lime mudstone intervals.

Jose Member. First described in the Cookes Range by Flower (1964), the Jose is an interval of dark (medium gray), oolitic packstone to grainstone with some beds containing small quantities of quartz sand that weathers in relief. Some of the thicker beds in the unit commonly also display fabric-selective dolomitization wherein only the ooid nuclei are dolomitized and stand out as small orange grains against the darker limestone around them. The Jose is the thinnest member of the formation, ranging in thickness from ~3 m to nearly 20 m in the sections that we have examined. Clemons (1991) reported greater thicknesses from some areas in his comprehensive analysis of the El Paso Group. These thicknesses probably are exaggerated by inclusion of strata that do not display the characteristics required for assignment to the Jose Member in the present study. This is definitely the case here in the Franklin Mountains, where previous authors (LeMone, 1996; Clemons, 1991; Hayes, 1975) equated the Jose with the Middle Sandy Zone of Harbour (1972) and the Chamizal Member of Lucia (1969). While the arenaceous interval identified as the Jose Formation by Flower and LeMone does indeed correspond almost precisely with the Middle Sandy Zone and the Chamizal, it is not equivalent to the oolitic package that has been traced across southern New Mexico as the Jose Member. The true Jose Member is present in the Franklin Mountains, but it is a characteristically thin (12–17 m) interval whose top lies ~18 m (or more) below the base of the sandy interval misidentified as the Jose in previous studies. The top of the highest dark grainstone bed is used in our lithostratigraphic scheme to define the base of the overlying McKelligon Formation. The interval between the top of the Jose Member and the top of the sandy interval, which is marked in the southern Franklin Mountains by a conspicuous orange-weathering sandstone bed 0.5–2 m thick, is reassigned to the base of McKelligon. That interval could be designated as a separate member of the formation below the Pistol Range Member of Flower (1964, 1969) and LeMone (1969, 1996), who used the aforementioned orange sandstone bed as the base of the McKelligon. A very strenuous climb is required to see the Pistol Range sandstone bed and the underlying dolomitic interval mistaken for the Jose Member here in McKelligon Canyon. The exposures along Scenic Drive (Stop 2) provide much easier access to these units and some remarkable reefs in the McKelligon Formation as well.

Influenced by the sandy and oolitic character of the unit, Clemons (1991) interpreted the Jose Member as the product of relative sea level fall that introduced shallower, higher energy conditions than those represented by the units above and below it. In contrast, we consider the Jose Member to be the result of submergence of the platform and the lithologies within the member to be the deepest water facies in the Hitt Canyon Formation (Taylor et

al., 2001). The oolite is dark, and well-developed cross-stratification is uncommon; most beds are either structureless or conspicuously burrowed. Additionally, thin intervals of ribbon limestone occur at or just above the base of the member in the Franklin and Hueco Mountains. These intervals of dark, very thin-bedded lime mudstone rhythmically interbedded with organic-rich shale might be the deepest water lithofacies in the El Paso Group, having formed in a deep shelf or upper slope environment seaward of the oolite. Alternatively, they might have accumulated as more proximal quiet water deposits that formed in the lee of a distal ramp oolitic shoal. The base of the Jose Member is defined by the lowest dark oolitic packstone to grainstone or the lowest dark ribbon limestone. Two ribbon limestone intervals are well exposed at the base of the Jose Member here at roughly 171–173 m into the section (Fig. 5). The same two recessive intervals are recognizable in the Police Academy section along Scenic Drive (Stop 2b; Fig. 6). The ribbon limestone is particularly important as a surrogate for the oolitic lithologies in the Hueco Mountains (Stop 4), where the overlying burrowed grainstone is not as dark as typical Jose, and oolitic fabrics have not yet been confirmed. The submergence event apparently also inhibited microbial reef growth. Although the Jose contains thrombolites in a few places, they are scarce, relatively small, and isolated in contrast to the large, laterally continuous buildups that characterize the bounding Cookes Member and McKelligon Formation. A thin (<15 m) interval of bioturbate lime mudstone to fine packstone separates the top of the Jose and the lowest McKelligon reef in many sections (including this one), making the contact less conspicuous than where reefs occur at the very base of the McKelligon.

The Jose Member is the most productive unit in the Hitt Canyon Formation for macrofossils in New Mexico, where it contains an abundant fauna dominated by asaphid trilobites (Loch et al., 2003). In contrast, it yields very few macrofossils in Texas, and most are mollusks of little biostratigraphic utility. A single collection of fewer than 20 trilobite specimens from the base of the Jose at Hitt Canyon in the northern Franklin Mountains consists entirely of hystricurid trilobites. Although the sample size is too small to allow rigorous statistical comparison, the contrast in generic composition is remarkably similar to that documented in Skullrockian faunas in the Tribes Hill Formation of New York by Westrop et al. (1993). As in the Tribes Hill, the trilobite faunas from the Jose suggest the existence of two biofacies: one dominated by mollusks and including hystricurid trilobites, and another dominated by asaphid trilobites with a much subordinate mollusk component.

Conodonts from the Jose assign it to the medial Stairsian *Macerodus dianae* Zone, allowing correlation with successions outside of Texas and New Mexico. The Jose submergence event is recorded in the central Appalachians, where an entire platform-to-slope transect is preserved, including intact shelfbreak and upper slope facies (Taylor et al., 1996). In that area, the highest member of the Grove Formation, recently named the Woodsboro Member by Brezinski (2004), records onlap and the replacement of shelfbreak microbial reefs and shelf-edge sands with dark, shaly upper slope facies during deposition of the *M. dianae* Zone

(Taylor et al., 1996). Farther west, in the Great Valley of Maryland, coeval strata of the "oolitic member" of the Rockdale Run Formation (Sando, 1957) record suppression of microbial reef growth and deposition of oolite, just as occurred in the deposition of the Jose Member in the Southwest.

The submergence event can also be tracked geographically using carbonate $\delta^{13}C$ profiles. The carbonate $\delta^{13}C$ stratigraphic profiles generated from multiple sections in the southwestern United States (Fig. 4) include a major (>2‰) negative excursion that coincides precisely with the base of the Jose Member. This "Jose Event" has proven useful in confirming the age-equivalence of the oolitic interval and associated ribbon limestones in the Franklin and Hueco Mountains to the Jose Member in southern New Mexico. The same negative excursion is also present just above the base of the San Juan Formation in the Argentine Precordillera (Buggisch et al., 2003), in the Malyi Karatau Range, Kazakhstan, and probably in the Arbuckle Mountains, Oklahoma (Gao and Land, 1991). Associated conodonts indicate that the event was virtually coincident with the appearance of the conodont *Paraoistodus proteus*, a species used to recognize the base of the British "Arenig." This linkage to the globally recognized Arenig transgression reinforces our interpretation of the base of the Jose as a product of sea level rise.

Drive out McKelligon Canyon Drive (east), turn right onto Alabama Street, then right onto Richmond Avenue, and continue to Scenic Drive. Turn right into the parking lot for the Police Academy (obtain permission in advance). Beware of traffic: Scenic Drive is narrow and heavily traveled.

STOP 2. McKelligon Formation and Upper Hitt Canyon Formation along Scenic Drive

Exit vehicles and walk southwest from the Police Academy along Scenic Drive, ascending stratigraphically to the conspicuous, orange-weathering sandstone at the base of the Pistol Range Member.

Stop 2a. McKelligon Formation—Pistol Range Member and Underlying Strata

The strata above the orange, dolomitic sandstone here at Scenic Drive constitute the type section of the McKelligon Formation (LeMone, 1969), which is well known for its prominent sedimentary cycles (Goldhammer et al., 1993) and biohermal complexes (Toomey and Ham, 1967; Toomey, 1970; Toomey and Nitecki, 1979; Rigby et al., 1999). The latter differ from the microbial buildups of the underlying Hitt Canyon Formation in the abundance of the demosponge *Archaeoscyphia* and the presence of the alga *Calathium*. The thin (1–1.5 m), orange-weathering, cross-bedded, dolomitic sandstone that occurs a short distance below the McKelligon reefal facies has played a prominent role in defining that formation boundary in all previous studies. Cloud and Barnes (1948) used this sandstone to define the base of their Unit B in dividing the El Paso Group into three parts, lettered from bottom to top. Flower (1964) and LeMone (1969, 1996)

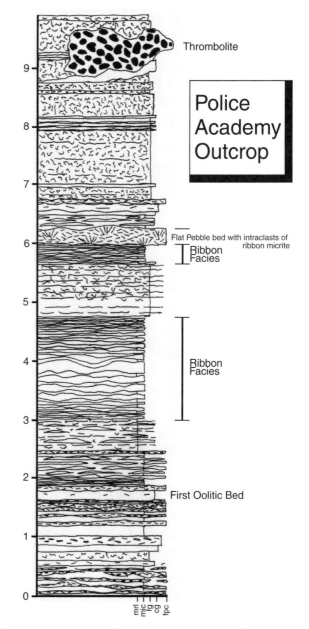

Figure 6. Detailed lithologic column of outcrop at the Police Academy (Stop 2b) showing the highest 1.7 m of the Cookes Member and the basal 8.3 m of the Jose Member of the Hitt Canyon Formation. Base of lowest oolitic bed defines base of Jose Member. See Figure 5 caption for lithologic abbreviations.

used the base of this sandstone to define the base of the McKelligon Formation, setting that bed and the overlying 21–24 m of dolomite apart as the basal Pistol Range Member of the McKelligon. Unfortunately, Flower and LeMone misidentified the 22 m of sandy dolomite directly below the Pistol Range Member as the Jose Formation, correlating it incorrectly with the dark oolitic limestone interval assigned that name in New Mexico, a miscorrelation perpetuated by Hayes (1975) and Clemons (1991).

One consequence of this miscorrelation is the interpretation of the base of the Pistol Range Member in the Franklin Mountains as an unconformity with a lacuna that includes the time represented by the Mud Springs Mountain Formation and Snake Hills Formation in New Mexico (Fig. 1, column 3). Flower (1964) erected those formations to accommodate 6–9 m of stromatolitic limestone with the gastropod *Bridgites* and as much as 18 m of overlying thinly bedded limestone (respectively) that occur directly above the (true) Jose in southern New Mexico (LeMone, 1983). Finding no evidence of these units or their faunas directly above what they believed to be the Jose in the Franklins, Flower (1969) and LeMone (1969) understandably concluded that an unconformity must exist to account for the absence of as much as 27 m of strata. Our discovery of a dark oolitic interval (the true Jose) some distance below the base of the Pistol Range reveals that the purportedly missing strata are present in the Franklin Mountains succession, but *below* the Pistol Range, rather than above it. The thickness of the interval between the base of the Pistol Range Member and the top of the true Jose Member along Scenic Drive could not be established with certainty owing to faulting that complicates that part of the Scenic Drive section. However, at Hitt Canyon, the top of our Jose Member is separated from the top of the sandy dolomitic interval mistaken for the Jose by 28 m of strata, including a 7 m stromatolitic reef interval. Similarly, the base of the Pistol Range Member lies at 220.5 m in the McKelligon Canyon section (Fig. 5), 35 m above the top of the true Jose Member, and stromatolitic reefs are common between those horizons. The reefs are all dolomitized, so it is not possible to say whether they originally contained *Bridgites* or any other diagnostic macrofossil. Nonetheless, the simplest interpretation is that the strata below the Pistol Range Member that we assign to the base of the McKelligon Formation (including the "Jose" and Chamizal Formations of previous authors) are the dolomitized equivalents of the Mud Springs Mountain and Snake Hills Formations in New Mexico. That being the case, there remains no evidence of an unconformity at the level of the Pistol Range sandstone.

Continue walking down-section, and return to the parking area for the Police Academy.

Stop 2b. Jose Member of the Hitt Canyon Formation at the Police Academy

The outcrop at the Police Academy exposes the highest few meters of the Cookes Member and the basal 8 m of the Jose Member (Fig. 6). The two ribbon limestone intervals seen at the base of the Jose Member at McKelligon Canyon are well developed in this section as well, although the base of the member is placed slightly lower at a thin bed of oolitic grainstone. Features seen here, but not seen in the Jose at McKelligon Canyon, include a prominent flat pebble conglomerate bed and an isolated thrombolite just under 1 m in thickness. Otherwise, the consistency in lithologic character between the two locations is remarkable. The lateral persistence of the Jose Member, not only along the length of the Franklin Mountains but across all of southern New Mexico,

justifies placement of the base of the McKelligon Formation at the top of this unique interval rather than at the base of the Pistol Range Member as defined by the sandstone bed higher in the section. Although it is well developed and conspicuous at the south end of the Franklin Mountains, the sandstone at the base of the Pistol Range thins northward to less than a meter at McKelligon Canyon and is represented at Hitt Canyon only by an inconspicuous interval of quartz sandy carbonate ~6 m thick. It has not been recognized in any sections in southern New Mexico.

Leave the Police Academy; turn right (west) to continue along Scenic Drive, then left onto Rim Road. Bear right at the stop sign onto Kerbey Street, then left onto Mesa (State Route 20); travel six blocks to West Schuster Avenue. Follow West Schuster to I-10; take I-10 west to Las Cruces, New Mexico. Follow I-25 north from Las Cruces to Truth or Consequences and take the northern exit (Exit 79). Pull into the Best Western Motor Lodge on the right. Driven distance is ~120 mi (193 km).

DAY 2—CABALLO MOUNTAINS, NEW MEXICO

Return to I-25 and travel south to Exit 29, Caballo Dam. Turn right at the end of exit the ramp, and go south for 2.9 miles on Route 187. Turn left (east) at a telecommunications tower; travel 1.2 mi, and turn left at three-way intersection, then travel 0.8 mi, and turn right (east) onto a dirt road. Go 0.3 mi to another three-way intersection, and turn left (north). Travel 1.0 mi, pass under I-25, then go 1.8 mi farther, with the river on your left, and bear right at the fork in the road; to the left is the east end of Caballo Dam. Travel 4.0 mi as the road winds, rises, and falls across the dissected bajada, and turn right onto a rocky two-track road (four-wheel drive required) that leads east toward the range. Driving distance: Motel to Exit 29, 21 mi (34 km); I-25 to the four-wheel drive road, 12 mi (19.3 km).

STOP 3. Cable Canyon/Bat Cave Gulch Measured Section

Stop 3a. Overview of the Sections

Compare the view to the east with Figure 7, which shows the locations of formation boundaries and measured sections in the footwall of the thrust fault that repeats the McKelligon Formation just north of Cable Canyon. Note the immense Bat Cave (BC) in the El Paso Group, barely visible at the left (north) edge of the figure, and the location of the old Sierrite Mine (SM) at the base of the Bliss on the north flank of the closer hill, capped by Hitt Canyon carbonates. The extremely glauconitic upper member of the Bliss Formation forms the conspicuous dark band on the hillside. These sections have provided the most continuous and highly resolved sequence of trilobite and conodont faunas, and the most complete $\delta^{13}C$ profile, yet recovered from the Upper Cambrian and Lower Ordovician of New Mexico. Collections from the Bliss and Hitt Canyon Formations reveal that these units span much or all of the Sunwaptan Stage, the entire Skullrockian Stage, and most or all of the Stairsian Stage (Fig. 3) in this area.

Figure 7. Distant view of Cable Canyon on the west side of the Caballo Mountains from the 4WD road in from the west (Stop 3a), showing the location of the Cable Canyon section (CC) of Taylor and Repetski (1995) just north of the Sierrrite Mine (SM) and the Bat Cave Gulch section (BCG) of the present study in the gully below the Bat Cave (BC).

Drive east for 2.1 mi (3.38 km) into Cable Canyon. Drive all the way to the cliffs on the north side of the canyon before turning left onto an old mine road that runs along the base of the steep slope supported by the Bliss Formation (the Jeep at the bottom of Fig. 8A is parked at that turnoff). Park where the road ends just south of small, old quarry in Precambrian basement. Hike to the quarry and ascend the gully upslope (east) of it, toward the cliffs. This is the Bat Cave Gulch section.

Stop 3b. Lower Segment of Bat Cave Gulch Section— Bliss Formation

Millardan Series—Steptoean and Sunwaptan Stages. The lower member of the Bliss is thin at this locality (Fig. 9), comprising a few meters of quartzitic sandstone at the base and 2 m of oolitic hematite at the top. This is the ore that was mined at the old Sierrite Mine on the south side of the canyon. The top of the oolite and its sharp contact with the lowest thin dolomite bed of the upper member are well exposed here. The upper member was measured and sampled in the cliffs and steep slopes on the south side of the gully, which ends at the base of the cliff supported by the Sierrite Member of the Hitt Canyon Formation, for which this locality is the designated type section.

The ages of the lower member and the basal strata of the upper member of the Bliss are poorly constrained. No diagnostic

trilobites of the medial Upper Cambrian Steptoean Stage were recovered here; however, Flower (1969) reported an *Elvinia* trilobite Zone fauna with the uppermost Steptoean genus *Camaraspis* from low in the Bliss at White Signal in southwestern New Mexico. The brachiopod *Eoorthis* was recovered in the present study from sandstone beds interstratified with oolitic hematite in the lower member at Lone Mountain and was reported from a few other locations in southern New Mexico by Flower (1969). This suggests a basal Sunwaptan age for the highest beds of the lower member. If that age is correct, the base of the Sunwaptan lies within the lower member, and the transgression recorded by the base of the upper member in New Mexico was a Sunwaptan event. The Sunwaptan strata in the upper member yield very few fossils, but conodont collections from dolomite beds at this locality verify the presence of faunas as old as the *Proconodontus tenuiserratus* Zone (or lowest part of the *P. posterocostatus* Zone) and as young as the uppermost Sunwaptan *Cambroistodus minutus* Subzone of the *Eoconodontus* Zone (Taylor and Repetski, 1995). Some of the trilobite species recovered from the Bat Cave Gulch section are shown in Figure 10. A small collection from the highest meter of Sunwapatan strata includes *Euptychaspis kirki* (Fig. 10P), a species restricted to the *Saukiella serotina* Subzone of *Saukia* Zone, corroborating the age indicated for these strata by the conodonts.

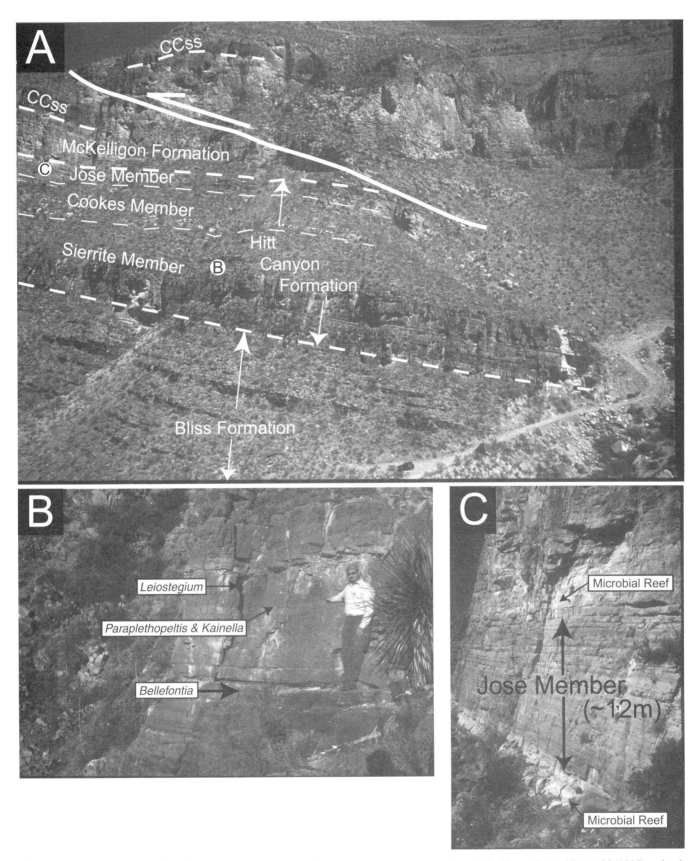

Figure 8. (A) View northeast (from Sierrite Mine) showing the Cable Canyon section in the Bliss studied by Taylor and Repetski (1995) and unit boundaries in the overlying El Paso Group used in the present study. CCss—Cable Canyon Sandstone. B and C in photo A mark the *stratigraphic* levels of the other photos; actual locations lie out of view to north. (B) Base of Stairsian Stage in the Bat Cave Gulch (BCG) section (Stop 3c); the bioclastic lag with lowest Stairsian fauna at the person's right hand; possible disconformity (top of lower Sierrite cliff) at parting just below knee. (C) Dark band formed by the Jose Member near the top of the BGC section.

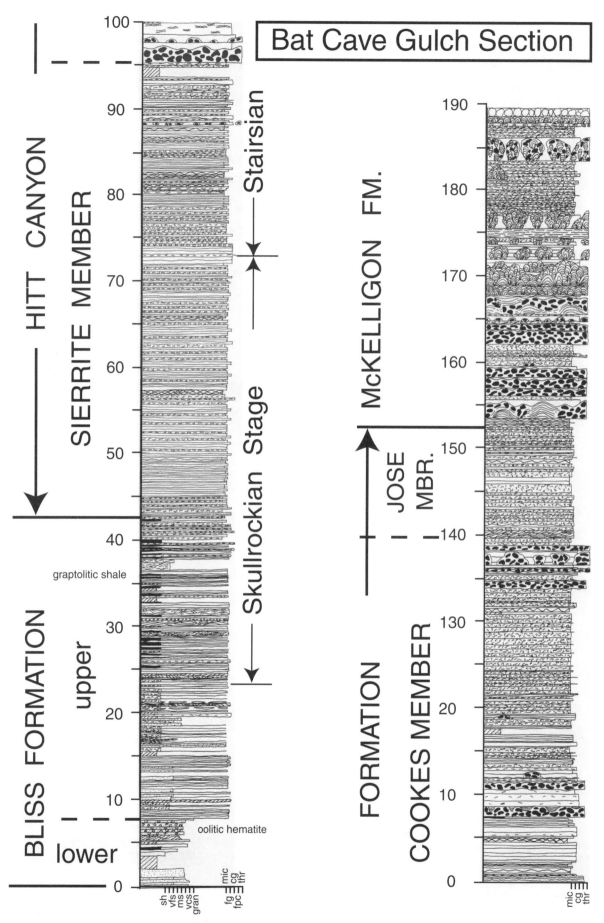

Figure 9. Lithologic column for the Bat Cave Gulch measured section in the Caballo Mountains, New Mexico. See Figure 5 caption for lithologic abbreviations.

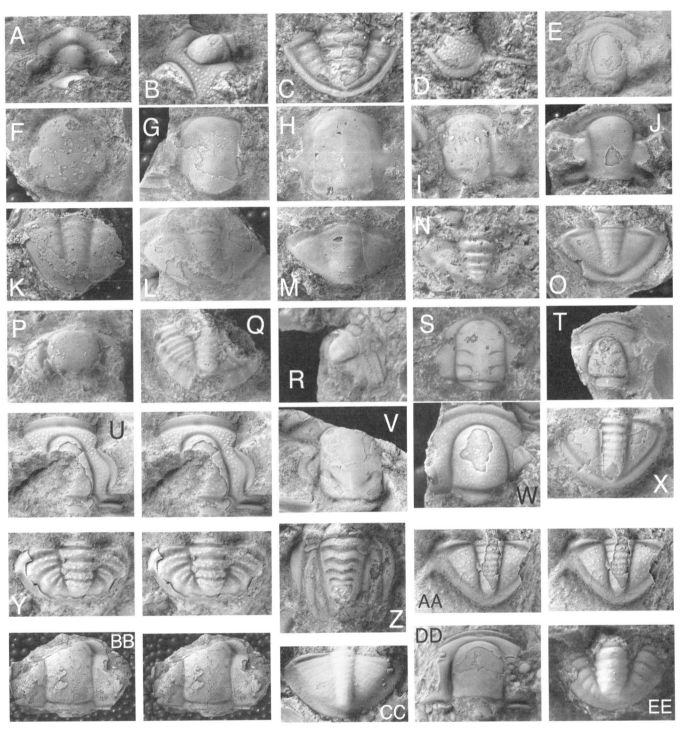

Figure 10. Selected trilobite cranidia (cr), pygidia (py) and librigenae (lib) from the Bliss and Hitt Canyon Formations. Dorsal view unless indicated otherwise. (A) *Highgatella cordilleri*, cr, X4.5, Bat Cave Gulch (BCG) 29.37 m. (B–D) *Politicurus* sp., BCG 71.45 m: (B) cr, X11.5, anterior oblique view; (C) py, X11, and (D) lib, X8. (E) *Paraplethopeltis* sp. 1, cr, X2.5, BCG 72.92 m. (F, K) *Symphysurina* n. sp. 6, BCG 37.85 m: (F) cr, X2.5; (K) py, X2. (G, L) *Symphysurina* n. sp. 1, BCG 31.34 m: (G) cr, X2; (L) py, X2. (H, M) *Symphysurina* n. sp. 5, BCG 36.63 m: (H) cr, X3.5; (M) py, X3.5. (I, N) *Jujuyaspis borealis*, BCG 36.63 m: (I) cr; (N) py, X5.5. (J, O) *Bellefontia* sp., BCG 66.54 m: (J, X2.5) cr; (O) py, X2. (P) *Euptychaspis kirki*, cr, X6, BCG 23 m. (Q) *Missisquoia depressa*, py, X6.5, BCG 24.08 m. (R) *Eurekia apopsis*, py, X8, BCG 23.6 m. (S) *Triarthropsis nitida*, cr, X8, BCG 23.6 m. (T) *Apoplanias rejectus*, cr, X2.5, BCG 25.45 m. (U, Y) Stereopairs of *Hystricurus* n. sp. A, BCG 135.6 m: (U) cr, X2.5; (Y) py, X4.5. (V, Z) *Kainella* sp.: (V) cr, X2, BCG 75 m; (Z) py, X3.5, BCG 72.92 m. (W) *Paraplethopeltis* sp. 2, cr, X5, BCG 72.92 m. (X) *Leiostegium manitouensis*, py, X4.5, BCG 75 m. (AA) Stereopair of *Perischodory* sp., py, X3.5, HUT1 32.22 m. (BB) Stereopair of *Leiostegium* sp., cr, X1, HUT 190.61. (CC) *Aulacoparia? huygenae*, py, X1.5, MCN 141 m. (DD, EE) *Jeffersonia* n. sp.: (DD) cr, X3, BCG 153.45m; (EE) py, X6, MCN 141 m.

Ibexian Series—Skullrockian Stage. Trilobites (Fig. 10) and conodonts from Bat Cave Gulch constrain the position of the base of the Skullrockian Stage to <0.5 m and the top of the stage to <1.5 m (Fig. 9). The stage is nearly 50 m thick, spanning the highest 19.3 m of the upper member of the Bliss Formation and the basal 29.95 m of the Sierrite Member of the Hitt Canyon Formation. Nearly all Skullrockian trilobite and conodont zones have been identified, although those within the Bliss are anomalously thin, reflecting the very low rock accumulation rates typical of glaucarenite-rich facies. The extinction horizon that defines the base of the Ibexian lies within an essentially monofacial interval of interbedded recessive glaucarenites and rippled, medium to coarse-grained bioclastic grainstone. Aside from a decrease in the degree of dolomitization of the grainstone beds, lithologies in uppermost Sunwaptan and basal Skullrockian strata are identical, providing no evidence to support models that invoke either a rise or fall in sea level as the cause of faunal change. Although condensed, the boundary interval at Cable Canyon is complete to the extent that the three thin trilobite subzones at the base of the Ibexian are all represented, including the basal *Eurekia apopsis* Subzone of the *Saukia* Zone, which was not discovered in previous work at this locality (Taylor and Repetski, 1995). The lowest *E. apopsis* collection (Bat Cave Gulch [BCG], 23.67 m) was recovered 40 cm above the top of a conspicuous, 20–30 cm, orange-weathering dolomite bed that serves as a useful marker for relocating the base of the Skullrockian on each visit and that is helpful in tying collections from the older Cable Canyon section into the newer BCG section. Closely spaced collections locate precisely the bases of the *Missisquoia depressa* (BCG 24.08 m) and *M. typicalis* (BCG 24.6 m) Subzones of the *Missisquoia* Zone. The base of the *M. typicalis* Subzone, which defines the top of the Ptychaspid Biomere, is marked as it is throughout Laurentian North America by coquinas of the olenid trilobite *Apoplanias rejectus* (Fig. 10T) and the brachiopod *Apheoorthis*.

The δ¹³C profile through the upper Sunwaptan and basal Ibexian at Cable Canyon does not approach the quality of those developed for this interval elsewhere (Ripperdan et al., 1992; Ripperdan and Miller, 1995), probably due to the extremely heterolithic character of the upper Bliss Formation at this locality. One feature that appears consistent with profiles developed elsewhere is the steady rise in values from −0.8‰ to +0.6‰ through the basal 5 m of the profile. That segment might represent the uppermost part of the strongly negative "HERB event" (see Ripperdan, 2002), which reached peak negative values just below the base of the *Cambrooistodus minutus* Subzone of the *Eoconodontus* Zone.

The appearance of *Highgatella cordilleri* (Fig. 10A) at 29.2 m marks the base of the *Symphysurina* trilobite Zone. Although many *Symphysurina*-rich collections were recovered from the upper Bliss, placement of the boundaries of the three subzones recognized in this zone in Oklahoma (Stitt, 1977, 1983) and Utah (Loch et al., 1999) has proven difficult. The defining species—*S. brevispicata*, *S. bulbosa*, and *S. woosteri* (in ascending order)—have not been identified with confidence, in part because the Bliss material does not adequately preserve prosopon

(surface texture), which is important in species diagnosis. However, even without prosopon, most of the *Symphysurina* material clearly represents undescribed species (e.g., Figure 10F–H and 10K–M). Of particular interest in the Bliss fauna are two cosmopolitan species that allow correlation with non-Laurentian successions: the olenid trilobite *Jujuyaspis borealis* and the dendroid graptolite *Rhabdinopora flabeliformis*. *R. flabeliformis*, which is abundant in two thin shale packages between 35 and 36 m, is noteworthy because this is the only known occurrence of this important Lower Ordovician index species anywhere within a paleoequatorial shelf sequence (Cooper et al., 1998). *J. borealis* (Fig. 10I and 10N), which occurs less than half a meter above the graptolitic shale, is a widely distributed species that allows correlation of a thin interval near the middle of the *Symphysurina* Zone from central Texas (Stitt and Miller, 1987) to Alberta (Norford, 1969). The presence of the graptolitic shale, typically a deepwater facies, in the Bliss is extraordinary. Interestingly, the associated conodonts represent the *Cordylodus angulatus* Zone, the interval in which evidence of a significant submergence (the Stonehenge Transgression) has been documented in platform sequences throughout North America (Taylor et al., 1992). The base of the *Bellefontia-Xenostegium* trilobite Zone lies very near the base of the Hitt Canyon Formation, and the top of the zone corresponds precisely with the top of the cliff (i.e., the top of the Sierrite Limestone as originally defined by Kelly and Silver [1952]).

Follow the base of the Sierrite cliff south for a short distance to the head of the next gully where a notch in the cliff provides access to the units above the Bliss. The climb up through the notch to the top of the cliff is not too difficult but is recommended only for those comfortable with heights.

Stop 3c. Upper Segment of Bat Cave Gulch Section—Hitt Canyon Formation

Skullrockian-Stairsian Stage boundary. Although macrofossil recovery is sparse through the Sierrite Member, the position of the Skullrockian-Stairsian boundary (the base of the *Leiostegium-Kainella* Zone) has been established to within <1.5 m in this section, 1.7 m in the Hueco Mountains, and <5 m at Hitt Canyon in the Franklin Mountains. This boundary, the top of the "Symphysurinid Biomere" of Stitt (1983), elsewhere is characterized by a thin zone dominated by the hystricurid genus *Paraplethopeltis*, separating the pre-extinction fauna of the underlying *Bellefontia-Xenostegium* Zone from the overlying replacement fauna of the *Leiostegium-Kainella* Zone. Here at Cable Canyon, the faunal turnover is accompanied by a lithologic change from the cliff-forming fine grainstone containing the pre-extinction *Bellefontia-Xenostegium* Zone fauna to a 1.7 m package of coarser bioclastic grainstone that includes centimeter to decimeter-scale skeletal lags of brachiopod and trilobite debris (Fig. 8B) similar to the coquinas that characterize biomere boundaries in the Upper Cambrian. The lowest post-extinction trilobite lag yielded a nearly monogeneric fauna dominated by *Paraplethopeltis* (Fig. 10E and 10W), suggesting that it might represent the *Paraplethopeltis* Zone. However, it also yielded two specimens of *Kainella* (Fig. 10V and 10Z) and is assigned accordingly to the *Leiostegium-Kainella* Zone.

It appears, therefore, that there is a small stratigraphic break in the Sierrite Member that omits the very highest beds of the Skullrockian Stage: the *Paraplethopeltis* Zone is missing and the *Leiostegium-Kainella* Zone rests unconformably directly on top of the *Bellefontia-Xenostegium* Zone. Hematitic staining of surfaces within the highest meter of the *Bellefontia-Xenostegium* Zone at Cable Canyon, and the occurrence of a thin interval with abundant quartz sand between the highest Skullrockian and lowest Stairsian collections in the Franklin and Hueco Mountain sections, support this interpretation. Some part of the *Bellefontia-Xenostegium* Zone might also be missing; however, the highest Skullrockian collection contains the hystricurid *Politicurus* (Fig. 10B–D), which is restricted to roughly the upper quarter of that zone in Utah (Adrain et al., 2003), suggesting that at least three quarters of the zone is present. Trilobites and conodonts recovered from the lowest Stairsian strata at Cable Canyon and Hitt Canyon indicate that little, if any, of the basal Stairsian is omitted. The presence of *Paraplethopeltis*, which ranges upward only a few meters into the *Leiostegium-Kainella* Zone in the Manitou Formation in Colorado (Berg and Ross, 1959), confirms that basal Stairsian strata are present in New Mexico and Texas. These strata also yield conodonts of the *Rossodus manitouensis* Zone, which barely extends upward into the Stairsian Stage, rather than those of the overlying Low Diversity Interval. The persistence across the study area of a 2‰ positive $\delta^{13}C$ excursion just above the base of the Stairsian, and its resultant utility as a datum on which to hang the isotopic profiles (Fig. 4), also suggests little if any loss of section from the base of the stage due to erosion or non-deposition. That excursion, which is the only level within the Stairsian where positive values are reached, is conspicuous in the middle of the La Silla Formation in the profile published by Buggisch et al. (2003) for the Argentine Precordillera.

The position of the base of the Low Diversity Interval at BCG 77 m, just a few meters above the base of the *Leiostegium-Kainella* Zone in the upper Sierrite Member (Fig. 9), is consistent with the relationship established for these zonal boundaries in previous studies (Ross et al., 1997; Miller et al., 2003). In contrast, the relative positions within the Hitt Canyon Formation of the next higher conodont and trilobite zones, the *Macerodus dianae* and *Tesselecauda* Zones, differs significantly from that reported from Utah, where Ross et al. (1997) reported that the base of the *M. dianae* Zone lies somewhere within the upper half to third of the *Tesselecauda* Zone. In Texas and New Mexico, the base of the *M. dianae* Zone was found consistently 15–20 m above the base of the Cookes Member. The only trilobite species recovered from the Hitt Canyon Formation that also occurs in Utah and Idaho, where the *Tesselecauda* Zone has been described, is *Hystricurus* sp. nov. A of Adrain et al. (2003) (Fig. 10U and 10Y). The occurrence of that species slightly more than 40 m above the base of the Cookes Member at Cable Canyon (BCG 135.6 m) places the base of the *Tesselecauda* Zone well *above* the base of the *M. dianae* Zone. The data currently available are insufficient to determine

whether the base of the *M. dianae* is older and/or the base of *Tesselecauda* is younger in the El Paso Group than in Utah and Idaho, or whether the discrepancy is an artifact of imprecision in the sampling or compositing of data in one or both areas.

What correlates very well across the southwest and into the Ibexian standard succession in Utah is the initiation of microbial reef growth during deposition of the Low Diversity Interval. The horizon that represents that event in the El Paso Group (base of the Cookes Member; Figs. 8A and 9) occurs consistently 15–20 m below the base of the *Macerodus dianae* Zone, supporting the assertion that the onset of reef growth was synchronous across the area. In Utah, prominent microbial reefs also appear in the Low Diversity Interval near the base of the Fillmore Formation (Evans et al., 2003), above a relatively thick succession of Skullrockian and basal Stairsian non-reefal carbonates. Another parallel between the reefs of the El Paso Group and the Fillmore Formation is a change in the structure and composition of the reefs up-section. In both areas, the alga *Calathium* is conspicuous and abundant in the higher reefs ("Church's Reef" in the Fillmore and reefs in the McKelligon Formation) but absent from those lower in the section ("Miller's Reef" and "Hintze's Reef" in the Fillmore and those in the Cookes Member). If the lower and upper reefs of the two areas are correlative as their similarities suggest, then any geochemical or lithologic evidence of the paleoceanographic event(s) that produced the Jose Member and associated isotopic excursion should be found in the Fillmore between "Hintze's Reef" and "Church's Reef." We propose Bat Cave Gulch as the type section for the Cookes Member owing to poor exposure and structural complications in the Cookes Range. Both contacts are well exposed at Cable Canyon: the base at the lowest meter-scale reef at BCG 95 m and the top (= base of the Jose Member) at the base of a prominent, dark, cross-stratified grainstone 139.75 m above the base of the section.

Continue upslope through the highest part of the Sierrite Member and intervals of well-preserved microbial reefs at the base and near the top of the Cookes Member. The dark band of non-reefal limestone at the base of the next set of tall cliffs (Fig. 8C) is the Jose Member.

Medial to upper Stairsian strata—Jose Member and McKelligon Formation. The sheer cliff at the top of the Bat Cave Gulch section exposes the entire Jose Member and all of a relatively thin McKelligon Formation, including the sharp contact with the resistant, brown-weathering Cable Canyon Sandstone at the base of the Montoya Group. The Jose lacks the ribbon limestone facies that it contains in Texas, but still has a deep subtidal aspect in the abundance of carbonate mud, much of it distributed as selectively dolomitized centimeter-scale patches in conspicuously burrow-mottled intervals of oolitic packstone. The dark color and absence of microbial reefs also reinforce the interpretation of this unit as representing deeper conditions than the units above and below it. Dominance of the trilobite fauna by an asaphid (*Aulacoparia*? *huygenae*; Fig. 10CC) also is consistent with a deeper water assignment. The unusual biofacies of the Jose Member has hampered attempts to assign a precise age to this unit based on its trilobites; however, the first thorough study of its trilobite fauna is in progress. Conodonts from the Jose assign it to the medial Stairsian *Macerodus dianae* Zone, but collections from the basal 5 m of the overlying McKelligon Formation might represent the overlying *Acodus deltatus* Zone, and it is possible that the very highest beds of the Jose do as well. Several collections from 15 or more meters above the base of the McKelligon contain *Oneotodus costatus* and are more confidently assigned to the *A. deltatus* Zone. Trilobites recovered from the McKelligon in the present study include the bathyurid trilobite genera "*Peltabellia*" (from this section, Scenic Drive, Cookes Peak, and Mescal Canyon), *Bolbocephalus* (Hitt Canyon, Mescal Canyon), *Petigurus* (Mescal Canyon), *Benthamaspis* (Scenic Drive, Mescal Canyon), and *Jeffersonia* (Scenic Drive). These genera are characteristic of the Jeffersonian Stage (Loch, 1995) of eastern Laurentia.

Allow enough time for a cautious climb back down to the vehicles while there is sufficient daylight to avoid the cacti. We will return to El Paso (135 mi, 217 km) via I-25 and I-10.

DAY 3—HUECO MOUNTAINS, TEXAS

From its intersection with Airway Boulevard, follow Montana Boulevard (U.S. Route 62) east for 19.5 miles, then turn right (southeast) off Route 62. Note the windmills on the southeastern skyline after ~4.5 mi; go straight through the four-way intersection 6.9 mi from Route 62, continue for another 3.4 mi, and turn left toward the windmills. Proceed 0.4 mi, turn right near the power poles, go 0.3 mi, turn left (upslope), and park. Distance from the Airway-Montana intersection: 30.5 mi (19 km).

STOP 3. Hueco Mountains Measured Section

Stop 3a. Bliss Formation

In contrast to the completed work in the sections to the west, what we present here in the Huecos is more of a work in progress. The Bliss is considerably thicker than it is to the west, with over 90 m of sandstone assigned to the lower member, much of it completely homogenized by burrowing. It is well exposed in the gully leading up to the notch in the cliffs through which the section was measured. The base of the upper member is marked by the appearance of recessive glaucarenite intervals just above a 7.65 m hematitic zone that caps that lower member. The upper member is a heterolithic mix of glaucarenite, quartz sandstone, and varied carbonate lithologies. The carbonates increase in relative abundance toward the intercalated contact with the Hitt Canyon Formation. Moldic specimens of *Symphysurina* were recovered from a thick sandstone bed low within the upper member of the Bliss, 95–100 m above the base of the section. This assigns those strata to either the *Symphysurina* or *Bellefontia-Xenostegium* Zone within the Skullrockian Stage. The biostratigraphic data confirm that the base of the Hitt Canyon Formation, again placed at the base of the pure, cliff-forming carbonate, is significantly younger than the base of the Hitt Canyon in New Mexico.

Recall that the cliff of grainstone at the base of the Sierrite Member at Cable Canyon consists entirely of upper Skullrockian strata of the *Bellefontia-Xenostegium* Zone; trilobites of the *Leiostegium-Kainella* Zone and the basal Stairsian positive δ^{13}C excursion were found just above the top of the cliff. The trilobites and δ^{13}C data recovered from this section in the Huecos place that faunal change and isotopic excursion at the *base* of the prominent cliff of Sierrite grainstone.

Stop 3b. Hitt Canyon Formation

The upper part of the Sierrite Member, above the cliff of fine grainstone, includes numerous meter-scale cycles, each of which shallows upward from lime mudstone at its base to a cap of coarse grainstone and/or microbial boundstone. The cyclic upper Sierrite is overlain by well-developed microbial reefs at the base of the Cookes Member, which include some remarkable stromatolites nearly 2 m tall. As elsewhere, the Cookes contains two prominent intervals of reefs and associated grainstone, one at the base and another high within the member. Thinly bedded lime mudstone with prominent burrows that often weather in relief dominates the intervals between the reefs. Isolated trilobite collections recovered so far from the Cookes offer promise of more recovery than has been typical of this member in sections to the west (e.g., Fig. 10AA and 10BB). The top of the Cookes is placed at the base of an 80 cm interval of dark ribbon limestone at the bottom of a cliff formed by the highest beds preserved at this locality. We interpret this recessive ribbon limestone as the base of the Jose Member, a correlation supported by the strongly negative (<−3‰) C-isotopic values acquired from this interval (Fig. 4). Except for a prominent 80 cm thrombolitic reef that sits directly on top of the ribbon limestone, the cliff is a thick package of non-reefal grainstone that displays light and dark banding on a 0.5 m scale. The banding is caused by alternation of dark medium to coarse grainstone with lighter, burrow-mottled intervals similar to those common in the Jose Member in New Mexico, although no oolitic textures have yet been found, and none of the grainstone is quite as dark as the typical Jose.

Allow sufficient time to descend the slope to be at the vehicles for departure at noon.

REFERENCES CITED

Adrain, J.M., Lee, D.-C., Westrop, S.R., Chatterton, B.D.E., and Landing, E., 2003, Classification of the trilobite subfamilies Hystricurinae and Hintzecurinae subfam. nov. with new genera from the Lower Ordovician (Ibexian) of Idaho and Utah: Memoirs of the Queensland Museum, v. 48, p. 559–592.

Berg, R.R., and Ross, R.J. Jr., 1959, Trilobites from the Peerless and Manitou formations, Colorado: Journal of Paleontology, v. 33, p. 106–119.

Brezinski, D.K., 2004, Stratigraphy of the Frederick Valley and its relationship to karst development: Maryland Geological Survey, Report of Investigations 75, 101 p.

Buggisch, W., Keller, M., and Lehnert, O., 2003, Carbon isotope record of Late Cambrian to Early Ordovician carbonates of the Argentine Precordillera: Palaeogeography, Palaeoclimatology, Palaeoecology, v. 195, p. 357–373, doi: 10.1016/S0031-0182(03)00365-1.

Clemons, R.E., 1991, Petrography and depositional environments of the Lower Ordovician El Paso Formation: New Mexico Bureau of Mines and Mineral Resources Memoir 125, 66 p.

Clemons, R.E., 1998, Geology of the Florida Mountains, southwestern New Mexico: New Mexico Bureau of Mines and Mineral Resources Memoir 43, 112 p.

Cloud, P.E. Jr., and Barnes, V.E., 1948, The Ellenburger Group of central Texas: University of Texas Publication 4621, 473 p.

Cooper, R.A., Maletz, H.W., and Erdtmann, B.-D., 1998, Taxonomy and evolution of earliest Ordovician graptoloids: Norsk Geologisk Tidsskrift, v. 78, p. 3–32.

Evans, K.R., Miller, J.F., and Datillo, B.F., 2003, Sequence stratigraphy of the Sauk Sequence: 40th anniversary field trip in western Utah, *in* Swanson, T.W., Western Cordillera and adjacent areas: Boulder, Colorado, Geological Society of America Field Guide 4, p. 17–35.

Flower, R.H., 1964, The nautiloid Order Ellesmeroceratida (Cephalopoda): New Mexico Bureau of Mines and Mineral Resources Memoir 12, 234 p.

Flower, R.H., 1968, Some El Paso guide fossils: New Mexico Bureau of Mines and Mineral Resources Memoir 22, no. 1, p. 3–19.

Flower, R.H., 1969, Early Paleozoic of New Mexico and the El Paso region, *in* Lemone, D.V., ed., The Ordovician Symposium: El Paso Geological Society and Society of Economic Mineralogists and Paleontologists, Permian Basin Section, p. 31–101.

Gao, G., and Land, L.S., 1991, Geochemistry of Cambro-Ordovician Arbuckle limestone, Oklahoma: Implications for diagenetic $\delta^{18}O$ alteration and secular $\delta^{13}C$ and $^{87}Sr/^{86}Sr$ variation: Geochimica et Cosmochimica Acta, v. 55, p. 2911–2920, doi: 10.1016/0016-7037(91)90456-F.

Goldhammer, R.K., Lehmann, P.J., and Dunn, P.A., 1993, The origin of high-frequency platform carbonate cycles and third-order sequences (Lower Ordovician El Paso Group, west Texas): Constraints from outcrop data and stratigraphic modeling: Journal of Sedimentary Petrology, v. 63, p. 318–359.

Harbour, R.L., 1972, Geology of the Northern Franklin Mountains, Texas and New Mexico: United States Geological Survey Bulletin 1298, 129 p.

Hayes, P.T., 1975, Cambrian and Ordovician rocks of southern Arizona and New Mexico and westernmost Texas: United States Geological Survey Professional Paper 873, 98 p.

Kelly, V.C., and Silver, C., 1952, Geology of the Caballo Mountains: New Mexico University Publications in Geology 4, 286 p.

LeMone, D.V., 1969, Cambrian-Ordovician in the El Paso Border Region, *in* Lemone, D.V., ed., The Ordovician Symposium: El Paso Geological Society and Society of Economic Mineralogists and Paleontologists, Permian Basin Section, p. 145–161.

LeMone, D.V., 1983, Stratigraphy of the Franklin Mountains, El Paso County, Texas, and Dona Ana County New Mexico, *in* Delaware Basin Guidebook: West Texas Geological Society Publication 82-76, p. 42–72.

LeMone, D.V., 1988, Precambrian and Paleozoic stratigraphy; Franklin Mountains, west Texas, *in* Haywood, O.T., ed., Centennial Field Guide: South-Central Section of the Geological Society of America, v. 4, p. 387–389.

LeMone, D.V., 1996, The Tobosa Basin–related stratigraphy of the Franklin Mountains, Texas and New Mexico, *in* Stoudt, E.L., ed., Precambrian-Devonian Geology of the Franklin Mountains, west Texas—Analogs for Exploration and Production in Ordovician and Silurian Karsted Reservoirs in the Permian Basin: West Texas Geological Society Publication 96-100, p. 47–70.

Loch, J.D., 1995, An affirmation of the Jeffersonian Stage (Ibexian) of North America and a proposed boundary stratotype, *in* Cooper, J.D., Droser, M.L., and Finney, S.C., eds., Ordovician Odyssey: Short papers for the Seventh International Symposium on the Ordovician System, Society for Sedimentary Geology (SEPM) Pacific Section Book 77, p. 45–48.

Loch, J.D., Stitt, J.H., and Miller, J.F., 1999, Trilobite biostratigraphy through the Cambrian-Ordovician boundary interval at Lawson Cove, Ibex, western Utah, U.S.A.: Acta Universitatis Carolinae-Geologica, v. 43, p. 13–16.

Loch, J.D., Owen, A.M., Taylor, J.F., and Myrow, P.M., 2003, The Lower Ordovician trilobite fauna from the Jose Formation (El Paso Group) of southern New Mexico: Geological Society of America Abstracts with Programs, v. 35, no. 2, p. 35.

Lucia, F.J., 1969, Sedimentation and paleogeography of the El Paso Group, *in* Lemone, D.V., ed., The Ordovician Symposium: El Paso Geological Society and Society of Economic Mineralogists and Paleontologists, Permian Basin Section, p. 110–133.

Miller, J.F., Evans, K.R., Loch, J.D., Ethington, R.L., Stitt, J.H., Holmer, L., and Popov, L.E., 2003, Stratigraphy of the Sauk III interval (Cambrian-Ordovician) in the Ibex area, western Millard County, Utah and central Texas: Brigham Young University Studies in Geology, v. 48, p. 23–118.

Myrow, P.M., Taylor, J.F., Miller, J.F., Ethington, R.E., Ripperdan, R.L., and Allen, J., 2003, Fallen arches: Dispelling myths concerning Cambrian

and Ordovician paleogeography of the Rocky Mountain region: Geological Society of America Bulletin, v. 115, p. 695–713, doi: 10.1130/0016-7606(2003)1152.0.CO;2.

Norford, B.S., 1969, The early Canadian (Tremadocian) trilobites *Clelandia* and *Jujuyaspis* from the southern Rocky Mountains of Canada: Geological Survey of Canada Bulletin 182, p. 1–15.

Rigby, J.K., Linford, C.B., and LeMone, D.V., 1999, Sponges from the Ibexian (Ordovician) McKelligon Canyon and Victorio Hills Formations in the southern Franklin Mountains, Texas: Brigham Young University Geology Studies, v. 44, p. 103–133.

Ripperdan, R.L., 2002, The HERB event: end of the Cambrian carbon cycle paradigm?: Geological Society of America, Abstracts with Programs, v. 34, no. 6, p. 413.

Ripperdan, R.L., and Miller, J.F., 1995, Carbon isotope ratios from the Cambrian-Ordovician boundary section at Lawson Cove, Ibex area, Utah, *in* Cooper, J.D., Droser, M.L., and Finney, S.C., eds., Ordovician Odyssey: Short papers for the Seventh International Symposium on the Ordovician System: Society for Sedimentary Geology (SEPM), Pacific Section, Book 77, p. 129–132.

Ripperdan, R.L., Magaritz, M., Nicoll, R.S., and Shergold, J.H., 1992, Simultaneous changes in carbon isotopes, sea level, and conodont biozones within the Cambrian-Ordovician boundary interval at Black Mountain, Australia: Geology, v. 20, p. 1039–1042, doi: 10.1130/0091-7613(1992)0202.3.CO;2.

Repetski, J.E., 1988, Ordovician conodonts from the Bliss Sandstone in its type area, west Texas: New Mexico Bureau of Mines and Mineral Resources Memoir 44, p. 123–127.

Ross, R.J. Jr., Hintze, L.F., Ethington, R.L., Miller, J.F., Taylor, M.E., and Repetski, J.E., 1997, The Ibexian, lowermost series in the North American Ordovician: United States Geological Survey Professional Paper 1579-A, 50 p.

Sando, W.J., 1957, Beekmantown Group (Lower Ordovician) of Maryland: Geological Society of America Memoir 68, 161 p.

Stitt, J.H., 1977, Late Cambrian and earliest Ordovician trilobites, Wichita Mountains area, Oklahoma: Oklahoma Geological Survey Bulletin 124, 79 p.

Stitt, J.H., 1983, Trilobites, biostratigraphy, and lithostratigraphy of the McKenzie Hill Limestone (Lower Ordovician), Wichita and Arbuckle Mountains, Oklahoma: Oklahoma Geological Survey Bulletin 134, 54 p.

Stitt, J.H., and Miller, J.F., 1987, *Jujuyaspis borealis* and associated trilobites and conodonts from the Lower Ordovician of Texas and Utah: Journal of Paleontology, v. 61, p. 112–121.

Taylor, J.F., and Repetski, J.E., 1995, High-resolution trilobite and conodont biostratigraphy across the Cambrian-Ordovician boundary in south-central New Mexico, *in* Cooper, J.D., Droser, M.L., and Finney, S.C., eds., Ordovician Odyssey: Short papers for the Seventh International Symposium on the Ordovician System: Society for Sedimentary Geology (SEPM), Pacific Section, Book 77, p. 133–136.

Taylor, J.F., Repetski, J.E., and Orndorff, R.C., 1992, The Stonehenge Transgression: a rapid submergence of the central Appalachian platform in the Early Ordovician, *in* Webby, B.D., and Laurie, J.F., eds., Global perspectives on Ordovician geology: Rotterdam, Balkema, p. 409–418.

Taylor, J.F., Repetski, J.E., and Roebuck, C.A., 1996, Stratigraphic significance of trilobite and conodont faunas from Cambrian-Ordovician shelfbreak facies in the Frederick Valley, Maryland, *in* Brezinski, D.K. and Reger, J.P., eds., Studies in Maryland Geology: Maryland Geological Survey Special Publication 3, p. 141–163.

Taylor, J.F., Loch, J.D., Repetski, J.E., Ethington, R.L., and Myrow, P.M., 2001, Evidence of a pronounced mid-Ibexian submergence event on the southern Laurentian platform: Geological Society of America Abstracts with Programs, v. 33, no. 6, p. A321.

Toomey, D.F., 1970, An unhurried look at a lower Ordovician mound horizon, southern Franklin Mountains, west Texas: Journal of Sedimentary Petrology, v. 40, p. 1318–1334.

Toomey, D.F., and Ham, W.E., 1967, *Pulchrilamina*, a new mound-building organism from Lower Ordovician rocks of west Texas and southern Oklahoma: Journal of Paleontology, v. 41, p. 981–987.

Toomey, D.F., and Nitecki, M.H., 1979, Organic buildups in the Lower Ordovician (Canadian) of Texas and Oklahoma: Fieldiana: Geology, new series, no. 2, 181 p.

Westrop, S.R., Knox, L.A., and Landing, E., 1993, Lower Ordovician (Ibexian) trilobites from the Tribes Hill formation, central Mohawk Valley, New York State: Canadian Journal of Earth Sciences, v. 30, p. 1618–1633.

Geological Society of America
Field Guide 5
2004

The consequences of living with geology: A model field trip for the general public (second edition)

David M. Abbott Jr.

Consulting Geologist, 2266 Forest Street, Denver, Colorado 80207, USA

David C. Noe

Colorado Geological Survey, 1313 Sherman Street, Room 715, Denver, Colorado 80203, USA

ABSTRACT

This field trip focuses on the impacts of geologic hazards, natural resources development, and other geologic features on human activity along the mountain front west of the Denver metropolitan area. The trip serves both as a trip for those in the Denver area and as a model of how common, technically oriented field trips can be converted into trips to educate the general public about such impacts. The many consequences of living with geology present questions about how geologic characteristics and processes should be recognized and mitigated. The questions' answers involve complex economic and political issues that must be answered on an individual or regional basis. The primary job of the geologic community is to educate the public about the impacts and their consequences so that informed public policy decisions can be made.

Keywords: impacts of geologic processes, general public, life style, field trips, outreach.

INTRODUCTION

The Concept of the Field Trip as a Model

This is a model for the type of field trip that we, as geologists, ought to be running far more frequently in each of our communities for the general public, service clubs, politicians, school groups, religious groups, and others. The geology of any particular area imposes constraints on what humans can do. These constraints are generally thought of as geologic hazards, but access to the natural resources we all use, from sand and gravel pits to water resources to oil and gas wells to large openpit mines to particularly scenic areas, may be local issues as well. The location of roads, bridges, reservoirs, landfills and other waste-disposal sites, etc., also have both geologic and publicpolicy consequences.

As a society, we can either plan for or fail to plan for these consequences. If we allow people to build homes in floodplains, they will become flood victims. If we prohibit building on flood plains, we have impacted the value of those lands. Whose ox gets gored, how, and when? Who pays? We don't offer answers to these questions; we will simply point out that these are issues that ought to be debated and that we, as geologists, can contribute technical information relevant to the debates. We believe that it is important that such field trips be as neutral as possible on the issues addressed. We can advocate either science or policy, but not both, without being viewed as having a conflict of interest (Abbott, 2000). The goal of these trips is general education, not advocacy of a particular policy.

Discussion points, which are presented in a different typeface, are included in the text to address specific consequences of living with geology and related policy questions. Use of a

different typeface helps separate the policy questions from scientific observations. We suggest that trips based on this one use a
similar method to separate geologic and policy discussions.

Almost the whole country is covered by existing technical
field trips that can be adapted for non-technical audiences and
decision-makers. The key adaptation is a shift in focus away from
attention to details of interest only to technical specialists to larger
topics that impact the public. The technical details, although
usually simplified for this type of audience, support the goal of
addressing how geology affects people at the site in question.

Stops are selected both as illustrations of particular issues
and, generally, for ease of access by those with limited physical
abilities. We have successfully run variations of this trip for various groups. Participants have included teenagers, who showed
real interest in some of the topics.

This text is the second edition of this trip; the first edition was
Abbott and Noe (2002). The 2002 edition was developed from
and incorporates portions of the original text from Noe et al.'s
1999 Geological Society of America field trip, "Bouncing boulders, rising rivers, and sneaky soils: A primer of geologic hazards
and engineering geology along Colorado's Front Range." Some
stops from that trip have been deleted and others have been added
to expand the scope of topics examined while keeping the trip
within a manageable time frame. Changes in this edition include
updating the text, including additional figures, deletion of the
Colorado School of Mines stop because of redevelopment—but
see the "in memory of a former stop" text—and the consequent
addition of different stops. The Red Rocks Amphitheater stop has
also changed due to completed renovation of the amphitheater.
These text changes reflect the continual changes resulting from
development, which require that guides be continually updated.

Introduction to this Particular Trip

An old commercial used the phrase "It's not nice to fool
Mother Nature." The Colorado Front Range is spectacular
because of its geology, but this geology has consequences for
those of us who live here. We love the views and beautiful river
canyons, and we also want cheap gasoline, electricity, building
materials, water, and building sites with gorgeous views, no
radiation, and no trash. This trip will focus on the realities of
these desires and the degree to which they can, and cannot, be
achieved. The trip will focus on various geologic characteristics
of the mountain front and how they impact land use planning
and our lives in general.

Although this trip is similar to the sort of field trips taken by
professional geologists, it is designed to inform non-geologists
of enough geology to understand why the issues presented are
important and to engage them in some reflection about the consequences of various policies (or lack thereof) in land use.

This road log contains the detailed directions from point-
to-point along with various observations and questions about
what we will see and talk about. The trip route is shown in
Figure 1. Have someone other than the driver in each car read

the road log as the trip progresses. Although the focus here is
on what we will see, the log contains various notes on similar
problems in well-known areas around the Denver metropolitan
area and the state.

Discussion Points

Points of discussion and related questions are included throughout
this field trip log. Individually, we may or may not agree on suggested
answers to the problems and questions raised by these points; nevertheless, we should think about them. Again, "It's not nice to fool Mother
Nature." If we try, we'll fail. Geologic events occur whether we plan for
them or not. Additional discussion points may occur to you; please share
them with the group on the trip.

It is useful to look at past geologic events as a means of planning
for future events and their impact on present and future development.
The Colorado Geological Survey (CGS) has a statutory mandate to
assist local governments with planning issues involving various aspects
of geology. Since the early 1970s, CGS geologists have reviewed plans
for thousands of subdivisions along the Front Range and elsewhere within
the state. The CGS has provided emergency assistance and advice for
incidents involving geologic hazards, in addition to research and mapping
of general geologic hazards and other geologic features (e.g., Soule et al.
[1976] on the Big Thompson flood; Soule [1978] on Douglas County; and
Noe [1997] and Noe and Dodson [1997] on heaving bedrock hazards).
Geologists from the U.S. Geological Survey, local colleges and universi-
ties, and private industry have all contributed to our knowledge of the
geology of the Front Range area.

There are thousands of pages of technical literature relating to
various aspects of the geology we will see during this trip. But that is too
much information, even for most professionals. If you want to know more,
various references are listed at the end of this log. One publication written
for the general public may be of particular interest: *The Citizens' Guide to
Geologic Hazards* (Nuhfer et al., 1993), which is available from the American Institute of Professional Geologists in Westminster, 303–412–6205,
or www.aipg.org under publications. State Geological Surveys frequently
have publications written for the general public on particular issues; for
example, Noe et al. (1997), *Heaving-bedrock hazards, mitigation and land-
use policy, Front Range Piedmont, Colorado.*

GENERAL GEOLOGIC SETTING

This trip extends from the hamlet of Marshall, just south of
Boulder, past the Turkey Creek water gap, where U.S. Highway
285 and Turkey Creek come through the Dakota Hogback, and
through a neighborhood to the southeast. The route roughly follows the topographic transition from the interior of the North
American continent to the Rocky Mountains. This topographic
transition reflects the geologic transition as well. To the east, the
landscape is generally flat all the way to the Appalachian Mountains and is underlain by sedimentary rocks that are more or less
flat-lying. To the west, the hard, crystalline rocks of the Front
Range rise dramatically.

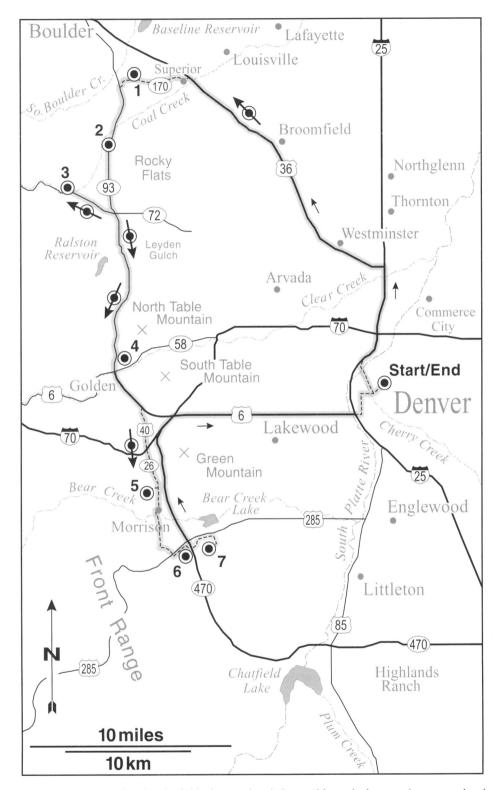

Figure 1. Index map showing the field-trip route in relation to cities and other prominent natural and cultural features in the Denver Metropolitan Area. Stops for the GSA Annual Meeting field trip are marked by numbered, bull's-eye dots; dots with arrows denote "roll-by" points of interest locations.

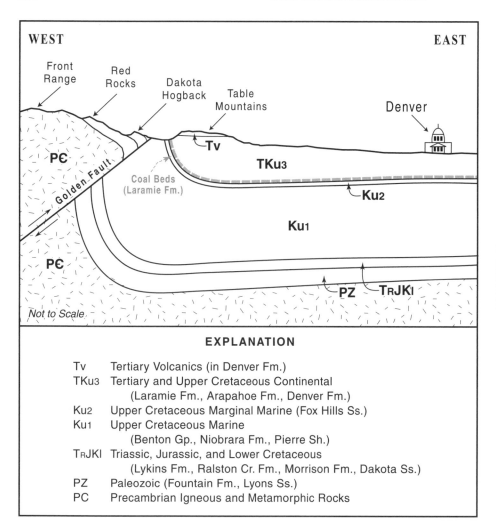

Figure 2. Schematic cross section through the Golden-Denver area showing the bedrock geology of the Denver Basin and the Colorado Piedmont (modified from Noe et al., 1999).

Along the mountain front, the sedimentary rocks have been folded and faulted from approximately horizontal to steeply angled (dipping) or vertical (Fig. 2) (Weimer and Ray, 1997). See Table 1 for a generalized stratigraphic section. Much is known about this transition and much could be said about how and when the various rock units formed and what has happened to them since. Continually acquired data answer some questions and raise more; there is still a great deal we don't know. However, today's trip focuses on geologic hazards and other geologic features affecting land use.

ROAD LOG

Trip Start

The trip for the GSA meeting begins at the Colorado Convention Center. However, for future use of this field guide, the zero point for mileage is the intersection of Marshall and Cherryvale Roads in Marshall[1] (39°57.29′ N, 105°13.40′ W), quite close to Stop 1.

Directions to Stop 1

Leave the Colorado Convention Center at 14th Street between Stout and Welton Streets by turning right on 14th Street. Proceed one block to Glenarm Street, and turn right (southwest). Proceed two blocks, and turn 45° to the right (west) onto Colfax Avenue. After one block, turn right onto Speer Boulevard (northwest), and continue to the I-25 north exit (1.7 mi [2.7 km] from the Convention Center). Continue on I-25 for 5.1 mi (8.2 km) to the exit onto U.S. Highway 36 (exit 217) to Boulder (west); the exit is from the left-hand lanes. Exit U.S. 36 at the Superior-Marshall (McCaslin Boulevard) exit, 21.0 mi (33.8 km) from the Colorado Convention Center. Turn left and cross over U.S. 36, then turn right on Marshall Road (Colorado Highway 170) at the Superior Market Place. Proceed 3.4 mi (5.5 km) to Cherryvale Road, in the hamlet

---

[1]The selection of the zero point reflects the fact that most subsequent users of this guide are unlikely to start from the Colorado Convention Center and the fact that the actual location of Stop 1 is private property. Most of the features described in Stop 1 can be observed from Marshall Road.

TABLE 1. GENERALIZED STRATIGRAPHIC SECTION FOR THE COLORADO PIEDMONT,
MORRISON QUADRANGLE (MODIFIED FROM SCOTT, 1972)

Section and description	Thickness	
	ft	(m)
Green Mountain Conglomerate (Paleocene)—Tgm	650	(198)
Conglomerate, sandstone, and shale. Contains andesite pebbles.		
Denver Formation (Paleocene and Upper Cretaceous)—TKd	950	(290)
Brown to olive-gray claystone, siltstone, sandstone, and conglomerate; contains three flows of potassium-rich basalt (Shoshonite); rich in andesite pebbles		
Arapahoe Formation (Upper Cretaceous)—Ka	400	(121)
White, gray, and yellow sandstone, siltstone, claystone, and conglomerate. Conglomerate clasts are sedimentary, igneous, and metamorphic rock.		
Laramie Formation (Upper Cretaceous)—Kl	550	(168)
Gray siltstone and claystone and yellow and white sandstone; coal in lower part.		
Fox Hills Sandstone (Upper Cretaceous)—Kfh	180	(55)
Olive to brown silty shale and yellowish-orange sandstone.		
Pierre Shale (Upper Cretaceous)—Kp	6200	(1890)
Olive-green shale, some beds of olive to gray sandstone, and limestone concretions.		
Niobrara Formation (Upper Cretaceous)—Kn		
Smoky Hill Shale Member:	140	(43)
Pale to yellowish-brown, thin-bedded, calcareous shale and thin-bedded limestone.		
Fort Hayes Limestone Member:	35	(11)
Yellowish-gray, dense limestone.		
Carlile Shale, Greenhorn Limestone, and Graneros Shale (Upper Cretaceous)—Kcgg	530	(162)
Gray claystone, siltstone, calcarenite, and hard limestone beds.		
Dakota Group (Lower Cretaceous)—Kd		
South Platte Formation:		
Yellowish-gray sandstone and dark gray shale.		
Lytle Formation:	300	(91)
Yellowish-brown sandstone and conglomerate.		
Morrison Formation (Upper Jurassic)—Jm	300	(91)
Red and green siltstone and claystone. Minor beds of brown sandstone and gray limestone.		
Ralston Creek Formation (Jurassic)—Jrc	90	(27)
Purplish-gray sandstone and siltstone, yellow silty sandstone.		
Lykins Formation (Triassic? and Permian)—TRPl	450	(137)
Maroon shale, sandy limestone, and maroon and green siltstone.		
Lyons Formation (Permian)—Ply	190	(58)
Yellowish-orange to yellowish-gray sandstone and conglomerate.		
Fountain Formation (Permian and Pennsylvanian)—Pf	1650	(502)
Maroon arkosic sandstone and conglomerate.		
Precambrian rocks—PC		
Igneous and metamorphic rocks.		

of Marshall (39°57.29′ N, 105°13.40′ W—this intersection is the zero point for subsequent mileage measurements). Turn right onto Cherryvale Road to private residence at 1670 Cherryvale Road (noted on mailbox) (0.15 mi [0.24 km]), and turn right into a parking area. This is Stop 1. *This is all private property; those on self-guided tours should examine the area from public roads or contact both property owners before entering this site.*

The mining area is reached by walking up the driveway across the street, to the northeast, and passing through a gate to the right of the house, onto another property.

Roll-By Points of Interest on the Way to Stop 1

Development Pressure along the Colorado Front Range

The Denver Metropolitan Area and other urban areas along the Colorado Front Range Piedmont were founded, and started to grow rapidly, following the initial gold rush in 1859. Denver, named after a Kansas territorial governor, was founded in 1859. The coming of the railroads greatly accelerated this growth, and a secondary service economy ensued and accelerated after the Civil War. William Jackson Palmer, a Civil War general, founded the Denver and Rio Grande Western Railroad (now part of the Union Pacific) and the city of Colorado Springs around 1870. Precious metal mining and the near-instant appearance of mining camps in the mountains, and coal mining, agriculture, and water development on the Piedmont were responsible for most of this early rapid growth. This continued unabated until the decline of the metal-mining industry following the silver crash of 1893. Many mansions, commercial buildings, and churches were built during the late 1880s. The Colorado State Capitol was started in 1890. Many of these buildings are built from Colorado native stone and are the focus of an interesting tour for a geologist (see Murphy, 1995). Colorado, the Centennial State, was admitted to the Union in 1876.

A century later, this area has seen rapid growth following World War II and during the regional "energy boom" years of the mid-1970s to early 1980s, a "bust" from ca. 1984 to the early 1990s, and another boom accompanying national trends that (arguably) continues until the present despite setbacks resulting from the dot-com crash of 2000–2001. The state of Colorado and the seven-county Denver metropolitan area grew at a similar rate of ~30% from 1990 to 2000 (U.S. Census Bureau). The regional population around Denver is now ~2.4 million, and the total Front Range regional population is ~3.6 million.[2]

The latest period of booming growth is characterized by rapid urbanization on the fringes and merging of all of the region's urban centers. This urbanization style is most apparent between Fort Collins in the north through the Denver metropolitan area (including Highlands Ranch and other communities in northern Douglas County and around Colorado Springs in the south).

According to *The Denver Post* (May 26, 2002, p. 1K), Colorado development currently occurs at the equivalent of a swath 1.5 mi (2.4 km) wide, stretching from Fort Collins to Colorado Springs (129 mi [208 km]), or 193.5 mi^2 (501 km^2) each year.

The predominant style of residential growth in this region is now single-family tract housing on lots ranging in size from ~3,500 to ~8,000 ft^2 (1/10–1/5 acre, or 325–743 m^2), with its attendant interspersed infrastructure. In recent years, the tendency has been toward construction on smaller lots with multi-story houses. This has probably slowed the per capita land consumption by development somewhat, but not its overall rate or amount. A tendency has been for infill of this higher density residential development among older, lower density residential and agricultural land uses along the urban fringes.

Discussion Points—Development Pressure

Debate over growth in and along the Front Range corridor has been raging for years. U.S. Hwy 36 between Sheridan Boulevard and

Boulder was all range and farmland (with the exception of a much smaller Broomfield—now the metro area's newest combined city and county) until ~20 yr ago when it started filling in. The pace has accelerated in the past 10–15 yr, and now little range and farmland is left and is rapidly disappearing into residential tracts, office campuses, and shopping centers. The formerly adequate four-lane highway, which had only the Broomfield exit between Denver and Boulder, is now overcrowded, and new exits have been and are being added. What you see along U.S. 36 is typical of the growth occurring between Fort Collins to the north and Colorado Springs to the south, particularly along Interstate 25.

The growth debate revolves around ways to preserve open space; at the same time, an expanding population, driven primarily by people moving into Colorado, demands affordable housing and convenient schools, grocery stores, other services, shopping, and nearby jobs. One side of the debate urges greater population densities requiring less acreage, increased reliance on public transportation, and other measures to control growth. However, land prices are climbing in the central urban areas. "Scrape offs," the practice of buying an older home, tearing it down, and building a new home that often occupies most of the buildable area of the lot, is increasingly common in the older, urban areas. Similarly, construction of row townhomes and apartment complexes changes the character of neighborhood. These changes are increasingly vigorously opposed by neighborhood groups. Neighborhood groups likewise oppose new or expanded transportation corridors, saying "Build these in someone else's neighborhood!"

Regulations added to building codes to increase health and safety are well intentioned but have the consequence of adding to building costs. Affordable housing for not just the traditionally "poor" groups but also teachers, police officers, firefighters, and an increasing percentage of the "middle" class is becoming an issue as housing costs increase. People are "voting" with their money and cars to move farther and farther away to where the cost of land is lower, resulting in lower building prices per square foot of living space. Most people still prefer to raise their families in single family residences surrounded by yards with grass.

Portland, Oregon, has been a leader in attempting to control growth by encouraging higher density housing in the urban core and developing light rail and other forms of mass transit. Despite apparent progress toward these goals, Portland reportedly finds itself with empty high-density housing as families continue to prefer suburban homes with yards.

Water availability is another serious issue. The Front Range urban corridor sits on the edge of the High Plains, in an area that explorer Stephen H. Long called the Great American desert (this was prior to penetration of the Basin and Range and Colorado Plateau). Colorado has had water courts for years debating water rights and their transfer. Growth exacerbates this problem. Downstream states and Mexico want their water as well. The current drought cycle adds pressure to the situation. Unfortunately, despite increasing public information and discussion about the drought, growth, and their interacting consequences, there is little agreement on solutions or on the need to change lifestyles and expectations, particularly in those areas where water is mined from aquifers.

Growth in the adjacent mountains is occurring along with growth on the plains. Along with increased problems of water availability and sewage disposal (wells and septic systems are not adequate solutions)

---

[2]The Front Range regional area or urban corridor encompasses the communities extending from north of Fort Collins and Greeley to south of Colorado Springs and from the bedroom communities in foothills to the west to the contiguous eastern suburbs. Cheyenne, Wyoming, on the north and Pueblo, Colorado, on the south will soon be included in the contiguous urban corridor if present growth rates continue.

comes the added **danger of forest fires**. Several large fires in recent years have burned homes and have threatened many more. Some of Colorado's worst fires burned during the summer of 2002, which set the single-year record for acreage burned prior to the "official" beginning of the fire season on June 15th. Major fires affected the westernmost Front Range suburbs during the summers of 2000 and 2002. Fires have occurred along the mountain front traversed in this trip, but as grass re-grows in the following year, the sites of these fires are obscured. Aside from the immediate dangers from burning, the burned-over areas become subject to flash flooding and mudflows due to incineration of the ground cover (Fig. 3). The mud and other debris washing into streams affect the fishery and contribute to the premature filling of water storage reservoirs.

STOP 1: COAL-MINE SUBSIDENCE AND FIRE, MARSHALL

Distance from zero point: 0.1 mi (0.16 km) north at Cherry-vale and Marshall (Hwy 170) Roads at 39°57.29′ N, 105°13.40′ W.

Marshall is the westernmost of a series of towns located along the Boulder-Weld coal field. These towns include Superior, Louisville, and Erie. Originally, these were coal mining towns rather than bedroom communities and, more recently, sites for various high-tech companies. Mernitz (1971) describes the history of Marshall and the environmental impacts of coal mining on the area. The coal occurs in several seams throughout the Laramie Formation. The large swales and ridges south of Marshall Road (Colorado Hwy 170) are related to both mining and coal mine fires. The mine fires, which reportedly started in the 1870s (Mernitz, 1971), are gradually burning out. In 1967, it was possible to smell the coal fires on a calm day along Marshall Road. Mernitz (1971) detected subtle evidence of continued burning in color infrared photographs. On cool, high-humidity days, steam can still be seen rising from cracks above currently burning areas.

The subsidence pattern in the open space area to the northeast of the intersection of Marshall and Cherryvale roads reflects the room-and-pillar mining pattern of the underground workings directly below (Fig. 4). The mine depth here is roughly 30–40 ft, and the coal seam was 5–6 ft thick. The piston-like nature of the sinkholes (depressions) reflects the very brittle, low tensile strength sandstone beds that overlie the coal (Fig. 5). Walking across the area, it is possible to detect steam and coal-gas vapors caused by a smoldering fire under this area. The irrigation ditch on the hillside has undergone significant structural strengthening through this section to maintain proper flowing grade.

The coal mine fire and related subsidence have imposed considerable land use constraints on the Marshall area. Houses have to be sited on a lot-by-lot basis, based on carefully locating unmined pillars and the limits of mining. Boulder County has acquired a good deal of land in the area for open space, thus avoiding future development problems. Significant areas of subsidence due to coal mining can be seen south of Marshall Road between Cherryvale Road and the hamlet of Marshall.

Figure 3. Aftermath of flash flooding in the 1996 Buffalo Creek fire area. The arrow points to a boulder in the crotch of a tree, indicating the depth of the debris flowing down this drainage during the flood. Photo by D. Noe.

Figure 4. Aerial photograph from a number of years ago of the collapsed coal mine near Marshall (from Knepper, 2002). "X" marks the intersection of Marshall and Cherryvale Roads. Room-and-pillar mining patterns are reflected in the subsidence features above these shallow workings. The arrow points to an area where an underground fire is currently burning within the coal seams.

Although mine fires are not a problem elsewhere in the Boulder-Weld coal field, subsidence problems exist in other areas. Louisville has been the site of extensive, detailed study (see the cover of Creath, 1996). Coal mine fires also occur in other parts of Colorado. Stressed vegetation and the lack

Figure 5. The piston-like character of collapse over a mined-out seam is shown by the addition of fencing in the collapsed area. The "regular" fence extends above Dave Noe's head and can be seen by the relative levels of the fence posts. Photo by D. Abbott.

of snowcover in the winter mark some of the fires in seams south of the Colorado River, between Glenwood Springs and Newcastle, in western Colorado. One of these seams initiated the Coal Seam Fire of 2002 that burned some homes on the western edge of Glenwood Springs. The brick red color of coal fire clinker (baked rocks overlying to coal; the color is due to the oxidation of iron) is visible near Walsenburg in southern Colorado and in the eastern Powder River Basin in northeastern Wyoming.

Discussion—Natural Resources and Electric Power

Coal is one of the chief sources of electric power. Colorado continues to mine and use substantial amounts of coal. Several coal trains, each carrying 100,000–110,000 tons, go through Denver every day. Coal only occurs in some areas, areas that may be valued for other reasons such as ranching, farming, and housing. Mining, particularly strip mining, disturbs the land.

But we all use electricity. The *Wall Street Journal* reported four years ago that our increasing use of computers and related technology is causing significant increases in the per capita consumption of electricity. Today's college students have computers, printers, TVs, phones, small refrigerators, and microwaves in their rooms. Not so long ago, an electric typewriter, clock radio, and stereo set constituted the major electrical appliances in dorm rooms.

We want the electricity. We don't want coal mining, nuclear power, oil and gas drilling, or hydroelectric dams. Solar and wind power are not sufficient to meet our demands for electricity, and do you really want to live next to a wind farm? Hydrogen fuel cells consume more energy than they produce in order to generate the hydrogen. We can't have all our wants. What will each of us, individually and collectively, do?

Directions to Stop 2

At 39°54.19′ N, 105°14.49′, turn around and return to Marshall Road, Colorado Hwy 170, and the zero point at the intersection of Cherryvale and Marshall Roads. Turn right on Marshall Road, and make a sharp left turn in 0.3 mi (0.48 km) to go up to the intersection with Colorado Hwy 93 (stop light). Turn left (south) on Hwy 93 (to Golden). Continue on Hwy 93 for 3.8 mi (6.2 km) to the parking lot on the right for the Rocky Flats Lounge; proceed to the south end of the lot. This is Stop 2.

Roll-By Points of Interest between Stops 1 and 2

Aggregate Quarry

Just south of the Jefferson County line near the crest of a hill, we will cross the conveyor belt for the shale-aggregate quarry. Prepare to turn right for Stop 2 in 0.6 mi (0.9 km). The quarry, which is well hidden from the road, is immediately behind you to the right (west).

STOP 2: ROCK AND AGGREGATE QUARRIES, NORTH ROCKY FLATS

Stop 2 is 39°54.19′ N, 105°14.49′ W; 4.4 mi (7.08 km) from our zero point.

When crossing the county line from Boulder County to Jefferson County, you will notice an increase in mining and industrial activities. On the right (west) side of Hwy 93 is a shale mining operation. The pit is several hundred feet deep and provides good exposure of the Pierre Shale, as well as an opportunity for fossil collecting (e.g., Cretaceous *Baculites* and other marine fossils). The Pierre Shale is mined and processed into a lightweight aggregate. Across the highway (east side), the shale is "baked" into a hard, light aggregate that is primarily used in the construction of buildings where the weight of concrete is an important consideration. The processed fine-sized material is also used as an alternative to sand on area roads during the winter. These materials create less dust than sand and help to mitigate some of Denver's PM_{10} air-quality index[3] concerns.

Just south of the shale-processing plant, the Rocky Flats Alluvium is mined for construction aggregate. The clays of the Laramie Formation are also mined for local brick production.

Looking again toward the mountains (west), we see a pronounced red scar along the base of the flatirons on Eldorado Mountain, ~2 mi (3.2 km) away. This is the Eldorado Canyon Quarry, which produced construction aggregate from the Lyons Sandstone. In the mid-1980s, due to lack of support by Boulder County residents, an application to expand the quarry was defeated. Boulder County Open Space has since purchased the site and attempted to reclaim it. The long, tan scars across the lower flank of the mountain are clay mines in the Dakota Group. Figure 7, which was taken further south, illustrates one of these clay mines. The refractory-grade, kaolinite-rich clay is used for industrial ceramics. Although these clay pits usually look abandoned, many of them are active mining sites, but they are not mined on daily basis. Mining only occurs when a supply of the particular clay in a quarry is needed. Clays vary in color and chemical quality, either or both of which may be needed for particular applications.

Figure 6. A mined-out portion of the Aggregate Industries Morrison Quarry, located south of the town of Morrison. This pit is hidden from public view by the high-wall on the near side of the photograph. The pit itself will be used as a water-storage reservoir by the town of Morrison. Lack of water storage for the town is currently limiting growth. Photo by D. Abbott.

Discussion—Quarries

Quarries are perceived as noisy, dusty, and ugly. But we want what comes out of them, the raw materials for concrete, for landscaping rock, for road and trails, for building stones, for clay and ceramic products, and kitty litter. Why is someone else's backyard a more appropriate location than yours? Some former aggregate quarries are now the ponds full of plants and wildlife lining the Platte River greenway, the area north of Chatfield Dam and C-470. Today's eyesore can be tomorrow's benefit (Figs. 6, 7, and 8).

Directions to Stop 3

Turn right (south) on Hwy 93 and proceed 2.3 mi (3.7 km) to the intersection with Hwy 72. Turn right (west) toward Coal

Figure 7. One of several clay pits along Colorado Hwy 93 north of Golden. Although this and other pits appear inactive, they are being mined as the particular clay available is needed for the blends used to make various varieties of brick and other ceramic products. The pit in this photo has changed since the picture was taken. Photo by D. Abbott.

---

[3]PM_{10} is a particulate matter index referring to particles <10 μ in size that is used by the Environmental Protection Agency as one of the measures of air quality. Particulates can be either solids or liquid drops. "Dirty air" is caused by particulates rather than most polluting gases, which are generally colorless.

Figure 8. Reclaimed aggregate quarry along the South Platte River north of Chatfield Dam. It is currently bordered by an upscale subdivision. Photo by D. Abbott.

Creek Canyon. After 1.7 mi (2.7 km), turn right on Plainview Road. Drive 0.2 mi (0.3 km), and pull over onto the right edge of the road. This is Stop 3 (39°52.61′ N, 105°16.15′ W).

Roll-By Points of Interest between Stops 2 and 3

Leyden Gulch

As we approach Stop 3, driving toward the mountains on Colorado Hwy 72, look to the left at the developing head of Leyden Gulch, which will be a topic of later discussion. Also, look

across the Gulch to the railroad track and the tan-painted railroad cars parked along a curve in the track (Fig. 9). These cars are permanently in place and are filled with rock. They provide a wind barrier to prevent cars on passing trains from being blown off the track. The Rocky Flats area is infamous for high winds. The high winds result from the fact that the low point in the Front Range at Devil's Thumb is on the east end of the Colorado River valley on the western slope is west of Coal Creek Canyon. Wind, like water, flows through the low points.

STOP 3: CANYON FLASH FLOODING, COAL CREEK CANYON

Stop 3 is 39°52.61′ N, 105°16.15′ W, 8.6 mi (13.84 km) from our zero point.

All larger drainages along the Front Range head in the mountains and fall rapidly from alpine elevations (above 9,500 ft [2900 m]) through steep-gradient canyons. These debouch at the mountain front onto the piedmont, where stream gradients lessen substantially. There is typically 6,000–8,500 ft (1830–2590 m) of relief from the highest mountain summits to the highest parts of the piedmont. The resulting orographic and meteorological effects on streamflow can be pronounced and extreme.

At this locality near the mouth of Coal Creek Canyon, we can see a sequence of flood deposits produced by events similar to the 1976 Big Thompson Flood. The Colorado Geological Survey (CGS) obtained a ^{14}C date from this locality of 955 ± 80 yr B.P. from what are interpreted to be the youngest of three flood deposits. CGS ^{14}C dates from similar materials at other Front Range localities range from ca. 10,000 yr B.P. to 300 yr B.P.

The greatest historical flood along the Colorado Front Range occurred in the drainage of the Big Thompson River, Larimer

Figure 9. The tan gondola cars above Leyden Gulch are filled with dirt and have been placed on the windward side of the rail line to help prevent derailments caused by high winds in this area. Photo by D. Abbott.

County, during the night of July 31–August 1, 1976. This flood can be attributed to one thunderstorm event, when a large cell remained nearly stationary over the middle part of the drainage basin for ~2.5 h. Nearly all of the tributary streams in this part of the basin were involved. East of Drake, Colorado, relatively little rain fell during the event. A peak discharge of the Big Thompson River of ~39,000 ft³/s was computed at the mountain front, ~4 mi west of Loveland. An estimated 139 people were killed during this event, and property loss of about $40 million resulted (1976 dollars). Overbank flooding downstream from the mountain front occurred through the city of Loveland to the river's confluence with the South Platte River. The Colorado and U.S. Geological Surveys documented the geomorphic effects of this event (Soule et al., 1976; McCain et al., 1979). These effects included debris avalanches and flows, deep scour of ephemeral streambeds, deep sheet erosion on hill slopes, and deposition of prodigious amounts of sediment on gentler slopes above normally active stream channels and on the river's piedmont floodplain. Clearly, being above the high water mark in the channel is not equivalent to safety for either people or structures.

Flood events of similar or lower magnitude have undoubtedly occurred in many Front Range drainages during the Holocene (Soule, 1999). Soule, Costa, and Jarrett of the Colorado Geological Survey studied paleoflood deposits in fourteen drainages along the Front Range during the late 1970s and early 1980s. This was done to support a broad research proposal to model Front Range paleofloods and modern floods. The study was never completed because of "fiscal malnutrition." Evidence for these occurrences include geomorphology of paleoflood deposits, discharge estimates based on flows necessary to move largest clasts (boulders) entrained in paleoflood deposits, depth of scour in streambeds, superposition of streamflows in channels, ¹⁴C dating, study of demonstrable glacier-related deposits, and meteorological computations. Jarrett (1987) offered field evidence for and proposed that the Big Thompson event has a recurrence interval of 10,000 yr and that storms of the magnitude to produce such an event must form below an altitude of 8,500 ft. Some debate about this continues, however.

Later in the trip we will be crossing several of these streams: Ralston Creek, Van Biber Creek (Stop 4), Clear Creek, Mount Vernon Creek, Bear Creek, and Turkey Creek. Ralston Creek is the only one with a dam upstream of where we will cross. Clear Creek has no flood control dams. The Bear Creek dam is well east of the Dakota Hogback at Morrison, leaving the town of Morrison unprotected from floods along Mount Vernon Creek and Bear Creek; both streams have flooded out Morrison within the lifetimes of some of the people on this trip. In 1969, Turkey Creek flooded and eroded out all four lanes of U.S. Highway 285 in one place and two lanes for a good portion of the distance between Tiny Town and the mountain front near The Fort Restaurant. In June 1965, the Plum Creek–South Platte flood eroded out I-25 at Castle Rock and collapsed or damaged most the of bridges south of Colfax Avenue. Although Chatfield Dam was built to prevent similar flooding in the Denver area, the areas upstream of Chatfield Dam remain unprotected and vulnerable.

Even very small drainages can flood and cause damage. In June 2004, a number of small storms around Golden precipitated 2 to 3 in (5–7 cm) of rain in short periods of time. One small drainage in north Golden flooded and destroyed a home built on the lower part of what was probably viewed as an insignificant gully. Significant amounts of mud and sand were deposited on Golden streets prior to reaching Clear Creek. Significant erosion also occurred in part of the Van Biber Creek channel (Stop 4).

The valley to the south of Highway 72 (up which we will drive) is the headwaters area for Leyden Gulch (Fig. 10). It is a fairly prominent valley that parallels the pronounced break in slope at the base of the foothills. The upper end of the gulch contains a steep channel that has eroded to within three-quarters of a mile (1.2 km) of the main channel of Coal Creek. It has migrated between 200 and 250 ft (61 and 76 m) northward in the past 30 yr. If this rate of advance continues, it is logical to expect that Leyden Creek will capture the upper (mountain) part of Coal Creek, an event known as stream piracy or stream capture.

Discussion—Flooding and Development

If a flood like the 1976 flood on the Big Thompson were to come down Coal Creek, the large boulders and other debris carried within the mountain canyon would drop out of the stream after the stream exits the

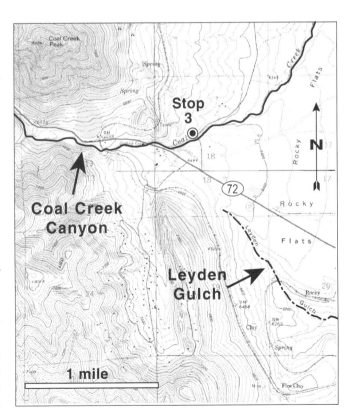

Figure 10. Topographic map showing the proximity of Leyden Gulch to the mouth of Coal Creek Canyon. The head of the gulch is eroding into soft Pierre Shale. How long until Leyden Gulch captures Coal Creek?

mountain front. This would tend to block the channel and cause water to spread out. If some of this water spread into Leyden Gulch, the stream capture process could take place rather rapidly (tens of minutes to a few hours?). If this were to occur, western Arvada would face an unplanned-for flood threat. If Coal Creek were not captured by Leyden Creek, flood damage to newly developed areas downstream on Coal Creek, especially near Superior and Louisville, could be significant.

Flood control dams are not popular with those favoring open rivers. They also do not protect those upstream from the dams. Furthermore, dams have been known to fail. Even small dam breaks can generate catastrophic results; for example, the Lawn Lake Dam failure in Rocky Mountain National Park that swept through Estes Park in 1982.

Even downstream of dams, developing housing in an area results in significant increases in surface runoff because roofs, driveways, walks, etc., are impermeable. They cannot absorb water or slow it down. As an area is converted from pasture or farm to urban houses and shops, the size and suddenness of flash flooding from the same-sized rainstorm can increase. As the small but high intensity storms in Golden in June 2004 demonstrated, even very small drainage basins can flood and cause significant damage property.

Figure 11 shows a small, tributary drainage to Coal Creek that could well be subject to damaged by a storm similar to those in Golden in June 2004 or as part of a 1976 Big Thompson–type event.

Figure 11. Looking up a small side drainage on the south side of Coal Creek at the mouth of Coal Creek Canyon near Stop 3. The homes in this drainage could be damaged by a storm event like the 1976 Big Thompson flood or an event like those that occurred in Golden in June 2004. Photo by D. Abbott.

Directions to Stop 4

Stop 4 is at 39°46.00′ N, 105°13.69′ W; turn around and return to Hwy 72. Turn left (east), and go to Hwy 93, then turn right (south) toward Golden. Once in Golden, turn left at the third stoplight, onto Iowa Street. The service station just off Hwy 93 and Iowa is a good place for a rest stop. Go three blocks on Iowa Street to the stop sign at Ford Street. Turn right on Ford Street and then almost immediately turn left into the parking lot of the Hillside Community Church. The lower parking lot on the east (far) side of the Church is Stop 4.

Roll-By Points of Interest between Stops 3 and 4

Devil's Backbone, Clay Pits, and Gas and Storage
Descending into Leyden Gulch on Hwy 93, we will travel along the base of a near-vertical exposure of lower Laramie Formation sandstone (locally know as the "Devil's Backbone"). Between these vertical sandstone ribs are the collapsed stopes of old, abandoned clay and coal mines and uranium prospects (Fig. 12). Coal was also mined east of the "Devil's Backbone" in the Leyden mines described in the following discussion points on the former gas storage, now water storage, facility.

Discussion Points—Natural Gas Supplies for Urban Areas
Keeping a supply of natural gas near urban areas is a continual problem for utility companies. These supplies are required to meet surges in demand during winter cold snaps and other events. On the east side

Figure 12. A former clay mine on the east side of the vertically-standing Devil's Backbone. Note the "window" over part of the pit. This area is now owl-nesting habitat. Photo by D. Abbott.

of the Devil's Backbone hogback, down Leyden Gulch, is the site of the former Leyden coal mine gas storage facility, formerly used to store 3.5 billion cubic feet of gas in an old coal mine. This facility was decommissioned following the discovery of leaking gas. Prior to the installation of the Leyden Mine facility, Public Service Company of Colorado, now part of Xcel Energy, used storage tanks like that pictured in Figure 13 for urban supplies. David Abbott remembers one of these tanks being located about a block from a junior high school. Clearly, this is not the type of facility that could be approved for construction in a developed area today. The Leyden facility replaced such storage tanks in the metro Denver area.

Closure of the Leyden facility has resulted in new techniques to provide the required surge-gas storage capacity. These techniques include the use of larger diameter, high-pressure mains and storage in abandoned gas reservoirs at distances not too far from the Front Range corridor. We are fortunate that the local geology contains such reservoirs in the Denver Basin. This would not work in other parts of the country such as New England. However, as demonstrated by the events in Hutchinson, Kansas, on and following January 17, 2001, where a sudden release of gas resulted in the destruction by fire of a number of businesses, using old fields for storage is not without problems. Lee Allison (2001), Kansas State Geologist, provides an excellent summary of the events.

The old Leyden Mine has been converted into a water storage facility. Underground water storage has the advantage of not being subject to evaporation losses. Because the coal extracted from the Leyden Mine was low in sulfur, acid resulting from oxidation of pyrite is not expected to be a problem in this facility. Figure 6 shows another example of a mine (quarry) being converted to water storage.

Figure 13. A former type of urban gas-storage tank used in the Denver area. The tank moved up and down on rollers as gas was added or withdrawn. Some of these tanks were located in residential areas. Photograph courtesy of Xcel Energy.

North Table Mountain

As we approach the town of Golden on Hwy 93, North Table Mountain comes into view (see Fig. 14). Both North and South Table Mountains are basalt-capped. The underlying sedimentary rocks are flat-lying and range in age from Cretaceous through Tertiary. The famous K-T boundary, the extinction event marking the end of the dinosaurs (except the birds), can be located on the slopes of North and South Table Mountains.[4]

Clay Pits in the Dakota Hogback

As you proceed south, clay pits in the Dakota Hogback can be seen on the right. The prominent, planar surfaces of tan-colored rock are bedding planes in the Dakota sandstone, which are steeply dipping. The clay-pit stopes also follow the bedding planes, and were typically dug downward until the water table was encountered.

STOP 4: ROCKFALL AND LANDSLIDES, NORTH TABLE MOUNTAIN, GOLDEN

Stop 4 is at 39°46.00′ N, 105°13.69′ W; 18.5 mi (29.8 km) from our zero point.

The rockfall and landslide hazard area around North Table Mountain was mapped by the U.S. Geological Survey during the 1970s (Simpson, 1973a, 1973b). Jefferson County has adopted the mapped hazard area as part of its geologic hazards overlay, and has considered it to be a "no-build" area. This was challenged in the 1980s by a developer, who staged an actual rock-rolling demonstration on the northwest flank of the mountain in order to prove the fallacy of the outer (distal) hazard-area boundaries. The demonstration was curtailed (and the subdivision application was subsequently denied) after several boulders rolled beyond the outer boundary, but not before one boulder had bounced at least 10 ft (3 m) into the air and knocked a cross-bar off a high-tension power-line tower at the base of the mountain (J. Hynes, 1999, personal commun.).

The rockfall and landslide hazard areas, as mapped by the U.S. Geological Survey, exist in both unincorporated Jefferson County and the incorporated city of Golden. Unfortunately, protection of the public from these hazards appears to stop at the jurisdictional boundary in this case (see Fig. 14). The houses on the hillside are located within the city of Golden, which has not adopted Simpson's maps. This subdivision was approved and built due the absence of any requirements for home-rule cities to follow the state-mandated, geologic-hazard review process for these subdivisions.

---

[4]Some controversy exists about whether the K-T boundary marked by impact debris and anomalously high iridium values caused or merely coincided with the mass extinction at the end of the Cretaceous. Most current evidence suggests the events coincided, but this does not necessarily imply a causal link. Likewise, whether birds should be classified within the dinosaurs is debated, although recent evidence from China strengthens the classification of birds as dinosaurs.

Building continues in this area of Golden. Figure 15 shows a "premium lots available" sign taken in late May 2004. The available lots are above existing homes. Figure 16 shows Dave Noe standing on the rocks excavated from a house under construction above existing homes in late May 2004. The source of these rocks is the basalt cliff on the skyline. Rockfall and landsliding have presumably transported the rocks to their present locations.

Discussion—Rockfalls and Landslides

Rockfall and landslide hazards are common in Colorado. Any area with cliffs and steep hillsides may have rockfall potential. The rimrock areas of Douglas County are subject to rockfalls, landslides, and debris flows. Some scars from the 1965 Plum Creek–South Platte River flood are still visible south of Castle Rock. Think about rockfalls the next time you drive through the Vail Valley. There are also some fascinating debris flow, mudflow, and landslide features in the Vail area.

Landslides on steep slopes can be a significant hazard. The destruction of several homes in one of the subdivisions on the north side of Green Mountain starting in 1998 is but one example. The existence of the landslides had been geologically mapped prior to development. The developer's geologic consultant recommended remedial actions that were subsequently ignored.

Several of the references listed at the at the end of this paper are maps published by the Colorado Geological Survey and the U.S. Geological Survey showing geologic hazards in various parts of the metro area (Chase and McConaghy, 1972; Scott, 1972a, 1972b; Simpson, 1973a, 1973b, 1973c; Spencer, 1961; Van Horn, 1957, 1972, 1976). Similar maps are available for other areas.

Why isn't there stronger geologic hazard zoning? Well, who is going to pay landowners for their lost property values? Think about how you would feel if your family ranch was put off-limits to development rather than providing the money for your children's and grandchildren's education. Somebody will suffer economic loss, individuals or taxpayers or both. The political debate involves whom and how much.

IN MEMORY OF A FORMER STOP: DIFFERENTIAL CLAY-PIT COMPACTION AND EXPANSIVE SOILS, CSM MARRIED HOUSING

Earlier versions of this trip stopped in the married student housing area of the Colorado School of Mines. After many years, the damaged buildings featured at this former stop have been torn down and new buildings built (maybe Mines got tired of being a "how not to do it" geologic hazards field trip stop). But we still have pictures that document what happened and so have left the relevant text in this field guide for your consideration.

This former stop looked at the challenges of multiple-sequential land use as related to clay mining operations and reclamation along the western side of Golden. Here, an array of open-stope clay pits along the east side of U.S. Highway 6 were reclaimed for several uses with a variety of problems and solutions, some more successful than others.

Figure 14. Photograph of North Table Mountain from the west. The capping rock of North and South Table Mountains is basalt. The flanks of this mesa are mantled with landslide and rockfall deposits. Photo by D. Abbott.

Figure 15. A "lots available" sign located above previously existing homes on the slopes of North Table Mountain in late May 2004. Photo by D. Abbott.

Figure 16. Dave Noe standing on rocks excavated from a home under construction above existing homes on the slopes of North Table Mountain in late May 2004. Photo by D. Abbott.

Figure 17. Heavily damaged married student housing at Colorado School of Mines. The building, which was razed during the summer of 2004, experienced differential settlement; part of it straddled a worked-out clay mine and landfill, while the other part was on a stable sandstone rib. The added lines along the roof segments were originally parallel and the highlighted former window was originally square. Photo by D. Abbott.

Figure 18. Cracks in the side of the building shown in Figure 17. The cracks are partially filled with calking used for earlier repairs, but the settling and resulting damage continued. Photo by D. Abbott.

Severe differential-settlement problems were experienced in the student housing at Colorado School of Mines, along Campus Road (Figs. 17 and 18). The fill used to reclaim this area settled, perhaps only a few percent of total volume, but the bounding sandstone ribs are stable. Major differential settlement and damage occurred where structures, flatwork, and roadways straddled these highly variable units. In 2004, the old student housing buildings were razed to make way for new buildings. Engineers working at the site were surprised when two sinkholes opened up overnight during site grading. Unless the conditions presented by the vertically dipping sandstone and clay strata are properly addressed in planning and appropriate mitigation taken, similar structural failures may occur with the new buildings.

South of the CSM campus, the clay pits were simultaneously used as landfills and to dispose of fly ash. Unresolved water quality issues led to the termination of this program, and the remaining reclamation is being performed with unregulated materials such as random fill. This may prove to be a future land-use issue for the city of Golden.

Some of these pits have been converted into the Fossil Trace Golf Course. In some of the exposed Upper Cretaceous sandstone exposures, dinosaur tracks are visible. The development of this golf course provoked some controversy over the number of trackways that would be preserved and the amount of public access available for those wishing to visit them without playing golf.

There are many reclaimed dumps—the older word for landfill—around the Denver area. Cherry Creek Shopping Center sits on one. The site of Mile High Stadium (recently demolished) and Invesco Field occupy an old Denver city dump.

Directions to Stop 5, Red Rocks Amphitheater

Stop 5 is at 39°40.00′ N, 105°12.40′ W: Exit the church parking lot by turning left (south) on Ford Street. Proceed downhill 0.7 mi (1.1 km) to 10th Street and turn right. Go two blocks to Washington Avenue. (The field-trip lunch stop will be at Parfet Park, located at 10th and Jackson, one block west of Ford.) Turn left on Washington Avenue. Cross Clear Creek, and continue through downtown Golden and up over a hill to 19th Street; turn right on 19th Street. At 6th Avenue (U.S. Hwy 6), turn left (south or east, depending on how you look at it). After 1.3 mi (2.1 km), turn right at the light onto Heritage Road. In another mile (1.6 km), merge on to U.S. 40 going south (or west) and continue under I-70, where the road designation changes from U.S. 40 to Colorado Hwy 26, and continue going south toward Red Rocks Park and Morrison. After 1.4 mi (2.3 km), turn right into Red Rocks Park (Gate 1). Follow the main road 1.5 mi (2.4 km) to the top of the amphitheater. This is Stop 5.

Roll-By Points of Interest between Stops 4 and 5

Clay pits along 6th Avenue

As you travel along 6th Avenue, there are various clay pits on the left in various states of being filled. One hopes that the lessons of Colorado School of Mines' experience with student housing have been learned here. The new Fossil Trace Golf Course occupies some of these pits.

I-70 Road Cut—Point of Geologic Interest

There are walkways along both sides of I-70 as it crosses the Dakota Hogback, which allow detailed examination of the exposed upper part of the Jurassic Morrison and Cretaceous Dakota Formations. LeRoy and Weimer (1971) published a detailed guide to both sides of the cut with pictures and detailed stratigraphic sections. Unfortunately, weathering and erosion in the >33 yr since the cut was first made obscure some of the detailed features of the stratigraphy. The Morrison Formation on the west side of the cut is variegated reds and greens and consists of sandstones and mudstones. It is the classic "Jurassic Park," from which skeletons of many well known dinosaurs have been excavated. It was originally described at and named for the town of Morrison, where some of the earliest excavations were made by Arthur Lakes. We'll discuss both the Morrison and the Dakota Formations in more detail at Stop 5.

Mount Vernon Creek Floods

After passing under I-70, the road follows Mount Vernon Creek down to Morrison. While Mount Vernon Creek is normally practically dry, it can flood. When it does, downtown Morrison could be in trouble. Water has flowed through the buildings in downtown Morrison in the past, and the increased construction of homes in Mount Vernon Canyon is probably exacerbating the flood problem by creating large, impermeable areas.

Alameda Road Cut and Red Rocks Park

The Alameda road cut and the north entrance to Red Rocks Park intersect Hwy 26 ~1.4 mi (2.3 km) south of I-70. The Alameda Road cut is the site of Dinosaur Ridge, a very worthwhile stop for those interested in dinosaurs. Red Rocks Park features some geologic exhibits and is an excellent place to examine some of the features of the mountain front.

STOP 5: BRINGING GEOLOGY TO THE PUBLIC: RED ROCKS AMPHITHEATER

Stop 5 is at 39°40.00′ N, 105°12.40′ W; 27, 1 mi (43.6 km) from our zero point.

The Red Rocks amphitheater is where many Coloradoans and visitors become aware of geology. The site is a marvel of natural rock formations, hewn stone, concrete, and wood (Fig. 19). Remnant hogbacks of the red Pennsylvanian-Permian Fountain Formation flank the seating areas and, along with the panoramic view of the Denver area, provide a dramatic backdrop for the stage. Built by 200 men from the Civilian Conservation Corps (CCC) and opened in 1941, the amphitheater is a tremendously popular attraction. It is visited by 750,000 people each year, including 350,000 who attend concerts given by national and international musicians. The Easter sunrise service at Red Rocks is a long-time favorite event for many Colorado families.

The amphitheater has recently undergone a major renovation to improve the facilities and infrastructure and to repair damage that has occurred over the years. In particular, the southeast part of the seating area was undergoing movement, deformation, and cracking. Was the amphitheater built on a landslide? Investigations by the Colorado Geological Survey and private geotechnical

Figure 19. Red Rocks amphitheater. Photo by D. Abbott.

firms determined that the damage was being caused by settlement and movement of fill, which was found to have been dumped into place as uncompacted fill by the CCC during construction (Dodson et al., 2003). Poor storm drainage had contributed to the problem, and large void areas had formed beneath the seats due to erosion and transport of the granular fill material. The area was remediated using compaction grouting, reinforced-soil slopes (RSS), and ground anchors. Lightweight, concrete flow-fill was pumped beneath the seats to fill the eroded voids.

Two of the more challenging aspects of the renovation were the site's historic designation, which required every removed item to be cataloged and reinstalled in its original position, and the need to design the improvements in harmony with the terrain and landscape. One of the highlights is a new, underground visitor center at the top of the amphitheater. The geologic constraints of this particular site necessitated the use of soil-nail walls and mechanically stabilized earth (MSE) walls up to 35 ft (10.7 m) high. The design has been form-fitted into the existing geology, exposing large boulders. The new visitor center includes a display of the geologic evolution of Red Rocks Park.

Discussion—Geology and Public Education

Over the years, the Rocky Mountain Association of Geologists has constructed a number of signs identifying formations and other explanatory plaques on the geologic features of Red Rocks Park. In the past few years, the Friends of Dinosaur Ridge have done a great deal of work preserving, making accessible, and providing explanations of Dinosaur Ridge along the Dakota Hogback immediately east of Red Rocks Park. These are commendable efforts at public education.

Natural history museums also assist in public education; however, museums tend to focus on mineral collections and fossil displays, which, while part of geoscience, are still only a small part of a broad field. Most geologists are neither mineralogists nor paleontologists, even though we know something of these subjects and may use them regularly. Do such exhibits contribute to the view expressed by some that geology is a form of stamp collecting?

How can we teach the public about broader aspects of geology? This trip is one method, but one that does not address the whole of the field either. Many of us became geologists because we took Geology 101, for whatever reason, and realized what a varied and fascinating field geology is. John McPhee's books, assembled in *Annals of the Former World* (1998), provide a sense of the coverage of the field and some of the things that we do. Again, think about how each of us can contribute to a broader understanding and appreciation of the variety of things that we collectively do for a living and that provide vital information and resources for society as a whole.

Directions to Stop 6

At 39°37.09′ N, 105°10.15′ W, turn around and drive out of Red Rocks Park, then turn right (south) on Hwy 26. When you reach the junction with Hwy 8 in downtown Morrison, turn right (west). Turn south at the second stop light, following the sign to U.S. 285 and crossing Bear Creek. Proceed south 1.6 mi (2.6 km), and turn left on Turkey Creek Road (at 39°37.89′ N, 105°11.58′ W). Proceed east for 1.6 mi (2.6 km) (watch for the sharp turn taking you under U.S. 285) to where you can pull off on the right side of the road near the far (east) end of the cut in the Dakota Hogback. This is Stop 6.

Roll-by Points of Interest between Stops 5 and 6

While driving south from Morrison, notice the lack of the type of suburban development observed along U.S. 36 between Denver and Marshall. Lack of water taps is part of the reason. Further south (south of U.S. 285 and Turkey Creek), suburban development begins again.

The access to Aggregate Industries' Morrison quarry is 1.2 mi (1.9 km) south of Morrison. Although not much more than the upper part of the crushing plant is visible from the road, Figure 6, the photo of the pit that will be turned into a water reservoir for Morrison, was taken in part of this quarry. This quarry has won various environmental awards for shielding its operations from view and for reclaiming those high-walls that can be seen from Denver (Fig. 20).

STOP 6: OIL RESERVOIR AND URANIUM DEPOSIT, TURKEY CREEK GAP

Stop 6 is at 39°37.09′ N, 105°10.15′ W, 35.5 mi (57.1 km) from our zero point.

Figure 20. Reclaimed high-wall in part of the Morrison aggregate quarry. The upper levels of the high-wall have been "painted" with a special mixture that creates a naturally weathered look. The benches have been backfilled and trees planted. The right-hand and lower parts of the quarry were part of the active operations when this picture was taken. Most people seeing this from the Denver metro area don't recognize this as a quarry high-wall. Photo by D. Abbott.

Oil Deposit

The tan sands at the top of the Dakota Formation contain a "fossil" oil deposit that can be seen as a distinct color change to a more greenish gray (see Fig. 21). The oil deposit is "fossil" because all of the lighter hydrocarbons have evaporated off, leaving only the heavy, asphaltic components. If you smell one of the oil-bearing rocks on a hot day, you may get a whiff of the oily smell. Breaking a piece of rock float open may help.

One of the main purposes of this stop is to examine what a hydrocarbon reservoir usually looks like. It is not an underground pool (there are very rare exceptions). Rather, it is a portion of a rock unit in which spaces or pores occur between the sand grains. The amount of this space, which can be filled by hydrocarbons or other fluids, is known as *porosity*, one of the two "million-dollar words" in the petroleum business. The other million-dollar word is *permeability*, which describes the interconnection of the pores, allowing fluid to flow.

A kitchen sponge and a Styrofoam coffee cup illustrate porosity and permeability. Both the sponge and the coffee cup are mostly pore space; they both have extremely high porosity. In the sponge, the pores are interconnected, allowing water to flow through; the sponge has very high permeability. The coffee cup has essentially the same porosity as the sponge, but the pores aren't interconnected, and the cup holds coffee. The coffee can't leak out unless the cup is cracked.

A well-completion technique known as hydrofracing (hydraulic fracturing has more syllables) artificially improves permeability. The process consists of pumping a fluid and a propping agent like sand into the rock from the well bore with sufficient pressure to fracture the rock. The propping agent remains in the newly formed fractures to help hold them open when the pressure is reduced and the fluid is pumped back out of the rock and the well.

The Dakota Formation sandstones exposed here are known as the D-J sandstones, which are important oil and gas reservoirs in the Denver Basin of northeastern Colorado.

Uranium Deposit

In addition to the fossil oil, the south side of the Turkey Creek gap contains a roll-front uranium deposit. This is most clearly seen as the "C" shape of the uphill side of the top part of the oil deposit (see Fig. 21). The shape reflects the cross-sectional shape of a tongue of groundwater moving through the rock. The groundwater picks up very small amounts of uranium and other elements weathering out of the mountain rocks. These elements are in an oxidized state. When the groundwater encounters reducing conditions (which can be caused by organic material like hydrocarbons) in the rocks through which the water is flowing, the uranium and other elements, like molybdenum (the "moly blue" in Fig. 22), precipitate in a reduced state. Roll-front uranium deposits represent one of the major types of uranium deposits in Wyoming and in western Colorado, southeastern Utah, and northwestern New Mexico.

Figure 21. Annotated picture of the south side of the U.S. Hwy 285 road cut showing the location of the fossil oil reservoir and the oxidized back of the roll-front uranium deposit superimposed on the reducing "fossil" oil. Photo by D. Abbott.

Figure 22. An annotated picture of a tongue of ilsemanite ("moly blue") illustrating the "C"-shaped nose of precipitation of minerals resulting from the movement of element-charged groundwater through the D-J sandstones that led to the formation of the roll-front uranium deposit. Photo by D. Abbott.

The problem with uranium deposits is the associated radiation. This is a Naturally Occurring Radioactive Material (NORM) site. An important point to remember is that radioactive elements are present throughout the environment in trace amounts. No place is radiation free.

New Hampshire is famous for its granite, and New Hampshire's granites are relatively high in uranium and other radioactive elements like thorium and potassium (^{40}K). People have been living in New Hampshire for over 200 yr without exhibiting abnormal health problems. This leads to the conclusion that the background radiation amounts in New Hampshire do not cause health problems. Although we cannot have zero radiation risk, there has been a long-term history demonstrating that normal, background amounts of radiation do not pose an abnormal health risk.

Nevertheless, there are areas where anomalously high amounts of radiation occur. The most common radiation problem is due to radon gas. Radon detectors provide very precise measurements of radon concentration. Because radon concentrations depend on atmospheric pressure and other weather-related phenomena, long-term measurement (several months) is required for accurate measurement of radon levels. If a radon problem is discovered, its solution is often not very expensive and involves providing improved ventilation of basement spaces.

Turkey Creek Flash Flooding Hazard

Turkey Creek, which normally doesn't have a particularly high flow, passes under this roadcut in a culvert. Turkey Creek has experienced flash flooding from time to time, and the potential exists for flood water and contained debris to cover and block U.S. 285, perhaps even damaging it. During heavy rains in early June 2004, Turkey Creek's flow was high enough to begin encroaching on the frontage road, although the road was not blocked.

Discussion—Natural Resources Are Where They Are

We all use natural resources, but most people object to mines and oil wells. They're not pretty. They can be messy. Natural resources are not located in places that we might view as convenient. The San Juan Mountains are scenic, in part because they were formed by processes that resulted in the deposition of a variety of metals. In *Geo-Logic*, Robert Frodeman (2003) provides an excellent discussion of the problems of acidic and metal-bearing waters in the Silverton Caldera in the southwestern San Juan Mountains. The area has been actively mined, but the hydrothermal alteration and mineralization that attracted the miners preexisted mining and generated acids and released metals naturally. Determining how much mining altered the natural system is difficult and debatable.

As this stop and the coal mines and clay quarries we've seen earlier show us, such deposits are where they are. If we want the products, we must exploit the resources. Even if we don't extract them, the deposits and the processes that form and modify them continue to work. It is no surprise that Leadville and Aspen have high concentrations of lead and other metals in the soil. Mother Nature put it there. Trace elements naturally occur in our water supplies. Denver's water is naturally high in fluorine and molybdenum, among other elements, because of the geology of the watersheds from which Denver gets its water.

Radiation occurs naturally. The amount of radiation in an area depends on the rock types present. Weathering and other geologic processes can concentrate uranium and other radioactive minerals in deposits like the one

at this stop. These are the sites of greatest radiation hazard. Published uranium exploration data are available that locate potentially hazardous areas, but this is not information readily available (both access to the documents and text comprehendible by the layman are a problem) to or consulted by land planners and developers. How should the available information and those who should be using it be brought together? The geologic information needs to be in a form the public understands, not in the form of typical technical reports. And those who could benefit from using the information must be made aware that the information is available for them to use. What can we, as geologists, do about this?

Roll-By Stop 7: Heaving Bedrock in Southern Jefferson County

This section contains narratives for a combined stop and roll-by tour. The field trip will visit areas of historical damage; additional areas may be visited as well.

Directions: continue along the frontage road to Quincy Avenue (stop light) and turn left (east). Descend a long hill, and turn right onto Simms Street (south) at the stoplight, then right onto Marlowe Avenue. The first damage-area tour consists of taking a right onto Union Street, left onto Urban Way, and left onto Tanforan Avenue back to Marlowe Avenue (east). Stop 7 is in the vacant lot at the corner of Marlowe, Union, and Tanforan.

This field trip mini-tour is designed to show the effects of differentially heaving bedrock on past, present, and future development areas of the Upper Cretaceous Pierre Shale outcrop belt. On Quincy Avenue, notice the long, parallel "speed bumps" that have severely deformed the pavement is several areas. We will see numerous examples of damage similar to that shown in Figure 23 on the Union-Urban-Tanforan loop.

Figure 23. A house damaged by heaving bedrock and displaying the typical "bat wings" at the bottom of the garage door caused by the heaving of the driveway and garage cement slabs. The lines drawn along the porch and garage gutters are not parallel, reflecting the structural damage caused by the swelling bedrock layers. Photo by D. Abbott.

Differential ground heaving has adversely affected development projects for over 30 yr in southern Jefferson County, in an area that is underlain by the Pierre Shale and other Upper Cretaceous formations. This geologic hazard is manifested by the progressive growth of long, somewhat parallel ridges separated by relatively inert swales. Over 2 ft (0.6 m) of differential movement has occurred in some cases within a few years following development. Certain neighborhoods in the Pierre Shale outcrop belt have performed well over the years, while others have sustained tremendous amounts of damage to structures, roads, and utilities. One 110-home neighborhood reportedly incurred more than $5 million worth of damage and mitigation costs after 15 yr, and the ground is continuing to heave.

Until around 1990, these problems were often attributed to the geological hazard of swelling soil. However, commonly used mitigative designs for swelling soil, such as drilled-pier foundations, floating-slab floors, and structural floors have been remarkably unsuccessful to date. Subsequent research by the U.S. Geological Survey and the Colorado Geological Survey (Nichols, 1992; Noe and Dodson, 1997) indicates that the differential movement occurs within near-vertical claystone bedrock beneath the ground, resulting from the combined effects of wetting and unloading surfaces. The hazard has been called "heaving bedrock" to alert engineers to a need to depart from standard "swelling soil" considerations. This area provides an excellent example of the problems with "best" or "standard" practices pointed out by Abbott (2003) and in Figure 24.

As a result of Colorado Geological Society education efforts, Jefferson County convened a task force of nearly 75 stakeholders in 1994 to develop minimum standards for site exploration, evaluation, and design to mitigate heaving bedrock. New land development regulations were enacted for the Dipping Bedrock Area in 1995 (see Noe, 1997). Trenching has been implemented to evaluate the geometric complexities of the dipping bedrock. Over-excavation and fill replacement to at least 10 ft (3 m) beneath foundations are now specified as minimum site construction standards if the geologic evaluation finds that the bedrock has a potential for differential heave. The new Jefferson County regulations have resulted in a significant reduction in damage to new homes and infrastructure.

Discussion—Expansive Soils

Expansive soils are a problem throughout much of the metro area. These are due to the presence of certain clay minerals, particularly bentonite[5], which can take water into their crystal structure and swell (which is how Kaopectate works) or later dry out. In the early 1960s, the flat-

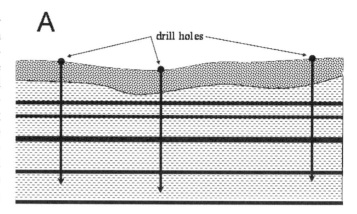

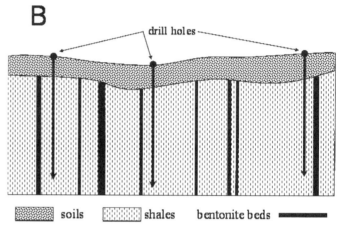

Figure 24. The upper diagram (A) illustrates the standard practice developed to test for swelling soils in the Denver metro area underlain by flat-lying rocks. This standard practice failed when applied to the steeply dipping to vertical rocks encountered in southwestern Jefferson County and other areas adjacent to the Front Range, as illustrated in the lower diagram (B). From Abbott, 2003.

lying swelling clay areas of southeast Denver were the subject of many newscasts and consternation as the foundations of then-new homes and other structures broke apart and cracks appeared in walls. The then-new Thomas Jefferson High School had significant swelling clay problems. During this time, soil testing for swelling soils, as shown in Figure 24A, came into use, and various engineering solutions were developed for areas in which flat-lying swelling soils were encountered.

In recent years, development along the hogbacks has revealed a whole new set of swelling clay problems where the clays are steeply dipping. The standard methods of drilling in order to test soils failed to alert developers to the problem—someone ignored the well-known geologic character of the mountain front. Again, the result is homes with significant structural damage. On the surface, the results look similar to what used to be seen at the Colorado School of Mines student housing. However, swelling soil damage involves differential heaving instead of the differential settlement experienced by the student housing. Although the symptoms look the same, the underlying cause, and therefore the solution to the problem, is very different.

[5]Bentonite is an older but common term for swelling clays that are now recognized as mixed layers of illite and smectite formed when volcanic ash falls into lakes or oceans. Charge imbalances in the clay mineral structure allow water to be absorbed into the mineral structure, causing the swelling that creates problems for foundations. Different bentonitic layers may have differing characteristics. "Bentonite" is still used for the clays used for such industrial mineral products as drilling mud, foundry casting binders, and kitty litter.

As with the rockfall and landslide hazards discussed earlier, geologists were aware that the sedimentary layers were turned on end near the mountain front. That information was not effectively communicated to developers and land-use planning agencies in a way that required action to account for differences in geology. The geologic information was ignored. To what degree is this failure to communicate the geologic profession's problem, and to what degree is it the problem of those who don't view geology as important or worth the bother? The result is economic losses for homeowners, developers, government agencies (the Jefferson County Public Schools had to condemn and demolish an elementary school), and taxpayers. The Colorado Geological Society has published a popular disclosure booklet for expansive soils (Noe et al., 1997) targeted at builders, homebuyers, and homeowners; over 210,000 copies have been distributed.

END OF TRIP

Directions back to the Colorado Convention Center: Take Marlowe Avenue to Simms Street and turn left (north). At Quincy turn left (west) and drive to the cloverleaf entrance ramp for Hwy C-470 westbound (it actually goes north here). Follow C-470 for 6.2 mi (10.0 km) and then exit onto I-70 eastbound. After 1.5 mi (2.4 km), exit onto the 6th Avenue Freeway (West 6th Avenue) eastbound. Take 6th Avenue all the way into Denver proper, where it loses its freeway configuration—stay in the left lane crossing I-25 and beyond. Turn left (north) at the second stop light, onto Santa Fe Drive. After ~1 mi (1.6 km) and crossing Colfax Avenue, the street will take a 45° turn to the right and become Stout Street. The Colorado Convention Center is two blocks beyond, at 14th and Stout Streets.

ACKNOWLEDGMENTS

We would like to thank Jim Soule, Jeff Hynes, and Karen Berry of the Colorado Geological Survey, who wrote parts of the original manuscript (Noe et al., 1999) that was used as a basis for this paper. Marilyn Dodson of Yeh and Associates provided information about the renovation project at Red Rocks Park. Curtis L. Johnson kindly used his sources at Xcel Energy to track down Figure 13, the picture of the old gas storage tank.

REFERENCES CITED

Abbott, D.M., 2000, Advocating science versus advocating policy: a conflict of interest, *in* Professional Ethics & Practices: The Professional Geologist, June 2000, p. 23–24.

Abbott, D.M., 2003, "Best Practices," a dangerous term: The Professional Geologist, November 2003, p. 14–16.

Abbott, D.M., and Noe, D.C., 2002, Consequences of Living with Geology: a model field trip for the general public, *in* Lageson, D., ed., Science at the Highest Level: Boulder, Colorado, Geological Society of America Field Guide 3, p. 1–16, http://www.gsajournals.org/gsaonline/?request=get-document&doi=10.1130%2F0-8137-0003-5(2002)003%3C0001:TCOLWG%3E2.0.CO%3B2 (accessed August 2004).

Allison, M.L., 2001, Hutchinson, Kansas: a geologic detective story: Geotimes, October 2001, American Geological Institute, http://www.agiweb.org/geotimes/oct01/feature_kansas.html (accessed August 2004).

Chase, G.H., and McConaghy, J.A., 1972, Generalized surficial geologic map of the Denver area, Colorado: U.S. Geological Survey Map I-731, scale 1:62,500.

Creath, W.B., 1996, Home buyers' guide to geologic hazards: American Institute of Professional Geologists, Westminster, Colorado, 303–412–6205, 30 p. (out of print).

Dodson, M., Arndt, B., and Andrew, R., 2003, Red Rocks Amphitheatre: 2001–2003 renovation retaining walls, *in* Boyer D.D., Santi, P.M., and Rogers, W.P., Engineering geology in Colorado—contributions, trends, and case histories: Association of Engineering Geologists Special Publication 15 and Colorado Geological Survey Special Publication 55, CD-ROM.

Frodeman, R., 2003, Geo-Logic: Breaking ground between philosophy and the earth sciences: State University of New York Press, 184 p.

Knepper, D.H. Jr., editor, 2002, Planning for the conservation and development of infrastructure resources in urban areas—Colorado Front Range Urban Corridor: U.S. Geological Survey Circular 1219, 27 p.

Jarrett, R.D., 1987, Flooding hydrology of foothill and mountain streams in Colorado [Ph.D. Dissertation]: Fort Collins, Colorado State University, 239 p.

LeRoy, L.W., and Weimer, R.J., 1971, Geology of the Interstate 70 Road Cut, Jefferson County, Colorado: Professional Contributions of the Colorado School of Mines Number 7, 1 sheet.

McCain, J.F., Hoxit, L.R., Maddox, R.A., Chappell, C.F., Caracena, F., Shroba, R.R., Schmidt, P.W., Crosby, E.J., and Hansen, W.R., 1979, Storm and flood of July 31-August 1, 1976, in the Big Thompson River and Cache la Poudre River basins, Larimer and Weld Counties, Colorado: Reston, Virginia: U.S. Geological Survey Professional Paper 1115, 152 p.

McPhee, J., 1998, Annals of the former world: New York, Farrar, Straus and Giroux, 696 p.

Mernitz, S., 1971, The impact of coal mining on Marshall, Colorado, and vicinity: an historical geography of environmental changes [M.A. Thesis]: Boulder, University of Colorado, 82 p.

Murphy, J.A., 1995, Geology tour of Denver's buildings and monuments: Historic Denver and the Denver Museum of Natural History, 95 p.

Nichols, T.C. Jr., 1992, Rebound in the Pierre Shale of South Dakota and Colorado—field and laboratory evidence of physical conditions related to the process of shale rebound: U.S. Geological Survey Open-File Report 92-440, 32 p.

Noe, D.C., 1997, Heaving-bedrock hazards, mitigation and land-use policy, Front Range Piedmont, Colorado: Environmental Geosciences, v. 4, no. 2, p. 48–57 (reprinted as Colorado Geological Survey Special Publication 45, 1997).

Noe, D.C., and Dodson, M.D., 1997, Heaving bedrock hazards associated with expansive, steeply dipping bedrock in Douglas County, Colorado: Denver, Colorado Geological Survey Special Publication 42, 80 p.

Noe, D.C., Jochim, C.L., and Rogers, W.P., 1997, A guide to swelling soils for Colorado homebuyers and homeowners: Colorado Geological Survey Special Publication 43, 76 p.

Noe, D.C., Soule, J.M., Hynes, J.L., and Berry, K.A., 1999, Bouncing boulders, rising rivers, and sneaky soils: A primer of geologic hazards and engineering geology along Colorado's Front Range, *in* Lageson, D.R., Lester, A.P., and Trudgill, B.D., eds., Colorado and adjacent areas: Boulder, Colorado, Geological Society of America Field Guide 1, p. 1–20.

Nuhfer, E.B., Proctor, R.J., and Moser, P.H., 1993, The citizens' guide to geologic hazards: American Institute of Professional Geologists, Westminster, Colorado, 303–412–6205, 134 p. (Spanish version also available).

Scott, G.R., 1972a, Geologic map of the Morrison quadrangle, Jefferson County, Colorado: U.S. Geological Survey Map I-790-A, scale 1:24,000.

Scott, G.R., 1972b, Map showing some points of geologic interest in the Morrison quadrangle, Jefferson County, Colorado: U.S. Geological Survey Map I-790-E, scale 1:24,000.

Simpson, H.E., 1973a, Map showing landslides in the Golden quadrangle, Jefferson County, Colorado: U.S. Geological Survey Map I-761-B, scale 1:24,000.

Simpson, H.E., 1973b, Map showing areas of potential rockfalls in the Golden quadrangle, Jefferson County, Colorado: U.S. Geological Survey Map I-761-C, scale 1:24,000.

Simpson, H.E., 1973c, Map showing earth materials that may compact and cause settlement in the Golden quadrangle, Jefferson County, Colorado: U.S. Geological Survey Map I-761-D, scale 1:24,000.

Soule, J.M., 1978, Geologic hazards of Douglas County, Colorado: Colorado Geological Survey Open File Report 78-5, 16 p.

Soule, J.M., 1999, Active surficial-geologic processes and related geologic hazards in Georgetown, Clear Creek County, Colorado: Colorado Geological Survey Open File Report 99-13, 6 p.

Soule, J.M., Rogers, W.P., and Shelton, D.C., 1976, Geologic hazards, geomorphic features, and land-use implications in the area of the 1976 Big Thompson flood, Larimer County, Colorado: Colorado Geological Survey, Environmental Geology 10.

Spencer, F.D., 1961, Bedrock geology of the Louisville quadrangle, Colorado: U.S. Geological Survey Map GQ-151, scale 1:24,000.

Van Horn, R., 1957, Bedrock geology of the Golden quadrangle, Colorado: U.S. Geological Survey Map GQ-103, scale 1:24,000.

Van Horn, R., 1972, Surficial and bedrock geologic map of the Golden quadrangle, Jefferson County, Colorado: U.S. Geological Survey Map I-761-A, scale 1:24,000.

Van Horn, R., 1976, Geology of the Golden quadrangle, Colorado: U.S. Geological Survey Professional Paper, v. 872, p. 116.

Weimer, R.J., and Ray, R.R., 1997, Laramide mountain flank deformation and the Golden Fault zone, Jefferson County, Colorado, *in* Bolyard, D.W., and Sonnenberg, S.A., eds., Geologic history of the Colorado Front Range, RMS-AAPG Field Trip #7: Denver, Rocky Mountain Association of Geologists, p. 49–64.

Geological Society of America
Field Guide 5
2004

Surface and underground geology of the world-class Henderson molybdenum porphyry mine, Colorado

James R. Shannon*
Eric P. Nelson
Department of Geology and Geological Engineering, Colorado School of Mines, Golden, Colorado 80401, USA

Robert J. Golden Jr.
Senior Geological Engineer, Climax Molybdenum Company

ABSTRACT

This field trip will visit a modern, large-scale underground block cave mining operation at the world-class Urad-Henderson porphyry molybdenum deposits on and beneath Red Mountain, in the historic Dailey-Jones Pass mining districts, Clear Creek County, Colorado. The underground tour summarizes the Henderson deposit geology and the current status of mining operations, and offers the opportunity to examine and collect rock specimens. The surface tour summarizes the regional and local geologic and structural setting of the deposits, and surface features that define and characterize the outer, peripheral parts of the intrusive-hydrothermal system. The mine is located in the northern Colorado Mineral Belt, in the Front Range of the Rocky Mountains, ~75 km west of Denver. The deposits consist of molybdenite-bearing, quartz vein stockworks at the cupola apices of highly evolved, silica-rich, subalkaline, leucorhyolite/leucogranite porphyry stocks. The system formed over ~3.0 m.y. between ca. 27 and 30 Ma by at least 23 intrusive events. Emplacement of the Red Mountain intrusive center and a second intrusive center at Woods Mountain may have been controlled by the NNE-trending Berthoud Pass–Vasquez Pass structural zone, a major Laramide-reactivated Precambrian shear/fault zone. A peripheral, 7.5 × 12.0 km, NNE-elongated, elliptical, pervasive chlorite alteration zone contains a well-developed system of radial quartz and base-precious metal veins.

Keywords: porphyry, Colorado Mineral Belt, molybdenum, alteration, vein.

INTRODUCTION

The Urad-Henderson porphyry molybdenum deposits are situated on and beneath Red Mountain, in the historic Dailey-Jones Pass mining districts, Clear Creek County, Colorado (Fig. 1). The Henderson Mine is a premier example of a world-class porphyry Mo deposit and also represents one of the finest examples of a modern, active, large-scale, underground block-cave mine. The mine is located in the northern Colorado Mineral Belt in the Front Range of the Rocky Mountains, ~75 km west of Denver. The Urad-Henderson deposits, together with sister deposits at Climax, Lake County, and Mount Emmons, Gunnison County, Colorado, are the classic deposits for defining the "Climax-type" of porphyry Mo deposit (White et al., 1981). The

*jshannon@mines.edu

Shannon, J.R., Nelson, E.P., and Golden, R.J. Jr., 2004, Surface and underground geology of the world-class Henderson molybdenum porphyry mine, Colorado, *in* Nelson, E.P. and Erslev, E.A., eds., Field trips in the southern Rocky Mountains, USA: Geological Society of America Field Guide 5, p. 207–218. For permission to copy, contact editing@geosociety.org. © 2004 Geological Society of America

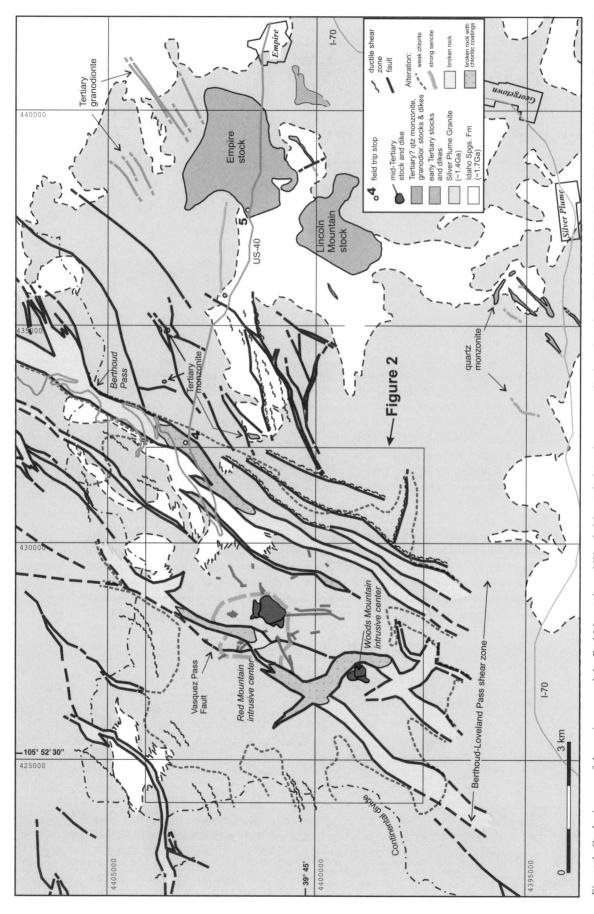

Figure 1. Geologic map of the region around the Red Mountain and Woods Mountain intrusive centers. Geology based on Theobald (1965), Theobald et al. (1983), and Shannon and Gunow (1983). The western, NE-trending broken rock zone within the Berthoud-Loveland Pass shear zone is the Vasquez Pass shear zone. Note that the younger Vasquez Pass Fault cuts obliquely across the Vasquez Pass shear zone just west of the Red Mountain intrusive center.

deposits consist of stockworks of molybdenite-bearing quartz veins at the apices of stock cupolas of highly evolved, silica-rich, subalkaline, leucorhyolite/leucogranite porphyry.

The main goal of the Henderson Mine field trip is to provide an opportunity to visit a world-class underground block-cave mining operation at a world-class porphyry Mo deposit. The mileage for the field trip log (Table 1) begins at the Morrison exit (Exit 259) on I-70 and is reset to 0 at the mine entrance gate. The underground portion (morning) of the tour will provide a summary of the Henderson deposit geology, an update on improvements at the mine and the current status of mining operations, and the opportunity to examine and collect rock specimens. The second part of the field trip (afternoon) will include an overview of the regional and local geologic and structural setting of the Urad-Henderson deposits and examination of surface features, which define and characterize the outer, peripheral parts of the intrusive-hydrothermal system. Some of the peripheral features will be examined underground and, weather permitting, on the surface.

BACKGROUND

Two main mid-Tertiary rhyolitic intrusive centers have been recognized in the Red Mountain area: (1) the Red Mountain intrusive center, and (2) the Woods Mountain intrusive center (Figs. 1 and 2). The Urad-Henderson deposits are associated with the Red Mountain intrusive center. The Urad orebody is the higher level system exposed at the surface. The first significant production was by the Primos Chemical Company in 1914. Most of the production was from 1967 to 1974 by Climax Molybdenum Company, a division of American Metal Climax, Inc. As of 1978, the total production from Urad is estimated at 13.7 mt (million tons) at 0.35% MoS_2 (Wallace et al., 1978). Molybdenite mineralization is predominantly structurally controlled by the "main fissure" and subsidiary structures, and is hosted in the Tungsten Slide Complex (a porphyry intrusion within the Red Mountain intrusive center) and Silver Plume Granite. The mineralization is interpreted to be associated with the Square quartz porphyry intrusion (Wallace et al., 1978).

The much larger Henderson orebody is located in the subsurface ~1,000 m below the summit of Red Mountain. The Henderson orebody was discovered in 1964, and production began in mid-1976. The discovery of the Henderson deposit was the direct result of intensive geologic studies conducted at the Climax mine during the late 1950s and early 1960s (Wallace, 1974). This research demonstrated the multiple intrusive nature of the deposit and the importance of the superposition of multiple ore shells in creating the economic molybdenum mineralization (Wallace et al., 1960, 1968). These concepts were applied to the Urad deposit and resulted in the discovery of the Henderson deposit. In 1978, the published combined proven and probable reserves were 303 mt at 0.49% MoS_2 with a 0.2% MoS_2 cutoff (Wallace et al., 1978). Total production at the end of 1986 was 65 mt at 0.44% MoS_2 (Carten et al., 1988a). The total production as of 31 December 2003 has been 172 mt at 0.39% MoS_2 (0.237%

Mo). The current reserves, as of 31 December 2003, are 166 mt at 0.352% MoS_2 (0.211% Mo) (Climax Molybdenum Company, a subsidiary of Phelps Dodge Corp.).

After the discovery of the Henderson orebody, continued intensive geological investigations resulted in a number of publications that further defined the intrusive and structural complexities of the deposit (e.g., Ranta et al., 1976; Wallace et al., 1978; White et al., 1981; and Carten et al., 1988a). Of great significance was the recognition of a suite of "igneous-hydrothermal" textures that appear to record the separation of hydrothermal fluids from the magma (Wallace et al., 1978; White et al., 1981; Shannon et al., 1982; Carten et al., 1988a, 1988b). These observations have provided one of the strongest cases for the derivation of metals, specifically molybdenum, directly from the magma for this type of porphyry deposit. Detailed studies of hydrothermal alteration have been ongoing and continue to make significant contributions to the understanding of porphyry molybdenum deposits (e.g., Carten et al., 1988a, 1988b; Seedorff, 1988; Seedorff and Einaudi, 2004a, 2004b).

The suites of multiple rhyolite-granite intrusions at both Red Mountain and Climax also include volumetrically minor biotite lamprophyres or kersantites. Field relations suggest that the lamprophyres (representing contaminated alkali-olivine basalt) are contemporaneous magmas that intimately commingled with the rhyolite magmas (Shannon et al., 1984; Bookstrom et al., 1988). These relations are particularly well developed in the radial dike swarm at Red Mountain and support a bimodal character to this mid-Tertiary magmatism. Age determination studies of the Urad-Henderson deposits have been ongoing (Schassberger, 1972; Naeser et al., 1973; Marvin et al., 1974; Bookstrom et al., 1987; various unpublished Climax Molybdenum Company reports). In summary, the K-Ar ages span 35.1–21.1 Ma, and the fission-track ages span 34.4–20.4 Ma (Table 2). More recent high precision $^{40}Ar/^{39}Ar$ age determinations (Geissman et al., 1992) have significantly reduced and refined the age span to 29.9–26.95 Ma (Table 3). This igneous activity is essentially contemporaneous with the initiation of Rio Grande rifting through opening of the upper Arkansas Valley graben between 29 and 28 Ma (Tweto, 1979). Various isotopic studies have been conducted and suggest that the Climax-type magmas were generated by low-percentage partial melting of biotite- or phlogopite-bearing, felsic, lower crustal Precambrian source rocks (Bookstrom et al., 1988; Stein, 1988). The geological and structural relations of intrusions and mineralization, together with paleomagnetic evidence, have suggested that the Red Mountain intrusive system has been tilted ~15°–25° to the southeast (Fig. 3; Geraghty et al., 1988; Geissman et al., 1992).

Over the years, Climax Molybdenum Company geologists have conducted a number of studies of the peripheral parts of the Urad-Henderson system, but most of this information has not been published (MacKenzie, 1970; Bookstrom, personal commun., 1978–1980; Shannon and Gunow, 1983, unpub.). MacKenzie (1970) stated that a strong propylitic alteration zone occurs as a large elliptical zone more than 2 mi

TABLE 1. HENDERSON MINE FIELD TRIP LOG

Time	Mileage	Site/stop
6:45 a.m.		Colorado Convention Center (Lobby B Cafeteria—GSA Field Trip Desk)
7:00 a.m.		Leave Convention Center—go west on I-70.
7:20 a.m.	0	Morrison Exit 259, through the Dakota Hogback and across Paleozoic disconformity with 1.7 Ga Idaho Springs Formation.
	5.4	Rocky Mountain overlook; view of the Continental Divide
	26.8	Empire Exit 232; take U.S. 40 west.
	29.1	Empire Post Office.
	35.6	The view from the highway looking WSW is of the east ridge of Red Mountain.
	36.1	Berthoud Falls.
	36.35	Henderson Mine turnoff (leave U.S. 40 before the lower hairpin turn).
8:00 a.m.	38.25	Henderson Mine main gate.
8:10 a.m.		Henderson Mine dry area (Administration Bldg).
		Video: Use of self rescuer.
		Video: Overview of Henderson Mine.
		Geology presentation: Henderson deposit.
9:30 a.m.		Henderson Mine—underground tour begins.
		Take the #2 shaft (personnel tunnel at 10,346 ft elevation) to the 7500 level.
		At the 7500 level, take the ramp to the 7700 level.
		7700 level; STOP U1: shop/warehouse walk-through.
		7700 level; STOP U2: production drift, drawpoints, orepass.
		Stockwork veining in Urad Porphyry.
		Examine Henderson Stock or Seriate Stock contact?
		Sample ore from active drawpoint.
		7065 level; STOP U3: underground primary crusher.
		At the 7065 level, take the ramp to the 7210 level.
		7210 level; possible STOP U4: active development heading in new production area. At the 7210 level, take the ramp to the 7437 level.
		7437 level; STOP U5: transfer station (PC1/PC2) on main conveyor belt. From the 7437 level, take the ramp to back to the 7500 level.
		7500 level; STOP U6: shaft area—examine fresh samples of Idaho Springs Formation and Silver Plume Granite. Examine bimodal lamprophyre-rhyolite porphyry dikes of the radial dike swarm. Return via the #2 shaft (from the 7500 level to the personnel tunnel at 10,346 ft elevation).
12:30 p.m.		Core shack—lunch.
		Overview of regional geologic and structural setting and surface geology. Examine rock/slab representative suites from the Urad and Henderson deposits and the surface.
1:45 p.m.	Reset to 0	Begin surface tour—Leave Henderson Mine main gate (1.9 mi from turnoff at US-40).
	0.5	STOP S1: Park at the south side of the pullout. Examine Precambrian Silver Plume Granite just north of the strong sericite alteration zone. Most of the outcrop is strong, pervasive chlorite altered with disseminated pyrite and some examples of the broken rock zone chlorite coatings. The west end of the outcrop has a narrow, structurally controlled quartz-sericite-pyrite zone with local advanced argillic alteration with Fe and Mn oxide staining and with irregular, thin quartz-galena-sphalerite veinlets.
	1.45	Turn off the pavement, cross the bridge, and park at the Forest Service gate. STOP S2: Scotia Mine. Walk ~100 m to a flagged path to Scotia Mine: there is a caved raise on the bench above Clear Creek and an open adit along the creek. Note the examples of vuggy quartz-fluorite veins and quartz-pyrite-galena-sphalerite veins with strong sericitic alteration hosted in Silver Plume Granite and pegmatite.
		Return to the pavement and regress west.
	1.55	Turn off the pavement (south) onto the Urad Mine access road.
	2.6	STOP S3: Urad tailings dam—overview of the Urad Mine and surface geology. Possible short hike to examine rhyolite porphyry type of a radial dike swarm.
	3.65	Return to the pavement and head east
	4.2	At U.S. 40, go east.
	5.55	STOP S4: Berthoud Pass shear zone. Park at the pull out on the south side of Hwy 40 at the trailer park. Carefully cross U.S. 40 to examine strong chloritized Silver Plume Granite with local broken rock zone chlorite coatings. There is also some localized strong argillic alteration.
	9.2	STOP S5: Optional stop to examine 62 Ma Empire Stock.
		Return to the Colorado Convention Center.

Note: The location of the surface stops are shown on Figures 1 and 2. Mine levels given in feet above sea level.

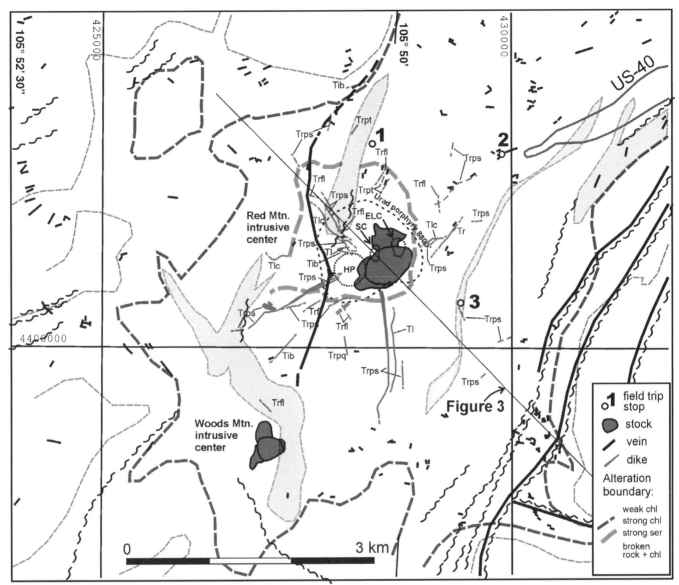

Figure 2. Geologic map of the immediate region around the Red Mountain intrusive center showing surface stops for the field trip and details of alteration boundaries, dikes, and veins in the outer, peripheral zone of the hydrothermal system. Gray areas from Fig. 1 are broken rock with chloritic coatings. Faults and shear zones taken from Fig. 1. Dike types: Trps—rhyolite porphyry, small K-spar; Trpt—rhyolite porphyry, tiny K-spar; Tl—lamprophyre (kersantite); Tib—igneo-fragmental; Trfl—flow-layered rhyolite; Tlc—composite lamprophyre/rhyolite porphyry, big K-spar; Tr—rhyolite; Trpq—quartz phenocryst rhyolite porphyry. Geology based on Theobald (1965), Theobald et al. (1983), and Shannon and Gunow (1983).

long in a NE-SW direction, centered on Red Mountain and the underlying Henderson orebody. His study was primarily based on detailed thin section observations, and his outer boundary of the propylitic zone was defined by the presence of some fresh biotite remaining in the rock. Bookstrom (1978–1980, personal commun.), based on reconnaissance traverses, suggested that the hydrothermal effects, especially chloritization of biotite, extend much farther away (up to 6.4 km) from the Red Mountain intrusive center than previously thought. He recommended a detailed study of the Red Mountain periphery. During the 1983 field season, the Climax Molybdenum

Company exploration group initiated a surface study (Shannon and Gunow, unpub., 1983). This study was not completed due to the shutdown or curtailment of Climax Exploration at the end of 1983. Some of the results of the 1983 field investigation are presented for the first time in this publication. They predominantly include a better understanding of the distribution and nature of faults, a subdivision of the radial dike swarm into subgroups, the distribution of peripheral hydrothermal features including veins and breccias, and the extent of sericitic and chloritic alteration. It should be noted that the study was entirely field based, and due to its premature termination,

TABLE 2. SUMMARY OF REGIONAL MAGMATIC EVENTS: RED MOUNTAIN AND VICINITY

Magmatic event	Method*	Source	Age (Ma)
Woods Mountain Center	K-Ar, Kspar	White, 1973	24.2 ± 0.9
Red Mountain Center	^{40}Ar/^{39}Ar	Geissman et al., 1992	30.4 to 28.4
	K-Ar, Bio-Kspar-Ser	Climax, unpub.	35.1 to 21.1
	FT, Zir	Marvin et al., 1974	34.4 to 20.4
		Bookstrom et al., 1987	
		Naeser et al., 1973	
		Climax Molybdenum Co, unpub.	
Hideaway Park Volcanics	K-Ar, San	Taylor et al., 1968	29.0 ± 3.0
Rabbit Ears Volcanics	K-Ar, San	Izett, 1966	33.0 ± 3.0
Mad Creek Stock (Empire)	FT, Zir	Bookstrom et al., 1987	40.5 ± 6.5
			39.4 ± 4.3
Montezuma Stock	Rb/Sr, Bio	Simmons and Hedge, 1978	39.0
	K-Ar, Bio	McDowell, 1971	39.6 ±1.2
	K-Ar, Bio	Marvin et al., 1989	37.9 ±1.4
	FT, Zir	Bookstrom et al., 1987	39.8 ± 4.2
	FT, Zir	Bookstrom et al., 1987	37.4 ± 3.0
	FT, Ap	Bookstrom et al., 1987	40.2 ± 8.8
Empire Stock	Rb-Sr	Simmons and Hedge, 1978	65.0
	FT, Sph	Cunningham et al., 1994	68.0 ± 6.2
	K-Ar, Bio	Marvin et al., 1989	61.6 ± 2.2
	FT, Ap	Cunningham et al., 1994	37.4 ± 2.8
Apex Stock	K-Ar, Hbde	Rice et al., 1982	79.3 ± 1.9
			76.7 ± 3.1
	FT, Zir	Cunningham et al., 1994	61.7 ± 6.3
	FT, Ap	Cunningham et al., 1994	71.3 ± 18.0
Eldora Stock	K-Ar, Hbde	Marvin et al., 1989	68.2 ± 4.1
	^{40}Ar/^{39}Ar	Berger, 1975	63.0
	FT, Zir	Cunningham et al., 1994	66.6 ± 6.4
	FT, Zir	Cunningham et al., 1994	61.8 ± 6.3
	FT, Ap	Cunningham et al., 1994	58.6 ± 9.4
Jamestown: Porphyry Mountain Stock	FT, Zir	Cunningham et al., 1994	44.8 ± 5.7
			45.1 ± 7.7
Granodiorite Stock	K-Ar	McDowell, 1971	79.6 ± 2.3
			73.6 ± 2.2
Silver Plume Granite	Rb-Sr	Peterman and Hedge, 1968	1,400
Idaho Springs Formation	Rb-Sr	Peterman and Hedge, 1968	1,700

*K-Ar—potassium-argon method (Kspar—K-feldspar; Hbde—hornblende; Bio—biotite; San—sanadine; Ser—sericite); FT—fission track method (Zir—zircon; Ap—apatite; Sph—sphene).

was not supported with follow-up laboratory work, especially petrographic observations.

GEOLOGIC AND STRUCTURAL SETTING

The host rocks for the Red Mountain and Woods Mountain intrusive centers predominantly consist of 1.4 Ga Silver Plume biotite-muscovite granite. The Silver Plume Granite in the Red Mountain area is the type locality for the Berthoud Plutonic Suite of Tweto (1987). It is an example of a Proterozoic monazite-sillimanite–bearing, two-mica granite representing a peraluminous portion of the anorogenic transcontinental province (Anderson and Thomas, 1985). Both intrusive centers also have minor pendants and inclusions of the 1.7 Ga Idaho Springs Formation (metasedimentary schist and gneiss). Large areas of

Idaho Springs Formation are exposed along the major Precambrian shear zones (Fig. 1).

In their classic paper on the Precambrian ancestry of the Colorado Mineral Belt, Tweto and Sims (1963) stressed the importance of northeast-trending Precambrian ductile shear zones in controlling the emplacement of Laramide and Tertiary igneous rocks and associated ore deposits. They also recognized that the Precambrian structures were reactivated during Laramide deformation and are characterized by a more brittle deformation style producing fault zones with breccia and gouge. The Berthoud Pass Fault was used as an example of a Precambrian shear zone that was reactivated during Laramide deformation. Previous workers have noted the proximity of the Red Mountain intrusive center to two major fault zones: the Berthoud Pass and Vasquez Pass faults (Ranta et al., 1976; Wallace et al., 1978). Theobald

TABLE 3. INTRUSIVE SEQUENCE AND ⁴⁰Ar/³⁹Ar AGES FOR THE RED MOUNTAIN INTRUSIVE SUITE

Event (youngest at top)	% of Mo orebody	Mineral*	^{40}Ar/^{39}Ar age[†] (Ma)
25° SE tilting—post-latest Oligocene (post instrusions)			
Ute	0	Or	28.4 ± 0.3
		Bio	27.6 ± 0.2
Vasquez	10	Or	28.71 ± 0.08
		Bio	27.6 ± 0.3
Nystrom	trace		
Ruby	1		
Seriate–Late Magnetite-Sericite Alt.		Mus-Ser	26.95 ± 0.08
		Mus-Ser	27.51 ± 0.03
Seriate	35	Or	27.1 ± 0.2
		Bio	27.6 ± 0.1
		Or	28.2 ± 0.2
		Bio	27.6 ± 0.2
		Bio	27.6 ± 0.2
East Lobe	1		
Arapahoe	trace		
Primos	11		
Henderson	27	Or	28.1 ± 0.2
		Bio	28.0 ± 0.2
Berthoud	trace		
Phantom	15		
Urad Porphyry	Small/localized	Or	28.0 ± 0.09
		Bio	28.6 ± 0.3
		Bio	28.3 ± 1.4
Peak Breccia			
Igneo-Fragmental Rock			
Concentric Dikes			
Red Mountain Porphyry		Or	29.9 ± 0.3
Rubble Rock Breccia			
Radial Dikes–Rhyolite		Or	29.4 ± 0.2
Radial Dikes–Kersantite Lamprophyre		Bio	29.81 ± 0.10
Square Quartz Porphyry	Most or all of Urad orebody		
East Knob Unit			
Tungsten Slide–Crowded Quartz Porphyry			
Tungsten Slide–Junk Rock Breccia			

*Or—orthoclase; Bio—biotite; Mus-Ser—muscovite-sericite.
[†]Geissman et al. (1992).

(1965) and Theobald et al. (1983) mapped discrete zones of broken and shattered rock that form complex anastomosing patterns of similar orientation as the Precambrian shear zones.

Detailed mapping by Theobald (1965), Theobald et al. (1983), and J. Shannon (Shannon and Gunow, 1983) showed the major structures in the area, and that there are discrete zones of fractured rock that form continuous, mappable zones (Fig. 1). These zones are characterized by areas of little or no outcrop and grassy, subdued topography. Most of the zones of fractured rock are cored by narrow mylonite zones ranging from 0.5 to 30.0 m thick. Mylonitic textures range from protomylonitic to ultramylonitic, the latter being more common, often cropping out as resistant ribs. In many cases it is the mylonitic core that enables the mapping of these dual structures. The mylonitic core of the structures are interpreted to be related to post–1.4 Ga Precambrian ductile shearing, and the fractured-broken rock envelopes are interpreted to be related to Laramide and mid-Tertiary brittle fault re-activation.

The orientation of mylonitic shears suggests the presence of three main structural orientations: (1) NNE with moderate to steep NW and SE dips, (e.g., Berthoud Pass and Vasquez Pass shear zones); (2) ENE with moderate to steep NW dips, (e.g., Bobtail Peak and Jones Pass shear zones); and (3) WNW with moderate NE dips. In the field, clear cross-cutting relations between the three shear zone orientations were not observed. However, the discontinuous nature of the WNW shears suggests that they are older than the NNE shears. The ENE shears are of similar orientation as the St. Louis Peak cataclastic zone, a major Precambrian shear zone to the north and northwest of Red Mountain. These ENE structures may predate (as

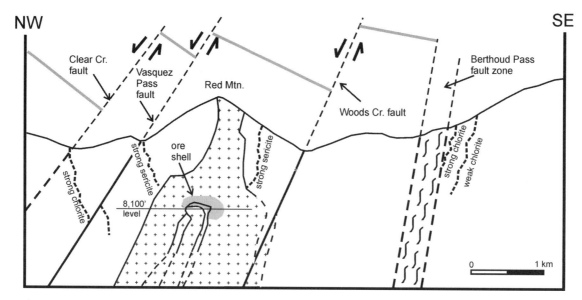

Figure 3. Schematic cross section modified from Geissman et al. (1992). Approximate cross section line shown on Figure 1. Gray lines above ground surface show a schematic marker illustrating the proposed offset of faults causing rotation of the Urad-Henderson system.

interpreted by Tweto and Sims, 1963) and/or be a conjugate shear set to the NNE shears.

Of the three shear zone orientations, the NNE structures are the most through-going and continuous. They include the two largest structures in the area: the Berthoud Pass and Vasquez Pass shear zones (oriented ~N30E). In the immediate vicinity of Red Mountain, the NNE shears form a broad, 5-km-wide structural corridor. The chlorite alteration zone is elongated along this structural corridor, and narrow zones of structurally-controlled chloritization extend to the NNE and SSW along the Berthoud Pass and Vasquez Pass shear zones. These relations support that the NNE Precambrian shears were reactivated prior to emplacement of the Red Mountain intrusive suite and that the reactivated fractured-broken zones were major fluid-flow paths for the Tertiary hydrothermal fluids.

The structural nomenclature in the study area is somewhat confused due to the local splaying off of the reactivated fractured zones. The best example is the Vasquez Pass shear zone and the Vasquez Pass Fault (Fig. 2). The Vasquez Pass shear zone is one of the major NNE trending Precambrian shear zones with a broad envelope of fractured rock interpreted to have been reactivated during the Laramide Orogeny. It passes through Vasquez Pass, crosses the lower northwest slope of Red Mountain, and passes through the saddle on the west side of Harrison Mountain. The Vasquez Pass Fault is a major brittle fault zone that was intersected in the #2 shaft and underground access and ventilation drifts of the Henderson Mine. The fault consists of ~20 m of breccia and gouge and is oriented ~N10E with a 65° dip to the northwest. In the immediate mine area, the Vasquez Pass Fault does not control mineralization or alteration and is interpreted to be predominantly

post-mineral. There is a suggestion that the reactivated portion of the Berthoud Pass shear zone (the Berthoud Pass Fault?) may also splay off of its master controlling Precambrian shear zone to the north, in the Fraser River valley south of Winter Park.

MIDDLE TERTIARY INTRUSIVE CENTERS

The Red Mountain and Woods Mountain intrusive centers were emplaced along the western portion of the Berthoud Pass/Vasquez Pass structural corridor (Fig. 1). The two intrusive centers are aligned along a N30E trend, suggesting that their emplacement was controlled by the Laramide-reactivated Precambrian shear zone. The Red Mountain intrusive suite includes a total of at least 23 intrusive events: 11 intrusive events associated with the shallower Urad system and at least 12 intrusive events associated with the deeper Henderson system (Table 3). Remarkably, the 23 intrusions were emplaced over an ~3 m.y. time span into a point source area, creating a bull's eye target. The bull's eye is further enhanced by the radial and concentric dikes that are common in the Urad system (MacKenzie, 1970; Wallace et al., 1978; Fig. 2). MacKenzie (1970) and Wallace et al. (1978) provided the first subdivision and relative ages of the surface dikes. A main group of surface dikes was referred to as the radial dike swarm and included a variety of dike types. Relative age relations suggest that, as a group, they are younger than the Tungsten Slide Complex (i.e., post-Urad orebody) and older than the Red Mountain Porphyry. Observations underground in the Henderson Mine, where the dikes are fresher, and on the surface in 1983, indicated that the radial dike swarm consists of at least five dike types (Fig. 2). These include small potassium feldspar (K-spar) rhyolite porphyry (RP) (Trps), big K-spar RP, lamprophyre (Tl), composite big K-spar/lamprophyre (Tlc), tiny K-spar RP (Trpt), and quartz phenocryst RP (Tr). The radial dikes extend up to ~2 km from the intrusive center. They are radial about the intrusive center but occur in three main orientations: NE, NW, and N-S. From detailed underground mapping of veins, Coe and Nelson (1997) showed that a number of vein sets are present in the orebody and consist mostly of radial and concentric veins.

The subsurface outline of the Urad Stock is shifted ~350–400 m NW relative to the center of the Urad system of intrusions on the surface (Fig. 2). The three main intrusion centers associated with the Henderson system (Henderson-Primos, Seriate, and Vasquez centers; see Carten et al., 1988a) occur along a N60E trend that is shifted ~100–150 m NW relative to the Urad system center. These relations are compatible with minor tilting of the system to the SE (Geraghty et al., 1988; Geissman et al., 1992). Interestingly, the N60E alignment of the Henderson intrusion centers is parallel to the orientation of the ENE-striking, Laramide-reactivated, Precambrian shear zones.

The Berthoud Pass/Vasquez Pass shear zones are inferred to have exerted a deep-seated structural control on emplacement of the Red Mountain and Woods Mountain intrusive centers. Local structures are inferred to have exerted a strong control on hydrothermal fluid flow associated with the Red Mountain center. However, it is unclear why these structures did not exert more control on the emplacement, shapes, and orientation of the mid-Tertiary intrusions. The strong radial and concentric dike pattern led Wallace et al. (1978) to suggest a near surface environment for the Urad system, whereas Bookstrom (1981) interpreted the well-developed radial dike (and vein) pattern to support a relatively atectonic transition between Laramide compression and Rio Grande rift extension at the time of Red Mountain igneous-hydrothermal activity.

A second mid-Tertiary intrusive center occurs ~2.5 km SSW of Red Mountain, along Woods Creek and on the low northwest slope of Woods Mountain (Wallace et al., 1978). The Woods Mountain intrusive center consists of at least four intrusions and exhibits a number of characteristics that strongly support a genetic relation with the Red Mountain intrusive center: (1) high-silica rhyolite composition; (2) a K-Ar K-spar age determination of 24.2 ± 0.9 Ma (recalculated from White, 1973, unpub.); (3) multiple intrusive character; (4) dominance of sericitic alteration with common occurrence of secondary fluorite; and (5) presence of intrusive and hydrothermal breccias (White, 1969; White, 1973, unpub.). By late 1980, three deep exploration drill holes tested the Woods Mountain intrusive center. Minor base metal (pyrite, galena, sphalerite ± rhodochrosite and fluorite) mineralization was encountered, but results were discouraging for the presence of deeper molybdenite mineralization.

MIDDLE TERTIARY HYDROTHERMAL ALTERATION

During the 1983 study of the Red Mountain periphery, two main types of hydrothermal alteration were mapped. There is a small, ~1.4 × 1.8 km zone of strong sericite alteration roughly centered on the peak of Red Mountain. The alteration is characterized by pervasive alteration of feldspars to sericite (+clay?) and removal of mafic minerals (biotite, chlorite, and magnetite). This alteration zone includes the bleached zone of MacKenzie (1970) but extends farther to the north and west. Most of the zone above 11,200 ft elevation is probably related to the Urad orebody. The northern limit of the sericite zone almost reaches the West Fork of Clear Creek at an elevation of 10,200 ft. In this area, the pervasive sericite alteration appears to be broad, coalescing halos on quartz-sericite-pyrite veins with associated base metals. The origin of this low-elevation northern extension of the sericite alteration zone is problematic. It is too low in elevation and too far away to be related to the Urad orebody. It may be an expression of small areas of molybdenite mineralization known to occur along the upper contact zone of the Urad stock or, less likely, an upward perturbation of the Henderson system phyllic alteration zone. An irregular linear tail of structurally controlled sericite alteration (probably related to the Urad system) extends to the southwest along the Vasquez Pass shear zone.

The chlorite alteration zone around the Red Mountain intrusive center is believed to be predominantly related to the Henderson orebody. It is a tremendous zone of pervasive hydrothermal alteration that forms a 7.5 × 12.0 km, elliptical, NNE-SSW elongated zone that extends from Berthoud Pass southwestward to Herman Gulch. The zone is characterized by the partial (distal) to complete (proximal) replacement of primary biotite by secondary chlorite. Based on MacKenzie's (1970) study, the propylitic alteration assemblage on Red Mountain also includes some epidote, carbonate, and disseminated pyrite. A genetic relationship between the chlorite alteration zone and the Red Mountain intrusive center is strongly supported by the central spatial position of the intrusive center and the spatial association of hydrothermal features, particularly veins with fluorite and rhodochrosite, and the geochemically anomalous molybdenum content of the base- and precious-metal mineralization within the chlorite zone.

The strong chlorite boundary generally outlines the area in which all biotite is megascopically replaced by chlorite. The outer weak chlorite boundary depicts where biotite in fresh rock becomes partially chloritized. A third type of chlorite alteration was recognized and mapped. It consists of localized zones along the reactivated fractured-broken rock envelopes where the fractured rock surfaces have unusually thick coatings of chlorite (up to 1 mm). Four main zones were recognized, and they occur within the strong chlorite alteration boundary (Figs. 1 and 2). They represent an interesting variant of the chlorite alteration because the overall zones are structurally controlled, and much of the chlorite is introduced along fractures in addition to that which is disseminated in the rock. In other words, the chloritic broken rock zones represent structurally controlled zones of chlorite veins. The chlorite-coated surfaces commonly have fluorite on them and also commonly display slickensided and/or striated surfaces supporting some post-alteration reactivated movement along the zones.

Limonite staining is common in the Red Mountain intrusive center periphery and represents a supergene alteration effect related to the hydrothermal alteration. Three general types of limonite staining were recognized and mapped. The first type of limonite staining is pervasive and occurs in the central area on Red Mountain. It includes a zone of strong limonite staining on the northeast side of Red Mountain that is coincident with the outcrop area of the East Knob porphyry, which is interpreted to be produced by the oxidation of disseminated pyrite in that unit (Wallace et al., 1978). A weaker zone of limonite staining is marginal to the strong limonite zone and is coincident with the zone of strong pervasive sericite alteration. The second type of limonite staining is structurally controlled and is related to oxidation of quartz-sericite-pyrite and base metal veins in the pervasive chlorite alteration zone.

The third type of limonite staining is subtler, has a broad regional distribution pattern, and is more problematic in its origin. This type of limonite staining occurs as a ring outside of the pervasive chlorite zone. It is generally recognized in the Silver Plume granite where light limonite coatings occur along fracture surfaces and along grain boundaries in the rock. In these zones, the Silver Plume Granite has fresh biotite and feldspars and few or no veins. The limonite does not appear to be a product of in situ breakdown of disseminated pyrite or mafic minerals. Two plausible interpretations for the limonite zone are (1) it is a regional limonite staining that is related to a Tertiary erosion-weathering surface (also suggested by P. Theobald, 1983, personal commun.); or (2) it is a weak, distal hydrothermal effect related to the Henderson orebody hydrothermal alteration. The weak, but pervasive, limonite zone tends to occur along the high ridges that surround the Red Mountain intrusive center, possibly supporting the first hypothesis. However, there is no direct evidence of a paleo-erosion surface in the immediate area of Red Mountain, and the limonite staining is not restricted to the high ridges only. Some specular hematite veins are present in the pervasive chlorite zone, and they extend out into the limonite stained zone. The best interpretation of the limonite-stained zone is that it is a distal alteration effect produced by the flushing of excess iron out of the peripheral alteration zones associated with the Henderson deposit. The limonite-stained zone is open to the NNE and SSW along the continuation of the Berthoud Pass–Vasquez Pass shear zone. If it is an outer, subtle hydrothermal effect related to the Henderson system, then it tremendously increases the area and volume of rock that has interacted with the hydrothermal fluids.

MIDDLE TERTIARY HYDROTHERMAL VEINS AND BRECCIAS

There are ~100 prospect pits and ~12–15 small mines located in the chlorite alteration halo around the Red Mountain intrusive center. The majority of the mines are on structurally-controlled quartz-pyrite-galena-sphalerite-chalcopyrite veins with or without associated fluorite, rhodochrosite, and calcite (Shannon and Gunow, 1983). A few of the mines have argentite and tetrahedrite.

Historic reports indicate that most of the mining and prospecting was for silver and gold. Total production of silver, gold, lead, and zinc was minimal, with ore assays in the range of 3–10 opt silver (Theobald et al., 1983). Molybdenite does not occur in the base metal veins but typically is anomalous in the 30–300 ppm range. The majority of the mines and prospects are situated just inside the strong, pervasive chlorite alteration boundary. The base metal veins are predominantly radial about the Red Mountain intrusive center (Fig. 2).

Two additional vein types are common in the Red Mountain chlorite zone. The first are vuggy quartz veins that are the most abundant vein type in the Henderson periphery. They are typically ≤5 cm thick and are lined with clear, 1–4-mm-long quartz crystals. Euhedral fluorite crystals sometimes occur in the open quartz vugs. The second type of veins is a peculiar chalcedonic breccia vein that commonly has been misidentified as rhyolite porphyry dikelets by previous workers. The chalcedonic breccia veins are light gray to cream-colored and are typically 1–40 cm thick and have moderate to strong sericite alteration halos. They typically contain from <1% up to 35% crystal and lithic fragments from 0.5 mm to 5.0 cm in diameter. The lithic fragments are generally subrounded to rounded clasts that were all derived from the Silver Plume granite wall rock. Crystal fragments include angular fragments of quartz and minor feldspar crystals. Some of the larger quartz crystals appear to be pieces of broken quartz vein material. The lithic and crystal fragments occur with rare limonite pseudomorphs after pyrite cubes, in a very fine-grained matrix of quartz and sericite. Some chalcedonic breccia veins are radial and occur along the same structures that controlled the base metal veins. However, most of the chalcedonic breccia veins occur in a strongly structurally-controlled (~N70E) vein zone that extends from Berthoud Pass to west of Jones Pass (at least 9 km). Vuggy quartz veins are commonly associated with both the radial base metal veins and the Berthoud Pass–Jones Pass vein zone.

The relative age and genetic relations between the Berthoud Pass–Jones Pass vein zone and the Red Mountain intrusive-hydrothermal center is not clear. The Berthoud Pass–Jones Pass vein zone is not centered on the Red Mountain system, and it extends outside of the pervasive chlorite alteration zone and over the Continental Divide, to the west. Observations by Theobald et al. (1983) suggest that the Berthoud Pass–Jones Pass vein zone may extend into the La Plata mining district and a mineralized-alteration zone near Steelman Creek. Climax Molybdenum Company prospected this mineralized zone during the early to mid-1980s. The system has rhyolite dikes and abundant fluorite mineralization associated with base and precious metal veins. Lack of significant molybdenite mineralization, a strong arsenic enrichment, and a general lack of highly evolved rhyolites indicated that the system may not be Climax type. These relations suggest that the Berthoud Pass–Jones Pass vein zone may be associated with a less evolved rhyolite-granite system (or systems) that extends, in the subsurface, from west of Jones Pass to the Stanley Mountain area near Berthoud Pass.

CONCLUSIONS

The Red Mountain intrusive center hosts the Urad and Henderson porphyry molybdenum deposits, one of the largest molybdenite occurrences in the world. The system had at least 23 intrusive events over an ~3.0 m.y. interval, all focused on a point source near the summit of Red Mountain. Emplacement of the Red Mountain intrusive center and a second intrusive center at Woods Mountain may have deep-seated structural control related to the Berthoud Pass–Vasquez Pass structural zone, a major Laramide-reactivated Precambrian shear/fault zone. The outer peripheral alteration effects produced a very large, 7.5 × 12.0 km, elliptical, NNE-SSW elongated, pervasive chlorite alteration zone. The NNE-SSW elongation of the chlorite zone and linear extensions along the major Precambrian shear structures suggest that the farther the hydrothermal fluids traveled away from the center, the more they were controlled by the older structures. The pervasive chlorite alteration zone contains a well-developed system of radial quartz and base-precious metal veins.

ACKNOWLEDGMENTS

The authors would like to thank Climax Molybdenum Company (both old and new) and Phelps Dodge Corporation for allowing publication of this paper and for approval of this field trip. The Climax Molybdenum Company is commended for its policy, from the 1950s through the 1980s, of supporting geologic research on these interesting deposits. Peter Faur and Eric Erslev are thanked for their helpful comments on the draft manuscript.

REFERENCES CITED

Anderson, J.L., and Thomas, W.M., 1985, Proterozoic anorogenic two-mica granites: Silver Plume and St. Vrain batholiths of Colorado: Geology, v. 13, p. 177–180.

Berger, G.W., 1975, ^{40}Ar/^{39}Ar step heating of thermally overprinted biotite, hornblende and potassium feldspar from Eldora, Colorado: Earth and Planetary Science Letters, v. 26, no. 3, p. 387–408, doi: 10.1016/0012-821X(75)90015-1.

Bookstrom, A.A., 1981, Tectonic setting and generation of Rocky Mountain porphyry molybdenum deposits deposits, *in* Dickinson, W.R. and Payne, W.D., eds., Relations of tectonics to ore deposits in the southern cordillera: Arizona Geological Society Digest, v. 14, p. 215–226.

Bookstrom, A.A., Naeser, C.W., and Shannon, J.R., 1987, Isotopic age determinations, unaltered and hydrothermally altered igneous rocks, north-central Colorado Mineral Belt: Isochron/West, no. 49, p. 13–20.

Bookstrom, A.A., Carten, R.B., Shannon, J.R., and Smith, R.P., 1988, Origins of bimodal leucogranite-lamprophyre suites, Climax and Red Mountain porphyry molybdenum systems, Colorado: Petrologic and Strontium isotopic evidence: Colorado School of Mines Quarterly, v. 83, no. 2, p. 1–24.

Carten, R.B., Geraghty, E.P., Walker, B.M., and Shannon, J.R., 1988a, Cyclic development of igneous features and their relationship to high-temperature hydrothermal features in the Henderson Porphyry Molybdenum deposit, Colorado: Economic Geology and the Bulletin of the Society of Economic Geologists, v. 83, p. 266–296.

Carten, R.B., Walker, B.M., Geraghty, E.P., and Gunow, A.J., 1988b, Comparison of field based studies of the Henderson porphyry molybdenum deposit, Colorado, with experimental and theoretical models of porphyry systems: Canadian Institute of Mining and Metallurgy Special Volume 39, p. 351–366.

Coe, J.A., and Nelson, E.P., 1997, Characterization of fracture networks using close-range photogrammetric mapping and GIS analysis, *in* Hoak, T.E.,

Klawitter, A.L. and Blomquist, P.K., eds., Fractured reservoirs: Characterization and modeling: Denver, Rocky Mountain Association of Geologists Guidebook, p. 43–55.

Cunningham, C.G., Naeser, C.W., Marvin, R.F., Luedke, R.G., and Wallace, A.R., 1994, Ages of selected intrusive rocks and associated ore deposits in the Colorado mineral belt: U.S. Geological Survey Bulletin 2109, p. 1–31.

Geissman, J.W., Snee, L.W., Graaskamp, G.W., Carten, R.B., and Geraghty, E.P., 1992, Deformation and age of the Red Mountain intrusive system (Urad-Henderson molybdenite deposits), Colorado: Evidence from paleomagnetic and ^{40}Ar/^{39}Ar data: Geological Society of America Bulletin, v. 104, p. 1031–1047, doi: 10.1130/0016-7606(1992)1042.3.CO;2.

Geraghty, E.P., Carten, R.B., and Walker, B.M., 1988, Tilting of Urad-Henderson and Climax porphyry molybdenum systems, central Colorado, as related to northern Rio Grande rift tectonics: Geological Society of America Bulletin, v. 100, p. 1780–1786, doi: 10.1130/0016-7606(1988)1002.3.CO;2.

Izett, G.A., 1966, Tertiary extrusive volcanic rocks in Middle Park, Grand County, Colorado, *in* U.S. Geological Survey research 1966: U.S. Geological Survey Professional Paper 550-B, p. B42–B46.

MacKenzie, W.B., 1970, Hydrothermal alteration associated with the Urad and Henderson molybdenite deposits, Clear Creek County, Colorado [Ph.D. Thesis]: Ann Arbor, University of Michigan, 208 p.

Marvin, R.F., Young, E.J., Mehnert, H.H., and Naeser, C.W., 1974, Summary of radiometric age determinations on Mesozoic and Cenozoic igneous rocks and Uranium and base metal deposits in Colorado: Isochron/West, no. 11, p. 1–16.

Marvin, R.F., Mehnert, H.H., Naeser, C.W., and Zartman, R.E., 1989, U.S. Geological Survey radiometric ages—compilation "C," part 5: Colorado, Montana, Utah, and Wyoming: Isochron/West, v. 53, p. 14–19.

McDowell, F.W., 1971, K-Ar ages of igneous rocks from the western United States: Isochron/West, no. 2, p. 1–4.

Naeser, C.W., Izett, G.A., and White, W.H., 1973, Zircon fission track ages from some middle Tertiary igneous rocks in northwestern Colorado: Geological Society of America Abstracts with Programs, v. 5, no. 6, p. 498.

Peterman, Z.E., and Hedge, C.E., 1968, Age of Precambrian events in the northeastern Front Range, Colorado: Journal of Geophysical Research, v. 73, no. 6, p. 2277–2296.

Ranta, D.E., White, W.H., Ward, A.D., Graichen, R.E., Ganster, M.W., and Stewart, D.R., 1976, Geology of the Urad and Henderson molybdenite deposits—A review: Colorado School of Mines Professional Contributions 8, p. 477–485.

Rice, C.M., Lux, D.R., and Macintyre, R.M., 1982, Timing of mineralization and related intrusive activity near Central City, Colorado: Economic Geology and the Bulletin of the Society of Economic Geologists, v. 77, p. 1655–1666.

Schassberger, H.T., 1972, K-Ar dates on intrusive rocks and alteration associated with Molybdenum mineralization at Climax and Urad, Colorado, and Questa, New Mexico: Isochron/West, no. 3, p. 29.

Seedorff, E., 1988, Cyclic development of hydrothermal mineral assemblages related to multiple intrusions at the Henderson porphyry molybdenum deposit, Colorado: Canadian Institute of Mining and Metallurgy Special Volume 39, p. 367–393.

Seedorff, E., and Einaudi, M.T., 2004a, Henderson porphyry molybdenum system, Colorado: I. Sequence and abundance of hydrothermal mineral assemblages, flow paths of evolving fluids, and evolutionary style: Economic Geology and the Bulletin of the Society of Economic Geologists, v. 99, p. 3–37.

Seedorff, E., and Einaudi, M.T., 2004b, Henderson porphyry molybdenum system, Colorado, II, Decoupling of introduction and deposition of metals during geochemical evolution of hydrothermal fluids: Economic Geology and the Bulletin of the Society of Economic Geologists, v. 99, p. 39–72.

Shannon, J.R., and Gunow, A.J., 1983, Red Mountain Periphery Study: Unpublished Climax Molybdenum Company Report, 113 p.

Shannon, J.R., Walker, B.M., Carten, R.B., and Geraghty, E.P., 1982, Unidirectional solidification textures and their significance in determining relative ages of intrusions at the Henderson mine, Colorado: Geology, v. 10, p. 293–297.

Shannon, J.R., Bookstrom, A.A., and Smith, R.P., 1984, Contemporaneous bimodal mafic-felsic magmatism at Red Mountain, Clear Creek County and Climax, Colorado: Geological Society of America Abstracts with Programs, v. 16, no. 4, p. 254.

Simmons, E.C., and Hedge, C.E., 1978, Minor-element and Sr-isotope geochemistry of Tertiary stocks, Colorado mineral belt: Contributions to Mineralogy and Petrology, v. 67, p. 379–396.

Stein, H.J., 1988, Genetic traits of Climax-type granites and molybdenum mineralization, Colorado mineral belt: Canadian Institute of Mining and Metallurgy Special, v. 39, p. 394–401.

Taylor, R.B., Theobald, P.K., and Izett, G.A., 1968, Mid-Tertiary volcanism in the Front Range, Colorado, *in* Epis, R.C., ed., Cenozoic volcanism in the southern Rocky Mountains: Quarterly of the Colorado School of Mines, v. 63, no. 3, p. 39–50.

Theobald, P.K., 1965, Preliminary geologic map of the Berthoud Pass quadrangle, Clear Creek and Grand Counties, Colorado: U.S. Geological Survey Miscellaneous Geologic Investigations Map I-443, scale 1:24,000.

Theobald, P.K., Bielski, A.M., Eppinger, R.G., Moss, C.K., Kreidler, T.J., and Barton, H.N., 1983, Mineral resource potential of the Vasquez Peak Wilderness Study Area, and the St. Louis Peak and Williams Fork roadless areas, Clear Creek, Grand, and Summit Counties, Colorado: U.S. Geological Survey Miscellaneous Field Studies Map MF-1588A, scale 1:50,000.

Tweto, O., 1979, The Rio Grande rift system in Colorado, *in* Riecker, R.E., ed., Rio Grande rift: tectonics and magmatism: Washington, D.C., American Geophysical Union, p. 33–56.

Tweto, O., 1987, Rock units of the Precambrian basement in Colorado: U.S. Geological Survey Professional Paper 1321-A, p. A1–A54.

Tweto, O., and Sims, P.K., 1963, Precambrian ancestry of the Colorado mineral belt: Geological Society of America Bulletin, v. 74, p. 991–1014.

Wallace, S.R., 1974, The Henderson ore body—elements of discovery, reflections: Society Mining Engineers American Institute Mining, Metallurgical Petroleum Engineers Transactions, v. 256, p. 216–227.

Wallace, S.R., Baker, R.C., Jonson, D.C., and Mackenzie, W.B., 1960, Geology of the Climax molybdenum deposit: A progress report: Geological Society of America Guidebook for Field Trips, Field Trip B3, p. 238–252.

Wallace, S.R., Muncaster, N.K., Jonson, D.C., MacKenzie, W.B., Bookstrom, A.A., and Surface, V.E., 1968, Multiple intrusion and mineralization at Climax, Colorado, *in* Ridge, J.D., ed., Ore deposits of the United States, 1933–1967 (Graton-Sales Volume): New York, American Institute of Mining, Metallurgical, and Petroleum Engineers, p. 606–640.

Wallace, S.R., MacKenzie, W.B., Blair, R.G., and Muncaster, N.K., 1978, Geology of the Urad and Henderson molybdenite deposits, Clear Creek County, Colorado, with a section on a comparison of these deposits with those at Climax, Colorado: Economic Geology and the Bulletin of the Society of Economic Geologists, v. 73, p. 325–368.

White, W.H., 1969, Geology of the Woods Mountain plug, Clear Creek County, Colorado: Unpublished American Metal Climax Inc. Report, 40 p.

White, W.H., 1973, Woods Mountain plug: Unpublished Climax Molybdenum Company Report, 3 p.

White, W.H., Bookstrom, A.A., Kamilli, R.J., Ganster, M.W., Smith, R.P., Ranta, D.E., and Steininger, R.C., 1981, Character and origin of Climax-type molybdenum deposits, *in* Skinner, B.J., ed., Economic Geology 75th Anniversary (1905–1980) Volume, p. 270–316.

Geological Society of America
Field Guide 5
2004

Walking with dinosaurs (and other extinct animals) along Colorado's Front Range: A field trip to Paleozoic and Mesozoic terrestrial localities

Joanna L. Wright*

University of Colorado—Denver, Department of Geography and Environmental Sciences, Campus Box 172, P.O. Box 173364, Denver, Colorado 80217-3364, USA

ABSTRACT

The Front Range of Colorado has been subjected to at least three major mountain-building episodes from Paleozoic time on; this field trip will examine some sedimentary deposits related to the first two of these. The first orogeny uplifted the late Paleozoic Ancestral Rocky Mountains and started a long period of terrestrial sedimentation; the second uplifted the late Mesozoic–early Tertiary Laramide Rocky Mountains. Faults in both cases strike approximately north to northwest; the Laramide Orogeny reactivated Paleozoic faults that may in turn have been reactivated Precambrian faults.

The earliest sedimentary rocks (Fountain Formation) deposited on the Precambrian basement in this area were deposited during the Pennsylvanian and Permian Periods under arid climatic conditions as alluvial fans along the eastern side of the Front Range arches, followed by braided rivers and sand seas (Lyons Sandstone). By the Late Jurassic, the climate was more humid and the area was a low-lying meandering river floodplain (Morrison Formation). During the mid-Cretaceous, with the advance of the Western Interior Seaway, the Denver area was beachfront property (Dakota Group) before fully marine conditions returned in the Late Cretaceous. By the end of the Cretaceous, with the initial phase of uplift of the Laramide Rocky Mountains, there was a return to terrestrial conditions (Laramie Formation), which have predominated ever since.

Both body and trace fossils of invertebrates and vertebrates have been preserved in several of these formations. In many cases, trace fossils are the only evidence of the animals that inhabited these areas.

Keywords: terrestrial, vertebrates, paleoenvironments, Front Range, sedimentology.

INTRODUCTION

During this field trip, we will visit several outcrops along the Front Range near Denver (Fig. 1) that show different terrestrial sedimentary environments. The localities will be visited in stratigraphic order; both sedimentological and trace fossil evidence will be considered in the discussion of possible paleoenvironments of the formations.

PROTEROZOIC BASEMENT

Proterozoic rocks were accreted to the Wyoming craton ca. 1.8–1.7 Ga (Condie, 1986). The main components of the Proterozoic rocks in this area are various types of gneiss, quartzite,

*jwright@carbon.cudenver.edu

Wright, J.L., Walking with dinosaurs (and other extinct animals) along Colorado's Front Range: A field trip to Paleozoic and Mesozoic terrestrial localities, *in* Nelson, E.P. and Erslev, E.A., eds., Field Trips in the Southern Rocky Mountains, USA: Geological Society of America Field Guide 5, p. 219–234. For permission to copy, contact editing@geosociety.org. © 2004 Geological Society of America iron formation, and several granitic intrusions ranging in age from 1.7 to 1.0 Ga. Near Boulder, the Proterozoic rock is the 1.7 Ga Boulder Creek granodiorite. The metamorphosed sedimentary rocks were originally shales and quartz-rich sandstones and conglomerates (Lytle, 2004). This suite of metamorphosed sedimentary and volcanic rocks is thought to have originally been formed in an island arc.

FOUNTAIN FORMATION

The Fountain Formation is an ~300-m-thick unit of red sandstone to conglomerate, with some thin beds of mudstone (Fig. 2). It is fairly resistant, and when tilted forms "flatirons," such as those near Boulder. This unit is famous for natural features such as Red Rocks Amphitheater, southwest of Denver, and

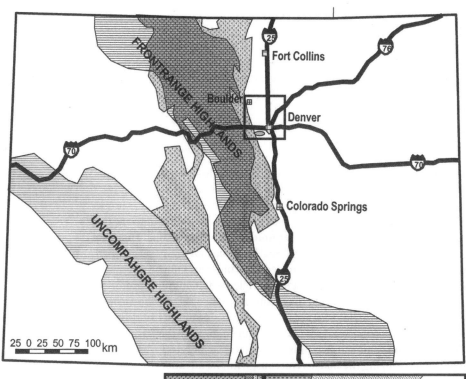

Figure 1. Geology of Colorado and the Front Range showing the location of the stops.

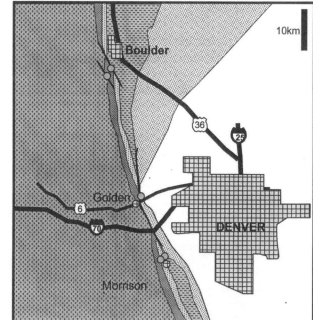

the Garden of the Gods, near Colorado Springs. The majority of the clasts in the sandstone and conglomerate are quartz, potassium feldspar, granite, and mudstone. The red color is more pronounced in the finer-grained beds than in the coarser-grained beds, and the paleo-transport direction ranges from northeast to southeast (present day coordinates). The Fountain Formation overlies and interfingers with the Casper Formation near the Wyoming border (Steidtman, 1976).

The Fountain Formation was formed during uplift of the Ancestral Rocky Mountains. The sediments were eroded from the granite mountains of Frontrangea and deposited as alluvial fans on the eastern flanks of the Ancestral Front Range (Figs. 1

and 3). The easterly transport directions indicate that the sediment source was to the west, and its wide range is consistent with the spread that would be expected from a series of alluvial fans. The lateral and stratigraphic relationship of the Fountain Formation with the Casper Formation in the north indicates that climatic conditions were likely similar but that the Fountain Formation was more proximal than the Casper Formation.

The contact between the Lyons Sandstone and the Fountain Formation is transitional. The base of the Lyons is placed at the base of a fine to medium-grained feldspathic sandstone that contains irregular patches of brown dolomite cement (Weimer and Erickson, 1976). The Lyons Sandstone is in general much lighter in color than the Fountain Formation and contains fewer shale beds.

LYONS SANDSTONE

The Lyons Sandstone is now considered to be one of the classic erg sandstones of the western United States. Yet, it was only in the early 1970s that research on modern dunes allowed detailed comparison with this formation. Prior to this, the Lyons Sandstone had been variously attributed to fluvial, coastal, or desert deposition. The classic work reinterpreting the Lyons Sandstone of the type area was published in 1972 (Walker and Harms), the same year that the Lyons Sandstone (near Golden and Morrison) was described as a fluvial deposit (Weimer and Land, 1972); the conclusions from both interpretations were soon combined into the current model (Weimer and Erickson, 1976).

The Lyons Sandstone has been used extensively as building stone in this area—examples of these flagstones face many of the buildings at the University of Colorado, Boulder. The Lyons Sandstone of the type area is very similar to the contemporaneous Coconino and De Chelly sandstones of Arizona, even to the types of animal tracks found preserved, although a much more varied ichnofauna is found in the Arizona deposits. The Lyons Sandstone further south is coarser grained and arkosic and bears a resemblance to the underlying Fountain Formation (Fig. 4).

Lithology

The Lyons Sandstone crops out along the Front Range from near the Wyoming border south to Colorado Springs and varies in thickness from 45 to 200 m (Hubert, 1960). It consists of two main lithologies: feldspathic quartz arenite and arkosic conglomerate. Micaceous siltstone and shale form a very minor component. At its type locality in Lyons, Colorado, north of Boulder, the Lyons Sandstone consists of well sorted, clay-free, fine-grained sandstone exhibiting large-scale, high-angle (up to 28°), tabular-planar cross bedding. Individual cross bed sets are up to 13 m thick and extend laterally for hundreds of meters (Walker and Harms, 1972). Medium-coarse sand grains are concentrated on some truncation surfaces but are not found within any of the sand beds. Some surfaces preserve low, paral-

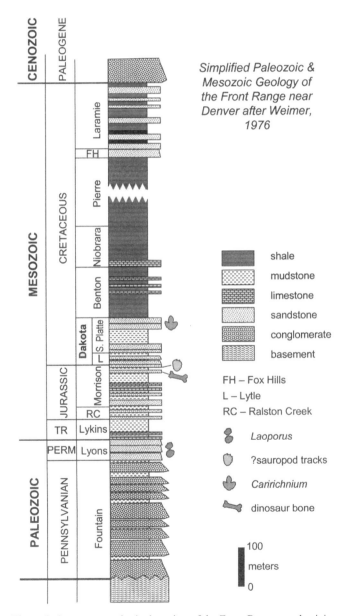

Figure 2. Summary geological section of the Front Range emphasizing the units seen on this trip.

lel, regularly-spaced, straight-crested ripples and, rarely, small circular impressions with raised rims. Tracks of vertebrates and invertebrates are rare and preserved in concave epirelief on foreset surfaces (Henderson, 1924). The vertebrate tracks show five sharp claws on both hands and feet and often also preserve low mounds behind the prints (Fig. 5). Unfortunately, the quarries in Lyons are closed now, but good examples of the formation crop out in various places in that area.

In the area around Morrison and Golden, the Lyons Formation is 50–80 m thick and developed as two main facies: a coarse-grained arkose and a fine-grained arenite (Weimer and Erickson, 1976). The coarse-grained facies dominates in the upper and lower sections, and the fine-grained facies dominates the middle section. Distribution of these facies varies throughout the area, with the coarse-grained facies more dominant in the north of the area near Golden and the middle, fine-grained facies better developed near Morrison and I-70, farther south (Weimer and Erickson, 1976). The coarse arkosic facies can be divided into fining-upward sequences, from conglomerate to sandstone, each 2–6 m thick, and these units are thicker toward the top of the formation. The base of each unit is scoured into the top of

the previous one (Weimer and Erickson, 1976). The transport direction, determined from trough axes, is northeast to east. The middle, fine-grained facies is very similar to the "typical" Lyons facies outlined above, with the exception that in the Golden area these strata consist of fining-upward packages and show only some trough cross-bedding and no tabular-planar cross-bedding.

Depositional Environment

The well sorted, clean, fine-grained sandstone layers with large, high-angle cross beds are all indicative of eolian deposition. The abundant, low, straight-crested, rounded ripples on lee slopes are characteristic of wind formation. The layers with concentrations of coarser grains are interpreted as lag deposits exposed and concentrated by eolian processes (Walker and Harms, 1972). The circular pits with raised rims are interpreted as raindrop impressions by comparison with similar impressions on modern dunes. Such impressions indicate a brief shower, because a downpour would obliterate all the prints, just as trampling obliterates individual footprints. Experiments on animal track formation indicate that the footprint fossils were made in dry sand (McKee,

Figure 3. Locality 1 (anticlockwise from top right): (A) the outcrop, (B) unweathered granite, (C) weathered granite, (D) the bleached zone, (E) the pink zone, and (F) contact with the Fountain.

1947). The invertebrate tracks have been attributed to scorpionids (ichnogenus *Paleohelcura lyonensis*), which are common in arid areas. The vertebrate tracks (*Laoporus coloradoensis*) were made by synapsids (mammal relatives), probably dicynodonts (squat, tusked herbivores).

The arkosic conglomerate facies was the result of fluvial deposition in a braided system. Each sedimentation unit is interpreted as the product of a single flood event. In the Golden area, it seems likely that the fine-grained sediment was blown in by wind but redeposited by water, whereas in the Morrison area, the middle fine-grained section was probably deposited in eolian dunes. Either sand supply for eolian processes was reduced, or the uplift of the mountains was reactivated in this area, because the fine-grained facies gave way to even thicker wadi (braided river) deposits at the top of the section (Weimer and Erickson, 1976).

Paleoenvironmental Setting

During depostion of the Lyons Sandstone, the Front Range from Boulder to Lyons, and probably into Wyoming (where the Casper Formation is of a similar facies and age), was a sand sea (erg). Farther south, near Morrison and Golden, the climate was probably still arid, but the dunes did not always extend that far south. There may have been a very long-lasting wadi channel in the Golden area because eolian deposition does not seem to have taken place there. Thin layers of windblown sand would be very susceptible to removal by periodic flash floods. Deposition around Morrison was from braided streams with an incursion of dunes in the middle of the section. Fossils and sedimentary structures such as raindrops are only found in the eolian deposits, but that does not mean that it did not rain or that the animals did not venture off the

Figure 4. (A) View toward Red Rocks Park: granite mountains in the background overlain by the pink Fountain Formation, then the pale Lyons Sandstone, and in the foreground the red muddy Lykins Formation. (B) Lyons Formation outcrop at Red Rocks Park (Stop 2b). (C) Arkosic cross-bedded sandstones in the Lyons Formation at Stop 2b. Scale bar is 15 cm.

dunes, but rather that the preservation potential of such sedimentary structures would be very low in coarse-grained sediment.

The Pennsylvanian and Permian Formations in the Front Range seem to have been deposited under arid climatic conditions along a narrow coastal plain with mountains to the west and an epicontinental sea to the east. The Casper Formation, to the north in Wyoming, is partially contemporaneous with the Fountain and Lyons formations and is interpreted to have been deposited under desert conditions, although over a somewhat wider area.

MORRISON FORMATION

The Morrison Formation, named after the town of Morrison, is one of the most famous and areally extensive dinosaur-bearing units in the United States. The Morrison Formation spans ~5–10 m.y. of the Late Jurassic Period, depending on the location (Peterson and Turner, 1998). Locally, it may extend right up to, or even into, the Cretaceous, but at Morrison, it stops short of that. There was an ~40 m.y. hiatus between the top of the Morrison and the base of the Dakota; in spite of that, the contact is extraordinarily difficult to place (Houck, 2001).

Lithology

The Late Jurassic Morrison Formation is 75–100 m thick in the area of Golden and Morrison. The type section is along Alameda Parkway at Dinosaur Ridge (Fig. 6) and consists of a 74-m-thick sequence of gray and red sandstone, mudstone, and claystone (Waldschmidt and LeRoy, 1944). There is now another very well-exposed section along I-70 north of Morrison that has a greater total thickness, although it does not preserve the topmost unit (Houck, 2001). The Morrison Formation is considered to be conformable with the underlying Ralston Creek Formation (Fig. 2), and the boundary is placed at the base of the first sandstone bed that is over 1 m thick (Peterson and Turner, 1998). The base of the Morrison is scoured into the underlying silt.

The Morrison is divided into six informal units (Fig. 7). In the type section, bones occur at the base of unit 4, and possible sauropod tracks occur at the base of unit 6. It is interesting to note that in the I-70 roadcut both bone fragments and tracks have been found in units 1 and 5, although not in the same beds (Houck, 2001). Both bones and tracks are found in sandstones in the Morrison area, whereas in many other places bones are preserved in the mudstones.

In the bottom half of the Morrison at the type section, thin layers of muddy limestone are fairly common. Fossils in some of these limestones include freshwater bivalves, ostracods, and charophytes. The associated mudstones and sandstones are commonly calcareous and sometimes contain carbonate nodules. It seems likely that Arthur Lakes' Quarries 5, 8, and 10 were located in this unit, the Grey Claystone and Limestone Unit.

The base of the overlying Lytle Formation (which marks the top of the Morrison) at the type section has been placed at four different stratigraphic positions over a 20 m interval by as many

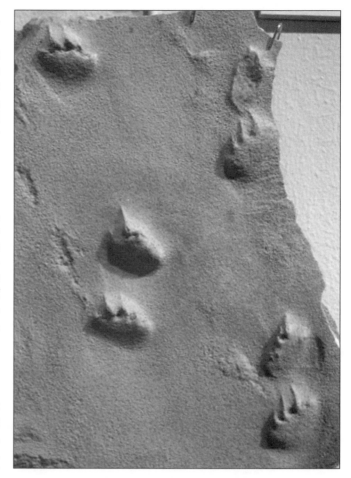

Figure 5. Tracks similar to those found in the Lyons Sandstone (University of Colorado at Denver collection). Tracks are ~40 mm long.

authors (Houck, 2001), based on such features as clay mineralogy or the presence or absence of clay in the sandstones.

Paleoenvironmental Setting

At the time of Morrison deposition, the Dinosaur Ridge area was at ~31°N along the western edge of Laurasia (Demko and Parrish, 1998), in a broad foreland basin with highlands to the west (Currie, 1998). Prevailing winds are thought to have been westerly, and thus, at least the western part of this large basin would have been in a rain shadow (Demko and Parrish, 1998). The Late Jurassic climate for the Colorado Plateau is interpreted to have been arid and warm with seasonal ranges between 10 and 40 °C (Demko and Parrish, 1998); however, in the Morrison area, there is evidence for perennial streams and freshwater lakes, and thus, the climate here seems to have been wetter (Houck, 2001). The Ancestral Front Range was probably still a topographic high to the west but would have consisted of low hills, probably vegetated, and would not have been a significant source of sediment into the Morrison basin (Demko and Parrish, 1998). The sediment

Figure 6. The Morrison Formation at Dinosaur Ridge. (A) Sandstone beds at Stop 3a. (B) Silicified dinosaur bone (sauropod rib head) at Stop 3a. (C) Basal sandstone of the Sandstone and Mudstone Unit (Stop 3b). (D) Deformed sandstone beds ("Brontosaur Bulges") at Stop 3b. Scale bars in A and D are ~1 m; in B, ~20 cm.

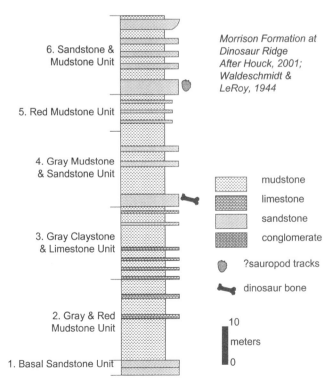

6. Sandstone & Mudstone Unit

5. Red Mudstone Unit

4. Gray Mudstone & Sandstone Unit

3. Gray Claystone & Limestone Unit

2. Gray & Red Mudstone Unit

1. Basal Sandstone Unit

Morrison Formation at Dinosaur Ridge After Houck, 2001; Waldeschmidt & LeRoy, 1944

mudstone
limestone
sandstone
conglomerate
?sauropod tracks
dinosaur bone

10
meters
0

Figure 7. Simple sedimentary section of Morrison along Alameda Parkway (after Houck, 2001) showing the location of the bones and the purported dinosaur tracks.

source for the sandstone is likely to have been to the southwest; Morrison paleocurrent directions show a generally eastward flow. This is consistent with the lithology of the sandstone, indicating a recycled orogen provenance (Currie, 1998).

The Morrison Formation is interpreted as the product of a meandering river system. The sediments were deposited in river channels, on floodplains, and in associated freshwater lakes (Hubert and Panish, 2000). At times, as represented especially in the lower half of the section, the floodplain is interpreted to have been wet enough that large, shallow lakes formed and existed long enough for the deposition of limestone around the edges (Dunagan, 1998). The Morrison is more sandy in the Morrison area than in many places, such as most of the Colorado Plateau and southeastern Wyoming. The lithology and bone preservation at Dinosaur Ridge are very similar to those seen at Dinosaur National Monument in northwestern Colorado. At times, the area was swathed in clouds of ash that settled onto the land and in the lakes, which likely made this area inhospitable for both plants and animals. This ash has now been altered to bentonite clay.

History of Dinosaur Finds

Dinosaurs were first found in this area on 26 March 1877 by Arthur Lakes, an English-born schoolteacher at the school of mines at Jarvis Hall, a prep school that would later become the Colorado School of Mines. Lakes found the bones while out "taking a geological section and measurements … near the little town of Morrison," and a few days later, on 2 April 1877, he wrote to Professor Othniel Charles Marsh of Yale College: "I discovered … some enormous bones apparently a vertebra and a humerus bone of some gigantic saurian." Lakes made the first collection of Morrison Formation dinosaur fossils for Marsh. Lakes developed several dinosaur quarries in this area. The locations of some of them have been lost, and some may have been destroyed in the development of Alameda Parkway. The bone outcrop at Dinosaur Ridge is thought to have been part of, or at least a lateral extension of, Lakes' Quarry 5 (Houck, 2001). By July 1877, Marsh had heard that similar giant bones had been found at a place named Como Bluff, near Laramie, Wyoming, and by the spring of 1879, he had sent his collectors, including Lakes, there. Lakes spent about a year excavating bones for Marsh at Morrison and another year at Como (Kohl and McIntosh, 1997). The bones in the Morrison Formation at Como were also well-preserved, silicified, and uncrushed, but they were preserved mainly in clays and so were easier to extract and were preserved across a wide area in flat-lying beds (regional dip ~6°) rather than in a confined scarp slope as at Morrison.

Arthur Lakes has been called the "Father of Colorado Geology." His discovery of dinosaurs in the west, and the couple of years he spent excavating them, was just a small part of his contribution to western geology. In March 1880, he returned to Colorado and taught at the now independent Colorado School of Mines, where he remained until his retirement in 1893. After his dinosaur excavation experiences, Lakes turned to hard rock geology, especially mining. He worked for the U.S. Geological Survey beginning in 1880, surveying the Mosquito Range and the Leadville district (Kohl and McIntosh, 1997).

DAKOTA GROUP

Lithology

The Dakota Group at Dinosaur Ridge consists of two formations, the lower Lytle Formation (24–37 m thick) and the upper South Platte Formation, which comprises the bulk of the group (Fig. 2). The Dakota Group is ~120 m thick and consists mainly of sandstone with some shale and mudstone (Fig. 8).

The contact of the Lytle Formation with the underlying Morrison Formation is difficult to place because the Lytle Formation differs from the Morrison Formation only in the proportion of sandstone (Weimer, 1970).

The Lytle Formation consists of medium to coarse-grained, trough cross-stratified sandstones, often with conglomeratic bases, interbedded with lenticular red, green, and gray siltstones and claystones. Current directions indicate an overall northeast transport.

The contact between the Lytle and the South Platte formations is placed at the change from oxidized mudstone and

sandstone (Lytle Formation) to reduced siltstone and sandstone (Plainview Member of the South Platte Formation).

The Plainview Member is ~9 m thick, light gray to tan, fine to medium-grained, cross stratified, friable sandstone with a few trace fossils. Transport direction was dominantly to the east (Weimer, 1970). It is overlain by 35 m of bioturbated sandy siltstone with thin black shale intervals (Skull Creek Interval), followed by 25 m of sandstone in which dinosaur tracks are preserved. This top sandstone interval—the "J" Sandstone—consists of buff quartzitic sandstone with lenses of gray siltstone and claystone. Bioturbation is common throughout the section.

The Dakota Group was deposited during the late Albian and early Cenomanian; the trackbearing layer is thought to have formed at ca. 98 Ma.

Paleoenvironmental Setting

The Lytle Formation is the product of fluvial deposition, much like the underlying Morrison Formation, although with less mudstone and less mature sandstone. The sandstones are

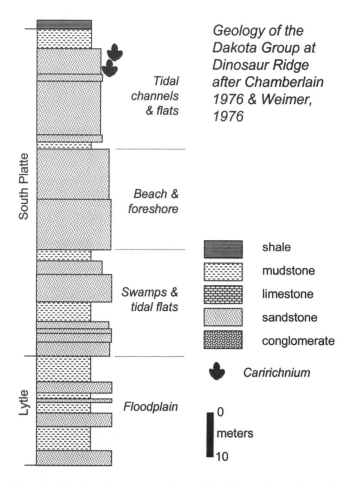

Figure 8. Simple sedimentary section of the Dakota Group along Alameda Parkway (after Chamberlain [1976] and Weimer [1976]).

interpreted to have been deposited in channels and the finer-grained deposits on floodplains (MacMillan and Weimer, 1976). The floodplain deposits are lenticular because of frequent cross-cutting and erosion by the channels (Weimer, 1970). These rivers drained northeast to a marine basin in Montana (Weimer, 1970).

The Plainview Member is interpreted as a transgressive sandstone formed as a result of the westward migration of the shoreline of the Western Interior Seaway (MacMillan and Weimer, 1976).

The depositional environment of the Skull Creek Interval is interpreted as a distributary bay, with coarsening upward strata at the Alameda Road section deposited in distributary channels and on beaches (Weimer, 1970). The "J" Sandstone was deposited in tidal sand flats and channels.

Invertebrate trace fossils of the South Platte Formation comprise a *Skolithos* assemblage, including *Diplocraterion*, *Skolithos*, *Teichichnium*, and *Ophiomorpha*. The first few meters of the South Platte Formation consist of several sandstone beds and represent a slow transgression. Overlying this, a more shaly interval containing *Arenicolites* indicates a deeper water depth of ~20 m. From here, the overall trend is one of shallowing upward, with the middle sandstone interval representing shoreline and foreshore deposition, grading upward into tidal and distributary channels and then at the top into oscillating shoreline deposits including such subenvironments as tidal channels and flats, mangrove swamps, and beaches. At the upper boundary with the Benton Shale, a *Rhizocorallium* assemblage indicates a rapid transgression to deep water (Chamberlain, 1976).

During the time the Dakota Group was being deposited, Denver, along with the rest of the Front Range, was beachfront property on the edge of the Cretaceous Interior Seaway. The shoreline fluctuated but seems always to have been very near Dinosaur Ridge. Dinosaur tracks help constrain water depths on some horizons.

LARAMIE FORMATION

Coal and clay from the Laramie Formation played an important role in the nineteenth century development of the Denver area. From the end of the nineteenth century until after the Second World War, the china clay (kaolinite) was mined for use in bricks and sewer pipes, tiles, and pottery. Once the clay had been mined out, the land was used for a landfill site. In 2001, the owner of the clay pits, Chip Parfet, donated some of the land to the city of Golden for reclamation for a golf course (Lockley and Wright, 2002). Part of this area, containing some of the best fossils, now forms the Parfet Prehistoric Preserve.

Lithology

The Laramie Formation was deposited at the end of the Cretaceous, during the Maastrichtian. It overlies the Fox Hills Sandstone and is overlain by the conglomeratic Arapahoe Formation. The lower half of the Laramie Formation consists of approximately equal amounts of sandstone, siltstone, and claystone (kaolinite) with minor coal beds, whereas the upper part is predominately claystone (Weimer, 1976). The thickness of the individual sandstone and claystone units is 6–12 m.

The sandstones have a scoured base and fine upward from coarse and medium grained to fine grained. They are poorly sorted and silty, buff to pale gray, and show trough cross stratification near the base and ripple laminations near the top (Weimer, 1976). Load casts are common at the base of the sandstone beds. Plant fossils are quite common in the sandstones—leaves, palm fronds, and logs 30 cm in diameter and tens of meters long are found concentrated in places at the base of the sandstone beds (Weimer, 1976). These logs indicate a transport direction of east-southeast. In the upper Laramie, the sandstone beds are thinner and finer grained (Weimer, 1976).

The claystones are generally structureless, light to medium gray kaolinite. Plant fossils also can be found in the claystone. The upper Laramie Formation does not contain any coal, and some of the claystone beds are laminated rather than blocky (Weimer, 1976).

Paleoenvironmental Setting

The Laramie Formation is interpreted as the product of a freshwater delta plain system (Weimer, 1976). The Laramie is the last set of sedimentary deposits formed before initial uplift of the Laramide Rocky Mountains. Weimer (1976) interpreted the section from the top of the Pierre Shale to the Arapahoe Formation as the product of a highly constructive delta in an unstable tectonic regime, with the Laramie Formation as a delta plain deposit.

The sandstone beds in the Laramie Formation were deposited in channels and crevasse splays. The claystones were deposited in swamps or bays or lakes. The channels flowed generally eastward and were flanked by low levees. Crevasse splays were formed during floods, and sandstones with concentrations of logs at the base were probably formed in this way (Weimer, 1976).

The pale, blocky claystone was deposited in well-drained restricted swamps, the dark, carbonaceous claystone in poorly drained swamps, and the laminated claystone of the upper Laramie Formation in lakes or bays. The upper Laramie Formation is interpreted as mainly a lacustrine or freshwater bay environment that experienced periodic incursions of sand (Weimer, 1976).

SUMMARY

The Paleozoic and Mesozoic terrestrial strata of the Front Range in the Denver area were deposited under a range of sedimentary environments. Only one of these formations contains significant osteological remains of animals that lived during these times, but most preserve fossil footprints, which, in most of these units, are the only evidence of the paleofauna and therefore valuable information in any discussion of the paleoenvironment. For instance, the dinosaur (and other vertebrate) tracks of the Dakota Group are the main evidence for subaerial exposure and/or very shallow water conditions in this unit. The ceratopsian tracks in the Laramie Formation are the only ones in the world and as such provide valuable information on the soft tissue structure of their feet and contribute to the debate over the way that these animals walked.

ROAD LOG

0.0   Set odometer to zero at the junction of northbound U.S. Highway 36 with Baseline Road in Boulder. Turn left (west) onto Baseline and remain on this road, out of the city and up a hill

2.1   First outcrop of the Fountain Formation on this road.

2.4   Panoramic viewpoint on the right (see below).

3.6   First outcrop of weathered granite.

4.1   STOP 1a. There is space for parking on the right, on the outside of the hairpin bend.

STOP 1a. Paleo Land Surface and Fountain Formation, Flagstaff Mountain, Boulder

The outcrop is ~30 m long with unweathered Boulder Creek granodiorite at its western end to the basal Fountain Formation at the east (Fig. 3). The base of the Fountain here is a granule conglomerate bed, ~20 cm thick, scoured into the underlying rock. This outcrop was first described in detail by Wahlstrom (1948).

The Boulder Creek granodiorite is a typical granodiorite, pale pink, containing phenocrysts of orthoclase and quartz with accessory amounts of biotite. The brown staining on some surfaces is the product of recent weathering. The granodiorite contains several veins that tend to have higher quartz content than the surrounding rock.

Walking east along the outcrop, the Paleozoic weathering of the granodiorite becomes obvious. The first signs are that the rock becomes very friable and progressively more bleached in the bottom half of the weathered zone. Above the bleached zone is a zone of red-stained rock, with the color intensity increasing up to the contact.

Even near the top of this weathered zone, however, the outlines of veins in the granodiorite are still clear. This means that the rock has not been mechanically disturbed in any way. Birkeland et al. (1996) questioned the presence of a true paleosol here and left open which chemical and mineralogical trends might be due to surface weathering versus due to diagenesis after burial.

This locality was a key outcrop for discussion of the origin of red color in sedimentary rocks. For a long time, it was assumed that the red was depositional (Wahlstrom, 1948), but Walker (1979) argued, based on observations of modern sediments in northern Africa, that the red in most desert sediments was diagenetic. This idea was very contentious for many years, but is now mainly accepted.

Return down the hill.

5.8   Panoramic viewpoint on the left hand side of the road.

If you pull off here and walk out to the viewpoint, which includes signs interpreting the geology of the landscape, you can see that here on Flagstaff Mountain, the rest of the Paleozoic and most of the Mesozoic rocks have been faulted out by a high angle reverse fault that separates the Fountain Formation from the Benton Shale. This is a very localized fault—looking north and south from this point, you can see that the complete sequence is preserved in the ridges and valleys just a few miles north and south of here.

Continue on down the hill.

6.5   Turn left on 6th Street; follow this road downhill (north); this street doglegs at Euclid and Rose Hill; take a left on Euclid, and then an immediate right to continue down 6th. Turn left (west) on Canyon Boulevard.

7.7   Turn right immediately after the Watershed School and immediately left into the parking lot for Settlers Park. Walk over the bridge and up the hill for ~100 m to Settlers Quarry.

STOP 1b. Fountain Formation and Lyons Sandstone Transition, Settlers Quarry, Boulder

This is just a quick stop to look at the Lyons Sandstone where it is similar to that in the type area. The quarries of the type area are now closed.

You are looking at the bedding planes at this locality. Note that the dip here is much steeper than at the previous locality. This is because of folding associated with the steeply dipping reverse fault that faulted out the Lyons Sandstone through the Niobrara Formation at Flagstaff Mountain. This area of the Front Range is riddled with similar faults; another example is the Golden Fault, which runs along the west side of the Dakota Hogback and has caused the beds in the Laramie Formation (see Stop 5) to be tilted up to vertical. The Lyons Sandstone here is also unusually highly indurated, again probably due to proximity to the fault.

This locality spans the contact between the Fountain and the Lyons Formations. At the west end of the outcrop, the typical granular arkosic conglomerate of the Fountain Formation is interbedded with the typical fine-grained sandstone facies of the Lyons Sandstone. The Lyons Sandstone here is bedded on a cm-dm scale, and the beds are similar in thickness and have planar contacts.

7.7   Leave the parking lot and turn left onto Canyon Boulevard, heading east.

9.4   Junction with 28th Street (which becomes U.S. 36). Turn right (south) onto 28th.

20.9   Junction of CO 121 with U.S. 36. Exit U.S. 36, and take 121 (Wadsworth Parkway) south.

31.1 Junction with I-70. Go west on I-70.

41.1 Morrison exit; turn left (signposted to Red Rocks Park) onto Hogback Road.

42.9 First exit for Red Rocks Park up West Alameda Parkway. Turn right and drive all the way up to the top circle parking lot for Red Rocks and the Visitors Center. The junctions are well signposted.

43.7 Parking lot on left for Red Rocks Trail; continue straight on.

44.1 Road goes through arch cut in Fountain Formation.

44.4 Stop 2a.

STOP 2a. Fountain Formation and Contact with the Idaho Springs Formation, Red Rocks Amphitheater

These typical outcrops of the Fountain Formation show thick beds of arkosic conglomerate interbedded with finer-grained laminated redbeds. Note the characteristic weathering of these sediments. Dark red, fine-grained beds between the paler conglomeratic beds are well exposed here. The contact is very different from the previous locality; the Precambrian basement rock here does not show deep weathering beneath the base of the Fountain Formation.

Historical Aside: Red Rocks Amphitheater

Red Rocks Amphitheater was built by the Civilian Conservation Corps (CCC) between 1936 and 1941. The CCC was set up by Franklin D. Roosevelt as part of the New Deal to provide employment during the Great Depression. Before construction, the area between the two flanking outcrops was strewn with boulders, which were blasted away. Red Rocks is not really a natural amphitheater, but the acoustics, as constructed and with help from geology, allows 20,000 people to clearly hear sound from the stage. A new Visitors Center has been built at the top parking lot that includes small historical and geological displays.

Cumulative	
(mi)	*Description*
44.4	Return back down the hill.
44.9	Turn left, signposted to "Geologic Marker."
45.2	The road terminates in a parking lot. Park and walk up the stone steps to the overlook. Here, the ridges of the Fountain, Lyons, and Dakota formations and the valley of the Lykins and Morrison formations can be seen below. Retrace your path almost to the Red Rocks turnoff.
45.5	Turn right back onto the road out of the park.
45.9	Marker on left for Fountain Formation.
46.1	Stop 2b.

STOP 2b. Lyons Sandstone, Morrison

This locality shows the coarser-grained facies of the Lyons Formation. The coarse sand and granule- to conglomerate-sized

grains indicate that this cannot be the result of eolian sedimentation (Fig. 4C).

Continue on down the hill.

Cumulative	
(mi)	*Description*
46.2	Marker on left for Lykins Formation.
46.5	Back at the junction of West Alameda Parkway with Hogback Road (where you turned off for Red Rocks Park), go straight over the main road and up the hill.
46.7	Stop 3a.

STOP 3a. Morrison Formation, Dinosaur Ridge

A small shelter and interpretive signs were erected here a few years ago for the benefit of the many visitors to this site, so this locality is easy to spot. One of the diagrams outlines the positions of some of the major fossil bones.

The dinosaur bones are preserved in the second of four stories of a compound sandstone at the base of the Grey Mudstone and Sandstone Unit (Waldschmidt and LeRoy, 1944). Framework grains in Morrison sandstones are 90% quartz; the remainder is feldspar (Hubert and Panish, 2000). The sandstone is well cemented; most of the cement is ferroan dolomite, and most of the rest is kaolinite—the total volume of cement is ~45% (Hubert and Panish, 2000). Transport direction in this bed was toward the southeast. Much of this bed was destroyed during the construction of Alameda Parkway, but the dinosaur bones also likely continue into the hillside. There are 109 bones visible in the outcrop, with many others in loose blocks along the road. This mass accumulation may have originally contained thousands of bones (Hubert and Panish, 2000).

The bone layer lies within a series of trough cross-bedded sets (Fig. 6A). Most of the cross bed sets are medium-grained sandstone, but there are some sets of pebbly coarse sandstone around the middle of the bone layer. Each story fines upwards. The elongate bones (e.g., Fig. 6B) seem to be bimodally distributed; most are parallel to the flow direction, and the rest are perpendicular; this is normal for a confined channel (Hubert and Panish, 2000). The bones disrupted the water flow and sometimes trapped coarse sand on their upstream sides.

The bone mineral is well-crystallized francolites (carbonate fluorapatite) (Hubert and Panish, 2000). The bones are radioactive.

Deposition of the Bone-Bearing Layer

The bone-bearing unit was deposited in the channel of a river that was meandering southward. Avulsion brought the river channel to this area of the floodplain, possibly as a result of heavy rainfall after a period of drought. The heavy rainfall caused a flood 1–2 m deep, but this is not thought to have killed the dinosaurs whose bones are found in this layer; rather, they were already dead and defleshed. Perhaps they had died of hunger or thirst during this putative drought, and their bones lay upon the floodplain only to be entrained in the flood, where they were rapidly buried in a trough cross-bedded coset by advancing subaqueous dunes. The bones do not show any evidence of surface cracking, so they cannot have been lying on the surface for very long.

Ribs, which are long and relatively light, tend to be oriented perpendicular to the flow direction. The larger bones rolled and slid in the current and are concentrated in what was the deeper part of the channel, where the flow was fastest. Smaller bones are not found in this deposit; they were probably swept much farther downstream (Hubert and Panish, 2000). The bones are all disarticulated, but it is not possible to determine how much of this occurred on the floodplain prior to entrainment and how much was as a result of transport.

There is a 2-m-thick lacustrine sequence directly on top of the bone unit, consisting of dolostone and gray calcareous shale, siltstone, and sandstone. This lake was saline due to the arid climate, and its presence probably sped up the cementation of the underlying sandstone and the diagenesis of the bones. The high volume of cement implies very early cementation as 40%–50% porosity is typical of freshly deposited river sands (Hubert and Panish, 2000).

The existence of the aforementioned drought is entirely conjecture. The drought is inferred because preservation of dinosaur bones is so rare in this part of the Morrison, yet channel sands and avulsion are fairly common. The bone accumulation bears some resemblance to bones at Dinosaur National Monument and at Bone Cabin Quarry, Wyoming. To form a mass accumulation of bones in a channel sandstone, several special circumstances seem to be necessary: (1) A large supply of bones in or near a dry river bed (drought is a handy way to achieve this); (2) A flood to scoop up the bones and deposit them in a channel where they are rapidly buried in sand; and (3) Early cementation will protect the bones from both decay and compaction (Hubert and Panish, 2000; Dodson et al., 1980). While here at Dinosaur Ridge the cement is dolomite, at Dinosaur National Monument it is silica from the alteration of volcanic ash.

STOP 3b. Morrison Formation "Brontosaur Bulges" (46.8 mi)

A little farther along from the bone outcrop, still on the scarp side of the hogback, some layers have been excavated away from underneath a sandstone layer (Fig. 6C and 6D). This is the basal sandstone of the topmost unit of the Morrison along West Alameda Parkway: the Sandstone and Mudstone Unit (Fig. 7). A number of bulges in this layer have been known for some time and were initially termed "load-pocket structures" (Weimer and Land, 1972). The cross section of these structures shows that several layers of sediment are deformed up to a depth of 50 cm. They are scattered randomly on the base of the bedding plane. They range in size from 30 to 90 cm in maximum dimension.

These load structures were reinterpreted as dinosaur tracks on the basis of their range in size, irregular distribution, and the fact that they do not occur at the boundaries of beds of different competencies, as would be expected in typical load structures.

The upper units of the Morrison Formation are thought to have been deposited by a meandering river system. The thin limestone beds do not occur here, and the sandstones and mudstones are less calcareous.

Cumulative (mi)	Description
46.9	Approximate boundary of the Morrison Formation with the overlying Dakota Group. There is also a panoramic viewpoint with interpretive signs of the geology of the Denver Basin.
47.1	Bedding plane covered with ripple marks.
47.3	Stop 4.

STOP 4. The Dinosaur Ridge Tracksite, Dakota Group

The footprints at Dinosaur Ridge are probably some of the most famous in the world. They were discovered in 1937 during the construction of Alameda Parkway, but remained an entirely local curiosity until the 1980s when this site, and similar ones nearby, started to be studied in detail (Fig. 9).

The dinosaur tracks are black because they have been colored in with charcoal (Fig. 10). This allows visitors to the site to be able to see tracks even on cloudy days or when the sun is shining directly on the bedding plane. It doesn't damage the footprints and washes off in about six months, so it is touched up regularly.

There are two main types of dinosaur footprints on the main horizon. Both types of footprints are tridactyl (have three toes). The thick-toed footprints were made by ornithopod dinosaurs—probably iguanodontids, although they could have been made by some very early hadrosaurs. The narrow-toed footprints were made by ornithomimids—often called ostrich dinosaurs because of their long necks and small heads. The tracks are attributed to these dinosaurs on the basis of comparisons with pedal morphology and on a temporal basis; i.e., what kinds of dinosaurs were around at that time.

If you look more closely at the footprint surface you can see that there are two sizes of the thick-toed tracks (Fig. 10B). The smaller ones are thought to be juvenile iguanodontids. The smaller ones seem to be better-preserved than the larger ones—they show more clearly-defined toes and heels. There is another morphological difference between the two sizes of iguanodontid tracks; many of the larger ones are associated with small impressions immediately in front of the middle toe (Fig. 10C). These are impressions of the trackmakers' front feet (Fig. 10D).

Some of the iguanodontid trackways are parallel, and this has been interpreted as evidence of gregarious behavior; however, there is no evidence that the makers of the parallel trackways were crossing the area as a group, so this conclusion is tentative.

These types of tracks are not the only ones known from the Dakota Group. At an outcrop near Golden, bird tracks were found associated with dinosaur tracks (Lockley and Hunt 1995). Tracks of crocodilians also have been found in this formation.

Continue on down the hill.

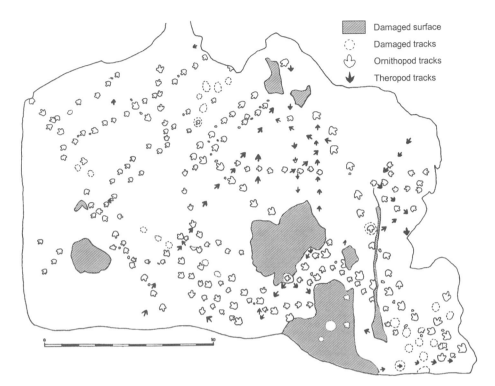

Figure 9. Main trackway surface at Dinosaur Ridge. Scale bar is in meters.

Damaged surface
Damaged tracks
Ornithopod tracks
Theropod tracks

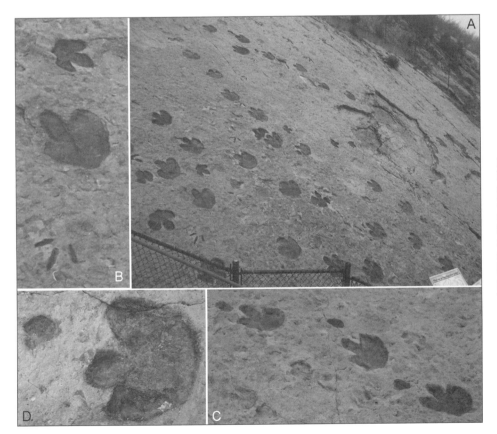

Figure 10. Dinosaur tracks at Dinosaur Ridge. (A) The main bedding plane. (B) The three different tracktypes at Dinosaur Ridge, from top to bottom: small ornithopod, large ornithopod, theropod—probably ornithomimids. (C) Quadrupedal trackway of large ornithopod. (D) Manus-pes pair of large ornithopod. Scale varies in these perspectives, but the large ornithopod tracks are ~50 cm long.

Cumulative (mi)	Description
47.4	Bedding plane covered with root impressions; an interpretive sign terms this a "mangrove swamp."
47.5	Approximate boundary of the Dakota Group with the overlying Benton Shale.
47.8	Turn left onto Rooney Road.
50.3	Turn right on to West Colfax Ave/I-70 Blvd/U.S. Hwy 40.
50.9	Turn sharp left onto U.S. Hwy 6 W/West 6th Avenue.
52.4	The junction with 19th Street is on the left. Park here on the corner. Walk ~100 m to the beginning of the paved trail. We are going to take the "Triceratops Trail" at the golf course and look at the dinosaur tracks and other fossils preserved there.

STOP 6. Fossil Trace Golf Course, Laramie Formation, Golden

When the owner of the clay pits, Chip Parfet, donated some of the land to the city of Golden for reclamation for a golf course, the city worked with the state archeologist to make sure that the dinosaur footprints were not damaged. Only one footprint had to be removed during the course of construction, and it is now on display at the golf course club house as part of an exhibit on the fossils of the area.

The Fossil Tracks

The dinosaur tracks are preserved as casts on the underside of the sandstone beds in the lower Laramie Formation (Lockley and Hunt, 1995; Lockley and Wright, 2002). Most of the tracks are very large (up to 50 cm long) and three, four, or five-toed (Fig. 11). The three-toed tracks fall into two categories: those with thick, rounded toes, and those with more elongate toes with pointed ends. The four and five-toed tracks are often associated in trackways, with the larger four-toed tracks falling behind and slightly inside the five-toed tracks. Some very small tracks, also preserved as casts, are just a few centimeters long; these are much rarer than the large tracks and have four or five long, narrow toes. Several large areas of bedding planes are covered with irregular bulges (Fig. 11A), and some bedding planes preserve small circular bulges on their underside.

Interpretation of the Tracks

The dinosaur tracks have been attributed to hadrosaurs (duckbills), theropods, and ceratopsians (horned dinosaurs). Hadrosaurs and theropods produced the tridactyl tracks, the hadrosaurs leaving those with rounded toes and the theropods those with longer, pointed toes (Fig. 11). The associated four and five-toed tracks were made by ceratopsians, based on the close correspondence of their foot skeletons with the tracks. The only other dinosaurs that could have made tracks like these are ankylosaurs (armored dinosaurs), and ankylosaurs have an unusually long pedal digit IV (Thulborn, 1990), which does not fit these pes (hind foot) impressions.

A couple of small trackways with tetradactyl manus (hand or fore foot) and pes impressions have been attributed to mammals (Lockley and Foster, 2003), and an isolated five-toed track (now buried) has been attributed to a champsosaur (a small, extinct, aquatic reptile). This is the only known champsosaur track in the world (Lockley and Hunt, 1995).

The bedding planes covered in irregular bulges are interpreted to be a result of large numbers of dinosaurs traveling over the bedding plane and trampling the ground underfoot so that no discrete trackways can be discerned.

The tracks of horned dinosaurs (e.g., *Triceratops*) are very rare. This is surprising because their skeletons, especially their skulls, are very common in the Late Cretaceous. This might mean that these dinosaurs lived in more upland or densely vegetated areas where tracks would be less likely to be preserved. This is still the only site in the world where ceratopsian tracks have been found. The tracks have been named *Ceratopsipes*, which means "horned dinosaur foot." These tracks are some of the youngest dinosaur fossils in the world.

This is the end of the field trip. The quickest route back to Denver is to head east on Highway 6, which goes all the way into the center of Denver (~14 mi).

ACKNOWLEDGMENTS

Thanks to Wes le Mesurier, who first suggested visiting the outcrop on Flagstaff Mountain. Thanks also to Peter Birkeland, who agreed to visit the above outcrop with me and explain what he knew about it; he also showed me the Lyons Sandstone outcrop in Boulder. This field guide was greatly improved by the comments of Eric Erslev and Eric Nelson.

REFERENCES CITED

Birkeland, P.W., Miller, D.C., Patterson, P.E., Price, A.B., and Shroba, R.R., 1996, Soil geomorphic relationships near Rocky Flats, Boulder and Golden, Colorado area, with a stop at the pre-Fountain paleosol of Wahlstrom (1948), in Thompson, R.A., Hudson, M.R., and Pillmore C.L., eds., Geologic excursions to the Rocky Mountains and beyond, Fieldtrip guidebook for the 1996 Annual Meeting of the Geological Society of America, Denver, Colorado, Oct 28–31, 1996: Colorado Geological Survey Special Publication 44, Field trip 27, 13 p.

Chamberlain, C.K., 1976, Field Guide to the trace fossils of the Cretaceous Dakota Hogback along Alameda Avenue, west of Denver, Colorado, in Epis, R.C., and Weimner, R.J., eds., Studies in Colorado Field Geology: Professional Contributions of Colorado School of Mines, v. 8, p. 242–250.

Condie, K.C., 1986, Geochemistry and tectonic setting of early Proterozoic supracrustal rocks in the southwestern U.S.: Journal of Geology, v. 94, p. 845–864.

Currie, B.S., 1998, Upper Jurassic–Lower Cretaceous Morrison and Cedar Mountain Formations NE Utah–NW Colorado: Relationships between nonmarine deposition and early Cordilleran foreland basin development: Journal of Sedimentary Research, v. 68, p. 632–652.

Demko, T.M., and Parrish, J.T., 1998, Paleoclimatic setting of the Upper Jurassic Morrison Formation: Modern Geology, v. 22, p. 283–297.

Dodson, P., Behrensmeyer, A.K., Bakker, R.T., and McIntosh, J.S., 1980, Taphonomy and paleoecology of the dinosaur beds of the Jurassic Morrison Formation: Paleobiology, v. 6, p. 208–232.

Dunagan, S.P., 1998, Lacustrine and palustrine carbonates from the Morrison Formation (Upper Jurassic), east-central Colorado, USA: Implications for

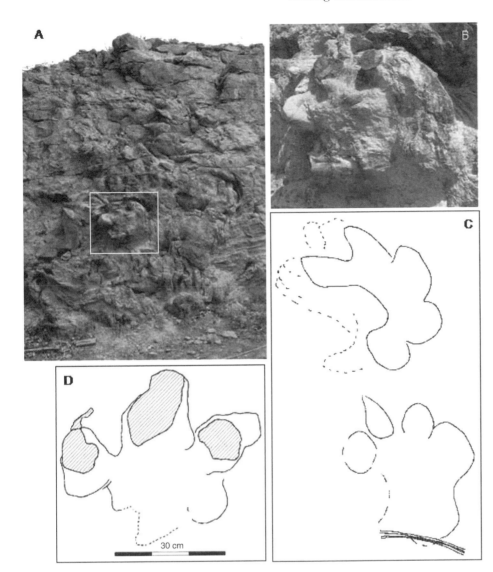

Figure 11. Tracks from the Laramie Formation at Fossil Trace Golf Course. (A) Trampled surface. (B) Close-up of a ceratopsian pes track from the same surface (track ~50 cm long). (C) Ceratopsian manus-pes pair (tracing; scale similar to B). (D) Hadrosaur track (tracing).

depositional patterns, paleoecology, paleohydrology and paleoclimatology [Ph.D. dissertation]: Knoxville, University of Tennessee, 276 p.

Henderson, J., 1924, Footprints in Pennsylvanian sandstones of Colorado: Journal of Geology, v. 32, p. 226–229.

Houck, K., 2001, Sedimentology and stratigraphy of the Morrison Formation in the Dinosaur Ridge area near Morrison, Colorado: The Mountain Geologist, v. 38, p. 97–110.

Hubert, J.F., 1960, Petrology of the Fountain and Lyons Formations, Front Range, Colorado: Colorado School of Mines Quarterly, v. 55, p. 1–242.

Hubert, J.F., and Panish, P.T., 2000, Sedimentology and diagenesis of the dinosaur bones exposed at Dinosaur Ridge along Alameda Parkway in the Morrison Formation (Upper Jurassic), Morrison, Colorado: The Mountain Geologist, v. 37, p. 73–90.

Kohl, M.F., and McIntosh, J.S., 1997, Discovering dinosaurs in the old West: Washington, D.C., Smithsonian Institution, 197 p.

Lockley, M.G., and Foster, J.R., 2003, Late Cretaceous mammal tracks from North America: Ichnos, v. 10, p. 269–276.

Lockley, M.G., and Hunt, A.P., 1995, Dinosaur tracks and other fossil footprints of the western United States: New York, Columbia University Press, 338 p.

Lockley, M.G., and Wright, J.L., 2002, Dinosaur tracks and plant fossils found in the Laramie Formation at the City of Golden recreation campus site: Report to Colorado Historical Society, Permit 2001-67, 20 p.

Lytle, L.R., 2004, Proterozoic of the Central Front Range—Central Colorado's beginnings as an island arc sequence: Boulder, Colorado, Colorado Scientific Society Proceedings of Conference on Colorado Geology, April 2–4, 2004, p. 20–21.

MacMillan, X., and Weimer, X., 1976, Stratigraphic model, delta plain sequence, J Sandstone (Lower Cretaceous) Turkey Creek area, Jefferson County, Colorado, *in* Epis, R.C., and Weimer, R.J., eds., Studies in Colorado Field Geology: Professional Contributions of Colorado School of Mines, v. 8, p. 228–240.

McKee, E.D., 1947, Experiments on the development of tracks in fine cross-bedded sand: Journal of Sedimentary Petrology, v. 17, p. 23–28.

Peterson, F., and Turner, C.E., 1998, Stratigraphy of the Ralston Creek and Morrison Formations (Upper Jurassic) near Denver, Colorado: Modern Geology, v. 22, p. 3–38.

Steidtman, J.R., 1976, Eolian origin of sandstone in the Casper Formation, southernmost Laramie Basin, Wyoming, *in* Epis, R.C. and Weimer, R.J., eds., Studies in Colorado field geology: Professional Contributions of Colorado School of Mines, v. 8, p. 86–95.

Thulborn, R.A., 1990, Dinosaur Tracks: Cambridge, UK, Cambridge University Press, 330 p.

Waldschmidt, W.A., and LeRoy, L.W., 1944, Reconsideration of the Morrison Formation in the type area, Jefferson County, Colorado: Geological Society of America Bulletin, v. 55, p. 1097–1114.

Wahlstrom, E.E., 1948, Pre-Fountain and recent weathering on Flagstaff Mountain near Boulder, Colorado: Geological Society of America Bulletin, v. 78, p. 353–368.

Walker, T.R., 1979, Red color in dune sand, *in* McKee, E.D., ed., A study of global sand seas: U.S. Geological Survey Professional Paper 1052, p. 61–82.

Walker, T.R., and Harms, J.C., 1972, Eolian origin of flagstone beds, Lyons Sandstone (Permian), type area, Boulder County, Colorado: The Mountain Geologist, v. 9, p. 279–288.

Weimer, R.J., 1970, Dakota Group (Cretaceous) stratigraphy, southern Front Range, South and Middle Parks, Colorado: The Mountain Geologist, v. 7, p. 157–183.

Weimer, R.J., 1976, Cretaceous stratigraphy, tectonics and energy resources, western Denver Basin, *in* Epis, R.C., and Weimer, R.J., eds., Studies in Colorado Field Geology: Professional Contributions of Colorado School of Mines, v. 8, p. 180–225.

Weimer, R.J., and Land, C.B., 1972, Lyons Formation (Permian), Jefferson County, Colorado: a fluvial deposit: The Mountain Geologist, v. 9, p. 289–297.

Weimer, R.J., and Erickson, R.A., 1976, Lyons Formation (Permian), Golden-Morrison area, Colorado, *in* Epis, R.C. and Weimer, R.J., eds., Studies in Colorado Field Geology: Professional Contributions of Colorado School of Mines, v. 8, p. 123–138.

Printed in the USA